Advances in Vehicle Design

To Ruth

Advances in Vehicle Design

by

John Fenton

Professional Engineering Publishing

Professional Engineering Publishing Limited
London and Bury St Edmunds, UK

First published 1999

ISBN 1 86058 181 1

A CIP catalogue record for this book is available from the British Library.

Printed and bound in Great Britain by St Edmundsbury Press Limted, Suffolk, UK

CONTENTS

Chapter 5: Vehicle development

Chapter 6: Systems development: powertrain/chassis

Chapter 7: System developments: body structure/systems

Preface

This literature survey is aimed at providing the vehicle design engineer with an update in vehicle and body systems. The author has scanned the considerable output of technical presentations during 1997– 9 to extract and distil developing technologies of particular import to the working designer. The easily digestible presentation, with unusually high dependence on diagrammatic presentation, continues the popular style used for the original handbooks that were compiled by the author, and published by Professional Engineering Publishing. These are listed on the Related Titles page overleaf. *Advances in Vehicle Design* serves both as an update to the earlier volumes and as a stand-alone volume. The referenced leads provided in the text are intended to help designers and engineers from whatever background discipline. Widespread availability of computing power to designers and engineers has created the possibility of considerably shortening the lead-times between design conception and prototype manufacture. Much of the material covered here will assist in establishing predictive techniques.

Advances in Vehicle Design is an update of vehicle and body systems design in that it provides readers with an insight into analytical methods given in a wide variety of published sources such as; technical journals, conference papers, and proceedings of engineering institutions, for which a comprehensive list of references is provided. The analyses are therefore not necessarily fully developed or rigorously evaluated. Recourse to the original references is necessary particularly in order to understand the limiting assumptions on which the analyses are based.

Much of the analytical work is centred around impending legislation and, where this is quoted in the text, it is for illustration only and it is, of course, important to examine the latest statutes in the locality concerned. The list of references given at the end of the volume is a key element of the publication, providing where possible a link to the original publication source. Where the original publication is not available in bookshops, many of the sources can be found in libraries such as those of the Institution of Mechanical Engineers, London, or the Motor Industry Research Association, in Nuneaton, UK, as well as the British Library. Other similar respositories of techincal information should be able to provide a selection of original source material. Where the source is a company announcement of techniques and systems, names, but not addresses, of the companies/consultancies are given. Most operate internationally and have different national locations, best found by enquiry in the country concerned. For the patent reviews in chapters six and seven, full specifications can be purchased from The Patent Office, Cardiff Road, Newport, NP9 1RH, UK.

JOHN FENTON

Related Titles

Title	Editor/Author	ISBN/ISSN
Multi-Body Dynamics	H Rahnejat	1 80058 122 6
Gasoline Engine Analysis	J Fenton	0 85298 634 3
Handbook of Automotive Body Construction and Design Analysis	J Fenton	1 86058 073 4
Cranes – Design, Practice and Maintenance	J Verschoof	1 86058 130 7
Handbook of Automotive Body Systems Design	J Fenton	1 86058 067 X
Handbook of Automotive Powertrain and Chassis Design	J Fenton	1 86058 075 0
Vehicle Handling Dynamics	J Ellis	0 85298 885 0
Handbook of Vehicle Design Analysis	J Fenton	0 85298 963 6
Automotive Braking: - Recent Developments	D Barton	1 86058 131 5
Brakes and Friction Materials	G A Harper	1 86058 127 7
Automotive Engineer Monthly Magazine	W Kimberley	0307/6490
Journal of Automotive Engineering (IMechE Proceedings Part D)		0954/4070

For full details of all Professional Engineering Publications please contact:

Sales Department
Professional Engineering Publishing
Nothgate Avenue
Bury St Edmunds
IP32 6BW
UK
Telephone: +44 (0)1284 763 277
Fax: +44 (0)1284 718 693
E-mail: sales@imeche.org.uk

Chapter 1: Materials and construction advances

The considerable comeback made by the steel industry in restating its case for structural superiority over the light alloys, and the ever moving goalposts of developing aluminium and polymer composites, open this chapter. The re-emergence of hybrid metal/plastic structures is also discussed as well as the creation of reinforced plastics with recycled-polymer matrices. The move to material property charts which lead to the creation of performance indices is next examined and the chapter concludes with the efforts to make self-pierce riveting a viable alternative to conventional welded joints in body construction.

Steel durability and structural efficiency

According to researchers at Lotus Engineering[1], market expectations for durability of vehicle body panels is typically seven years or 100 000 miles, with an expectation of 10 years corrosion-free life. British Steel engineers[2] have shown the main influences on durability with the corrosion triangle of Fig 1. In product design the traditional approach has been to remove moisture traps, allow better penetration of paint and evolve design with greater resistance to stone chipping. There had also been gradual adoption of one-side zinc-coated steel panels, offering protection on the inside and good paintability on the outside. More recently there had been a move from hot dip galvanized to electrozinc panels, spurred by Japanese manufacturers; these offer a range of ductilities, increased weldability and the ability to alloy the coating to provide various coating thicknesses and corrosion resistance levels. Now double-sided coated steels are favoured with differential coating weights: typical thicknesses are now 45–60 g/m^2. Future quality improvements are promised by development work in surface treatment and formability. Permanent organic based topcoats or phosphate coatings are providing improved performance in both weldability and corrosion prevention. Another new initiative is a drive by the OEMs to replace electroplated products by 'galvanneal' ones for outside parts, in the interests of production economy. These have the ability to provide good paintability.

In a joint venture between British Steel Strip Products and Rover Group, trial parts were prepared for the Rover 600, including front fender, front door skin, rear door skin and bonnet. In 1997 all new models were produced with two-sided galvanneal full-finished panels. In the longer term, the authors see the development of non-bake hardening, higher strength steel substrates with enhanced formability for use with galvanneal coatings, and the introduction of extra high strength transformation induced plasticity (TRIP) steels as substrates for zinc type coating which will offer yield strength approaching 1500 MPa, alongside high ductility. Prepainted body finished sheet steel, which obviate the electropheritic primer coat are also expected, as is a possible new coating process based on vapour deposition of zinc in a vacuum.

Car body weight reduction of 25%, without cost penalty, plus a 20% reduction in part count, has been the result of the final phase of the USLAB project for light-weighting steel automotive structures carried out by an international consortium of steel companies supervised by Porsche Engineering Services. The objective of a feasible design, using commercially available materials and manufacturing processes, has also been met. An 80% gain in torsional rigidity, and 52% in bending, has also been recorded and estimated body-shell cost of $947 compares with $980 for a year 2000 comparison figure for

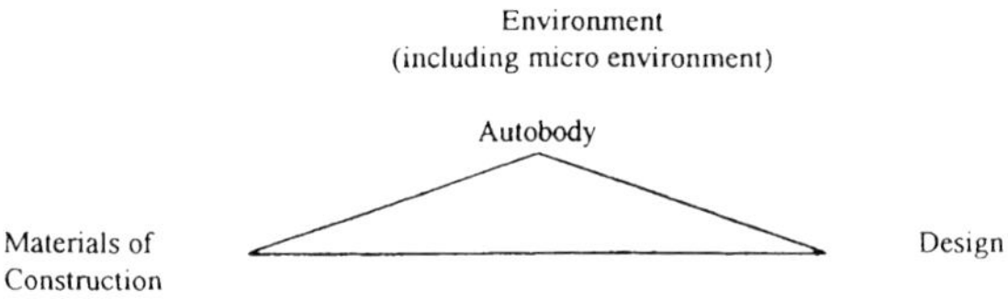

Fig 1: Triangle of factors affecting corrosion

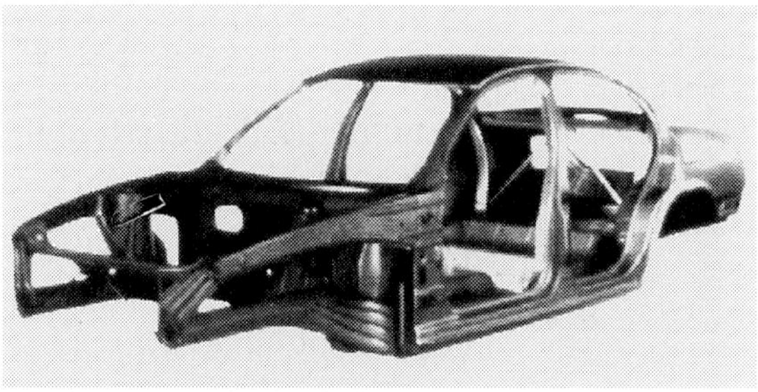

Fig 2: ULSAB body shell

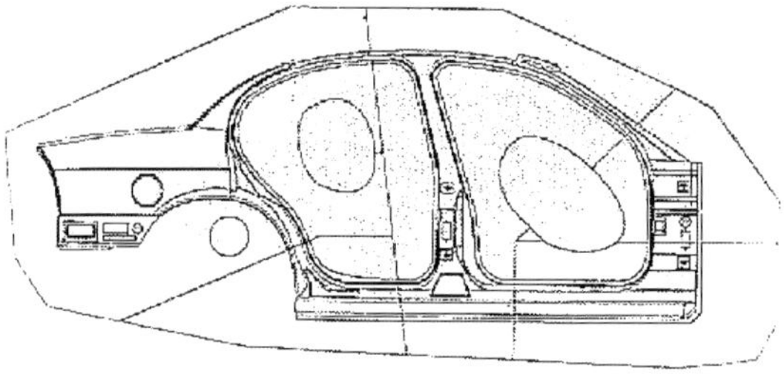

Fig 3: Monoside panel

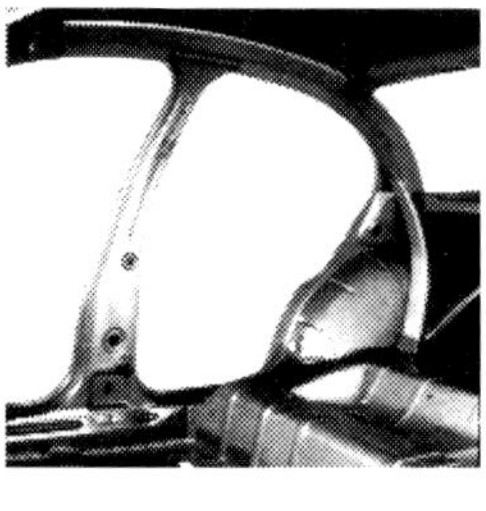

Fig 4 : Side roof rail

a conventionally constructed body to a similar specification. The 203 kg body shell (Fig 2), of 2.7 metres wheelbase, has a torsional rigidity of 20,800 Nm/deg and a first torsional vibration mode of 60 Hz, using the well-documented construction techniques.

A supercomputer analysis of crashworthiness has shown that frontal NCAP and rear moving-barrier levels have been achieved at 17% higher than the required speed. The body has also satisfied 55 kph/50% offset AMS, European side impact and roof crush requirements. The 35 mph frontal NCAP showed a peak deceleration of 31g, considered satisfactory in that stiffer body sides are required to meet 50% AMS offset requirements. The offset AMS frontal impact deceleration showed a peak at 35g, considered a good result in relation to the severity of the test. The simulation was carried out at 1350 kg kerb weight, plus 113 kg luggage and 149 kg for two occupants. At the final count, high strength steel was used for 90% of the structure, ranging from 210 to 550 MPa yield, in gauges from 0.65 to 2 mm. Around 45% of the structure involved laser-welded tailored blanks, including the monoside panels, Fig 3, which ranged in gauge from 0.7 to 1.7 mm and was made up of steel elements having yield strengths from 210 to 350 MPa. Fig 4 shows the hydroformed side roof rail.

To the casual observer the elements of the complete body shell 'look' more structurally efficient but they still seem a far cry from a fully optimized shape for the steel 'backbone', as much of the material seems to be still used in relatively low stress areas where its function is as a 'cover' rather than a fully working structural element. Perhaps the next stage in weight reduction should be in designing a highly efficient steel monocoque, then moulding over plastic elements for the non-structural covering and closure shaping details, using the hybrid metal/plastics techniques discussed later in this chapter.

Vigorous development of light alloys

Phase 2 of Audi's aluminium alloy body construction programme was unveiled at a recent Motor Show in the form of the Al_2 concept car, Fig 5. The 3.76 metre long times 1.56 metre high car weighs just 750 kg in 1.2 litre engined form, some 250 kg less than a conventional steel body vehicle, it was argued. The number of cast nodes has been reduced compared with the Phase-one aluminium alloy structure of the Audi A8. Most of the nodes are now produced by butt-welding the extruded sections. High level seating is provided over a sandwich-construction floor. Use of internal high pressure reshaping techniques has reduced the number of shaping and cutting operations required

The 1997 International Magnesium Association Design Award for applications went to the die-cast instrument panel support beam, Fig 6, on the GM G-van. It is a one-piece design weighing, at 12.2 kg, 5.9 kg less than the welded steel tubular structure it replaced. Proponents of magnesium alloy are pointing out that as die-castings in the material solidify from the outside in, they enjoy a dense chilled skin together with a relatively course-grained interior. Skin thickness is said to be relatively constant regardless of total wall thickness so that reductions in section size can often be made when the material is used as a substitute. The AZ91D (9% aluminium, 1% zinc) alloy is now finding application in removable rear seats for minivans. In the seat systems shown in Fig 7, Delphi Automotive have achieved 60% weight reduction over conventional construction by adopting a cast magnesium cushion frame and an extruded

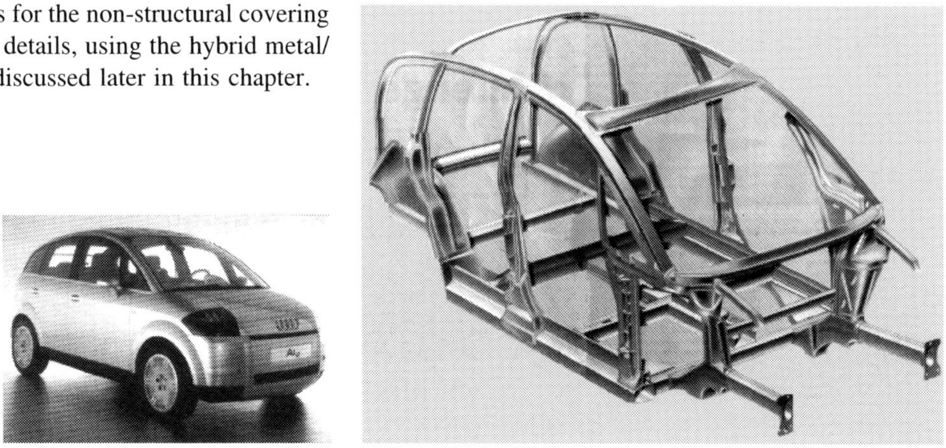

Fig 5 Al_2 concept car and structure

aluminium back frame. Other magnesium grades are being used in combination with engineering plastics for composite instrument panels. VW have also demonstrated a Polo door in pressure die-cast magnesium with a carbon fibre reinforced outer panel, which offers over 40% weight reduction over conventional steel construction.

Hybrid metal/plastic systems

Metal/plastic composites have been promoted again by Bayer in a recent presentation to body engineers[3]. A compelling argument is made for using simplified thin-wall structures in high strength metals which are stabilized by plastic composites which can also cut down on welding operations when combined as an in-mould assembly. The complexity of CV-cab and car bodies has often ruled against the effective use of lightweight metal systems on their own.

In the latest version of the technique, the processes of inserting, where metal parts such as bushes are included in a polymer moulding, and/or outserting, where various functions in plastic are moulded on to a metal baseplate, are taken a stage further. Cross-sectional distortion of thin-walled metal beams can be prevented by relatively small forces applied in the new process by the presence of moulded plastic supports in the form of x-pattern ribbing.

The deformation behaviour of various section profiles under bending and torsion is shown in Fig 8 a and b. Interconnecting points between plastic and metal are preformed in the metal part before it enters the plastics mould. Either corrosion-protected steel or aluminium alloy is the normal choice of metal with glass-fibre reinforced, impact modified, polyamide-6 (Bayer's Durethan BKV) being the plastic choice. Bayer say it is not always desirable to have the metal 'preform' in one piece, separate sections being joined by moulding resin around the prefabricated inter-

Fig 9: Front end structure in hybrid metal/plastic

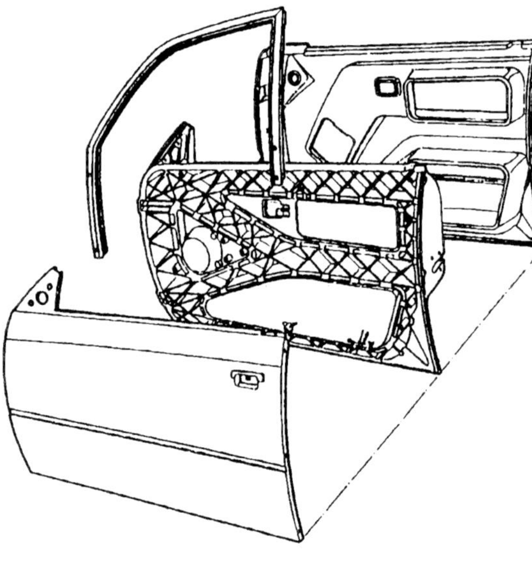

Fig 10: Metal/plastic hybrid door structure

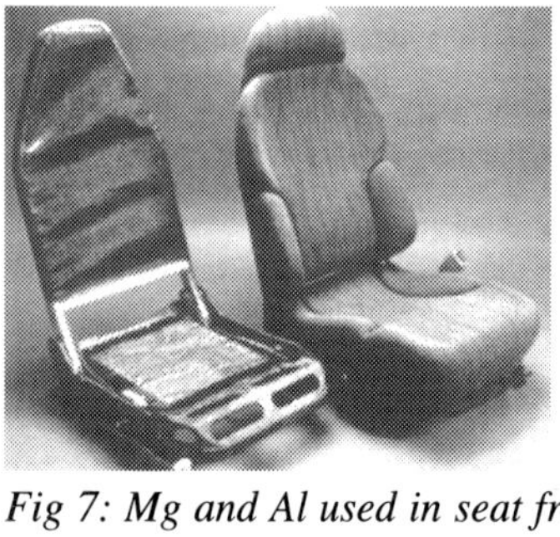

Fig 7: Mg and Al used in seat frame

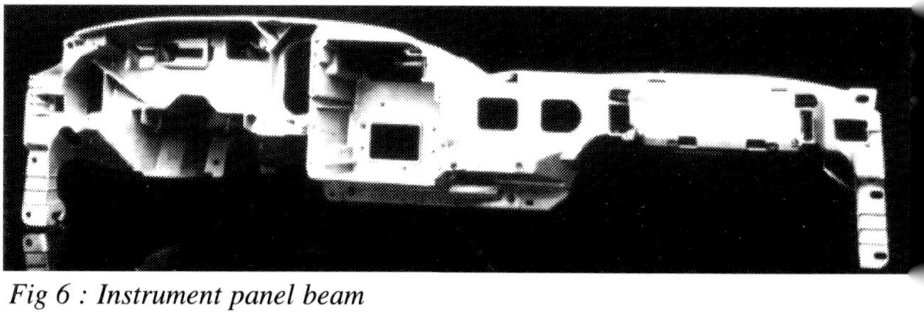

Fig 6 : Instrument panel beam

locking points or by means of clinching integrated into the mould. It is said that for recycling purposes, it takes only a few seconds to break a metal/plastic composite, using a hammer mill, and that the resin element has properties akin to the virgin material, on reuse.

Sample applications include instrument panel supports. In the cross-car beam co-developed by GM's Delphi a 40% reduction in weight was obtained against the replaced assembly, along with a 10% reduction in component and investment cost, for the same performance capability. Car front-ends with integrated bumper systems are further good candidates, Fig 9. A research project has also been carried out on car side doors having sufficient structural integrity to transmit impact forces from A- to B-post in the closed position. Fig 10 shows a sample door, requiring no further framework to support wing mirror, lock or other door 'furniture'. Seat frames with integral belt-anchorages have also been made

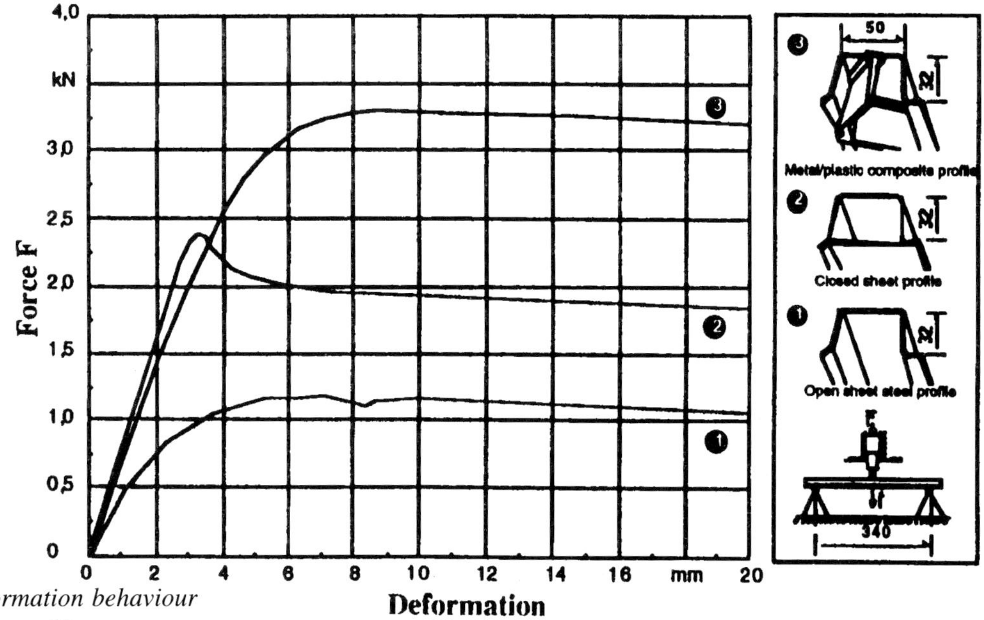

Fig 8a: Deformation behaviour

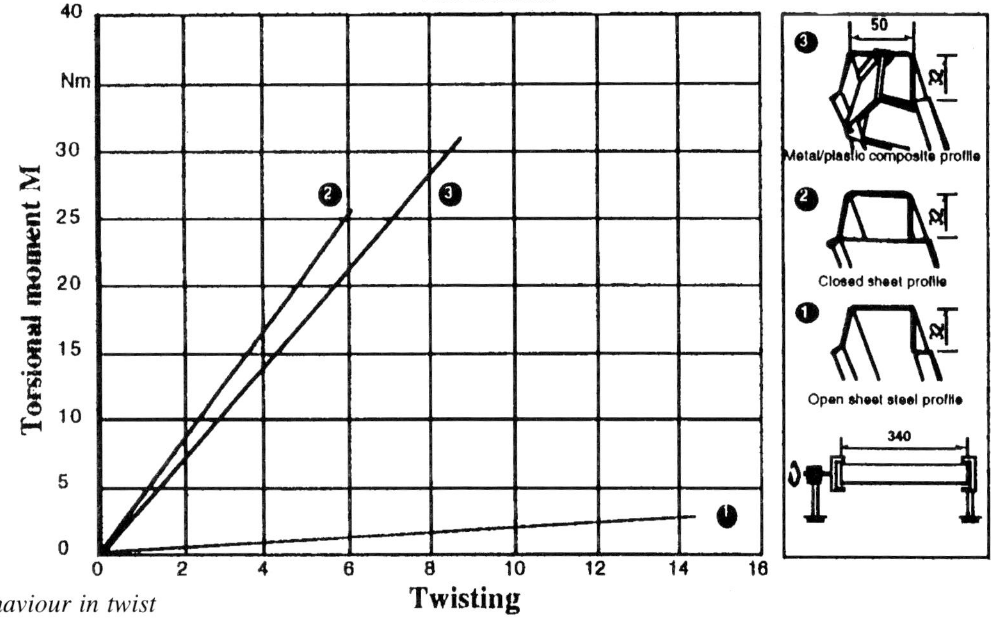

Fig 8b: Behaviour in twist

for the Mercedes-Benz Viano minibus in the process. The first volume car to incorporate the technology is the Audi A6. The front-end design was developed in association with the French ECIA company. The part is injection moulded in one piece and incorporates engine mountings, together with support for radiator and headlights.

Bayer have also been active in a joint-venture glazing project with GE Plastics. A $40 million development programme is under way on abrasion-resistant, coated polycarbonate vehicle windows as an alternative to glass systems. An abrasion-resistance equal to that of glass is targeted and high-volume production of windows is foreseen. Some 40% weight saving is forecast against equivalent glass systems and greater design freedom plus higher impact resistance is also claimed. Better resistance to forced entry, and reduced injuries in side-impact and roll-over accidents, have also been mooted.

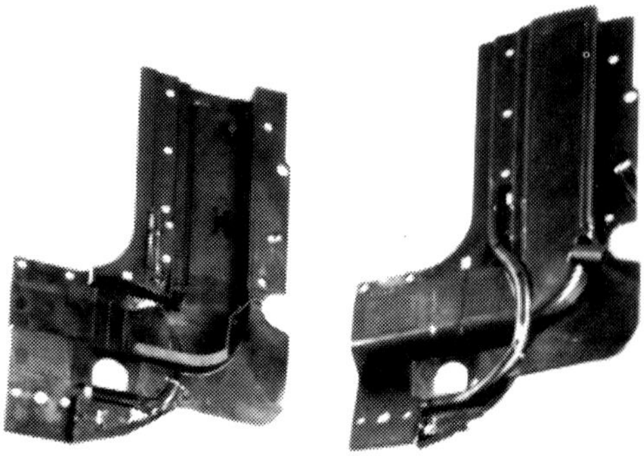

Fig 13: Audi A6 sun-roof

Structural design in polymer composites

Work reported from Delphi Interior and Lighting[4] has shown the possibility of constructing instrument panel crossbeam and display panel assemblies in injection-moulded plastics. The structural crossbeam provides both structural integrity, to the assembly, and an air passage for the primary air vents; it attaches to the A-pillars at each of its ends and operates as a simply supported beam. In one concept by the company, Fig 11, the beam supports the display panel but is not integral with it. In the second concept, Fig 12, the panel and beam form an integral structure. The advantage of this second concept is the allowance of further integration of components into the single total assembly.

Both concepts were built for retrofitting into a current production vehicle so that proper in-service evaluation could be obtained. In the CAD analysis

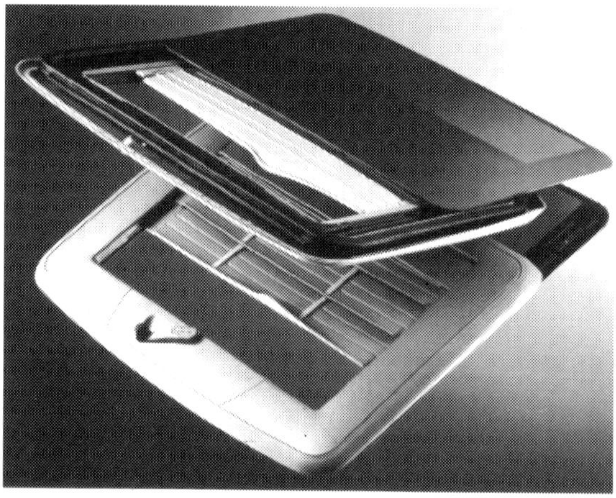

Fig 14: Sun-roof with venetian blind

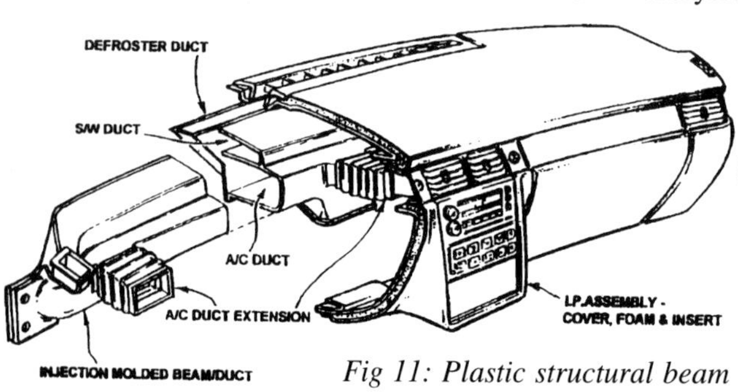

Fig 11: Plastic structural beam

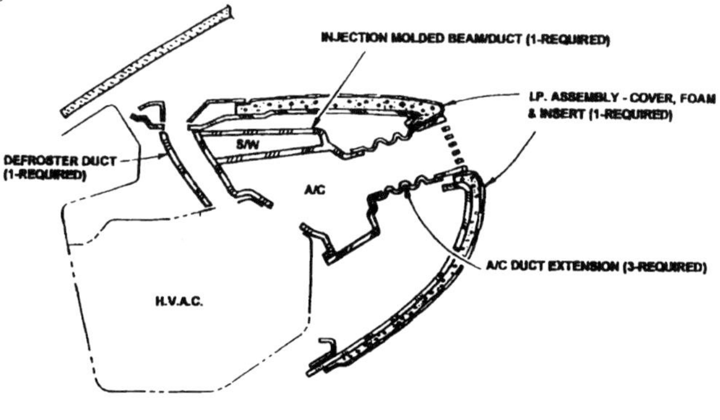

the initial objective was to develop a design for the plastic structures that would provide static stiffness equivalent to that of the baseline steel structure. The first natural frequency of the beam was also matched to the steel baseline. Both objectives were met without resort to wall thicknesses that would have compromised conventional injection-mould cycle times.

The next requirement was to meet FMVSS impact energy absorption levels resulting from input loads to the knee bolsters supported by the system in a 30 mph crash. It was found that materials with strain rates as low as 2% could absorb this energy without overload to the occupant's femur. Finally the structure had to meet these performance levels under specified temperature and vibrational environments, necessitating an evaluation of creep and fatigue capability. Temperature range of –30 to +190 F and fatigue criteria were based on subjection to service loads measured under real road conditions.

Retention of properties USCAR/EWCAP Class 3
Temperature/Humidity Testing

	Standard 15% GR PBT	Crastin• HR5015F	Standard 30% GR PBT	Crastin• HR5030F	Competitive HR PBT (30% Glass)
Tensile strength, Mpa					
Initial	109	74	145	99	114
After 40 cycles*	57	75	90	104	106
Retention, %	52	101	62	105	93
Elongation, %					
Initial	3.0	3.5	2.8	3.3	3.3
After 40 cycles*	1.5	2.9	1.3	2.8	2.2
Retention, %	50	83	50	79	67

* Each cycle consists of 6 hr. at 90°C/95% Rh followed by 2 Hr. at 125°C with humidity vented

Recycled PET, and prime PBT, for sun-roof parts

Following the recent interest shown in Chrysler's plan for using low-cost, glass-reinforced polyethylene terephthalate (PET) for its China car body construction, comes an announcement from DuPont that it is supplying recycled PET for corner mouldings of the Webasto sun-roof of the Audi A6, Fig 13. The mouldings hold the sliding roof frame's aluminium rails together and incorporate guides for the cable that moves the roof and channels off rain-water. The material has withstood 30,000 open–close cycles in trials of the sun-roof. In another sun-roof application, DuPont's polybutylene terephthalate (PBT) resin by Westmont Technik of Dusseldorf is used in a new design of sliding roof incorporating a built-in venetian blind, Fig 14. The PBT slats of the blind can be adjusted to deflect cooling air into the vehicle in hot weather. The company has also introduced a PBT grade, Crastin 5000, for electrical connectors and parts. This has improved hydrolysis resistance for high temperature and high humidity operation. The first variants produced, for connectors, have 30% glass reinforcement and properties are seen in the table, *left*.

Further details have also been released on the Chrysler China car which is now called the Composite Concept Vehicle. Its shell structure is made up of four major elements, Fig 15, to fit over a steel chassis. Target cost of the PET-based GRP is less than £2 per kg, compared with £6.50 – £8.50 for SMC and SRIM.

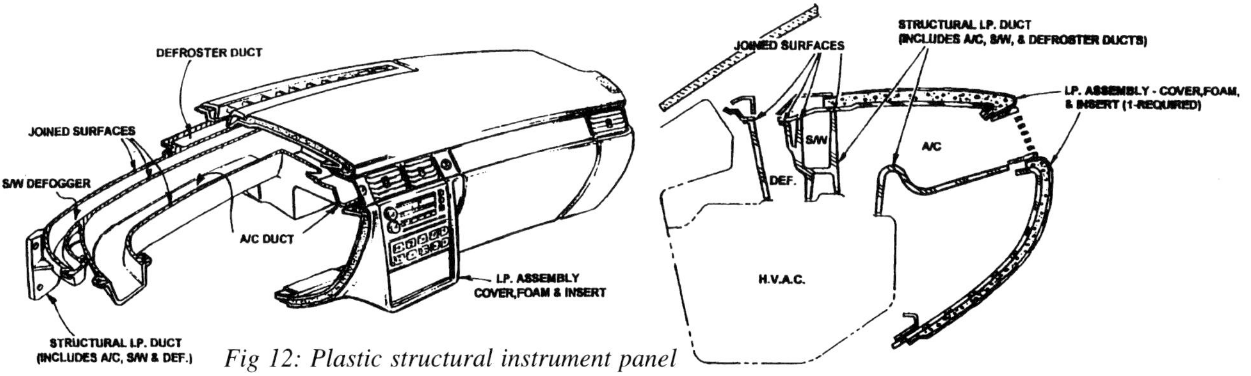

Fig 12: Plastic structural instrument panel

When the four shell elements are bonded together enough torsional stiffness is available without the need for further reinforcement. Ashland Pliogrip fast-cure adhesive is used and cycle time for making the four main mouldings is three minutes, seven minutes faster than for equivalent sized SRIM mouldings. The low cycle time has been achieved by a sequence of gate openings and gas injections into the mould to help guide the process, which was devised using computer-aided mould-flow techniques. An 8500 tonne machine injects the PET resins into the 145 tonne moulds, which each measure a massive 4.5 X 2.5 X 2 metres.

Structural–coloured fibre has body trim potential

Nissan has announced the development of what is claimed to be the world's first structural-coloured fibre, in a joint venture with Teijin Limited and Tanaka Kikinzoku KK, Fig 16. The colour of the fibre is produced by its structure and does not involve the use of dyes. The principles of this technology are based on research into biometrics, an engineering practice that looks at prominent functions of living things and applies them in practical applications. One of the findings of this research is the principle behind the iridescence of the wings of Morpho butterflies native to South America. Research has been conducted to apply this knowledge to the production of coloured fibres for interior and exterior trim materials, including seat fabric. The structural-coloured fibre has been developed by applying the theory of multilayered thin-film interference to a cross section of the fibre. This is the principle behind the iridescent scales of Morpho butterfly wings. The fibre displays a high metallic sheen and a clear colour shade under exposure to different types of light, including natural, fluorescent and incandescent. The colour properties and tint also vary depending on the light source used and the viewing angle. As the fibre does not use any dyes, the production of the fibre reduces environmental pollution due to waste dye solutions. It also prevents any worries about skin rashes or other allergic reactions to dyes.

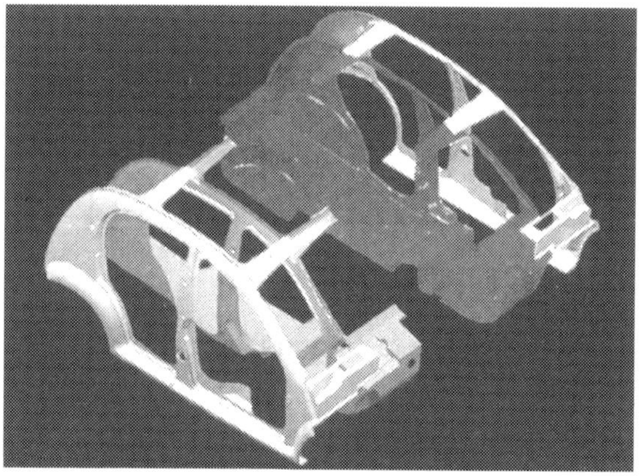

Fig 15: Composite Concept Vehicle shell

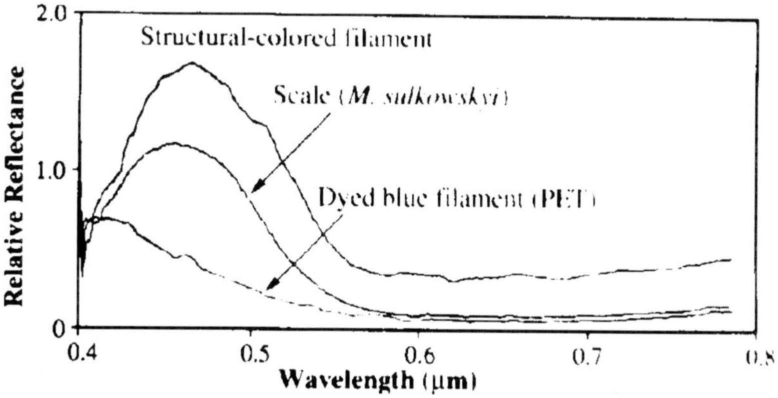

Fig 16: a: right*, Reflective spectra of some samples; b,* opposite*, the technology explained*

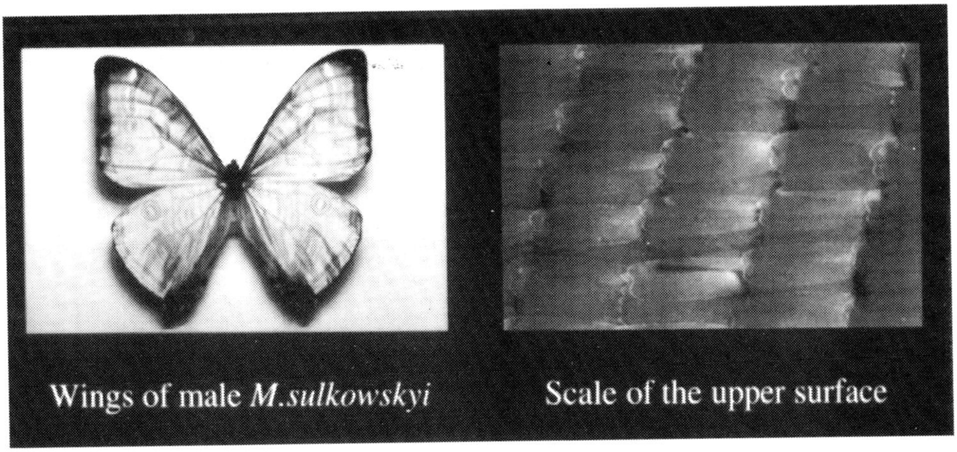
Wings of male *M.sulkowskyi*
Scale of the upper surface

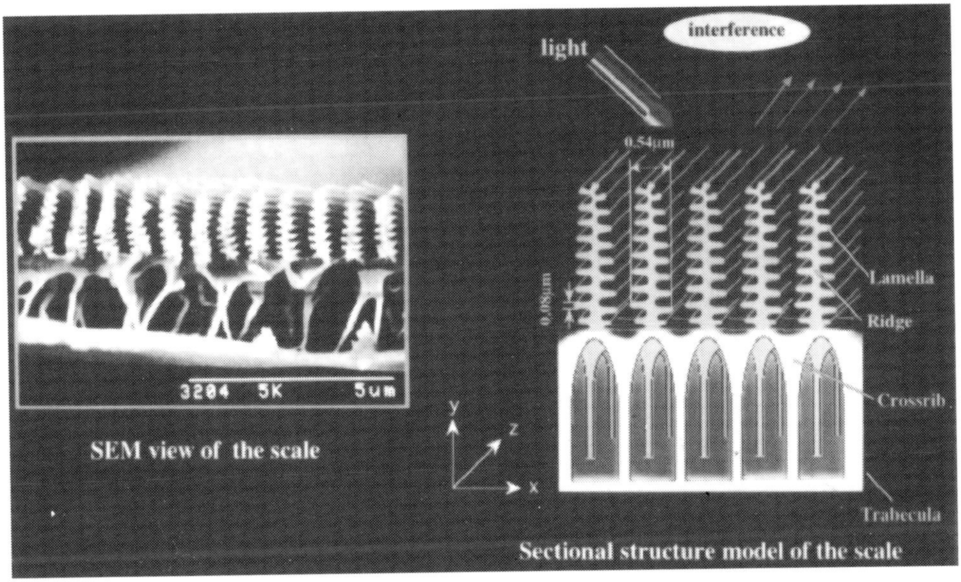
interference
light
0.54μm
Lamella
Ridge
Crossrib
Trabecula
3204 5K 5um
SEM view of the scale
y
z
x
Sectional structure model of the scale

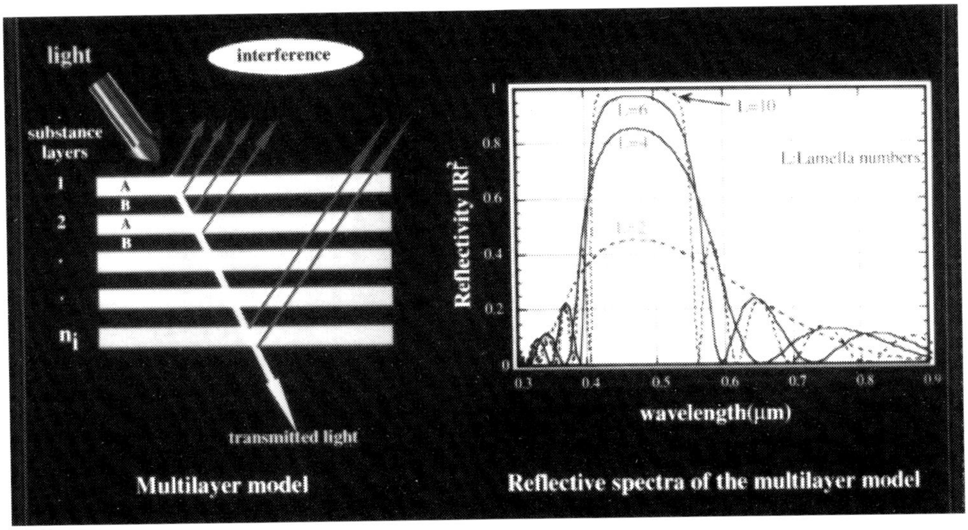
light
interference
substance layers
1
A
B
2
A
B
n_i
transmitted light
Multilayer model
L=10
L=6
L=4
L=2
L:Lamella numbers
Reflectivity $|R|^2$
wavelength(μm)
Reflective spectra of the multilayer model

Material property charts and performance indices

A rigorous method for evaluating the choice of materials for new designs has been made by a researcher[5] of the Cambridge University Engineering Design Centre which avoids the 'hunch' or 'guesstimating' approach so common in making these decisions.

The author points out that it is seldom that the performance of a component depends on just one property of its constituent material. So by plotting one property against another (stiffness and density, say, for lightweight components), and mapping out the fields of materials in each material class, rapid assessment of material suitability is possible.

Fig 17 shows how, on a log scale plot of elastic modulus and density, the data for a particular class of materials clusters together in envelopes. The sloping lines show how other derived properties, based on the two basic ones, can also be plotted on the graph. The elongated envelopes refer to material classes in which there is structure-sensitivity, in that properties can be altered significantly by such techniques as micro-alloying and heat treatment. While the modulus of a solid is a relatively well defined quantity, the author explains, strength is not and the envelopes take up different shapes, Fig 18. The shapes alter again when fracture toughness and other properties are plotted against density and the author provides explanations for these in the book, based on the behaviour of the material's molecular structure.

In general the charts show the range of any given property accessible to the designer and identify the material class associated with segments of that range.

Logical design involves asking such questions as 'what does the component do; what is to be maximised and minimised; what non-negotiable conditions must be met and what negotiable and desirable conditions . . .', defining function, objective and constraints. Property limits can be derived from these together with indices that will optimize material selection. Performance indices are groupings of the material properties which can maximize some performance aspect of the component, ratio of modulus to density being the classic one of specific stiffness, which would be a key consideration in the design of a light, stiff tie-rod. Many such indices exist each characterizing a particular combination of function, objective and constraint, as in the beam case, Fig 19.

In this 'index and chart' material selection procedure, the index isolates the combination of material properties and shape information that maximiszes performance prior to using the chart for selecting the appropriate material as described above. Fig 20 shows how the index derived by isolating the key parameters from the equations describes the objective function after the constraints have eliminated the free variables.

Design of a mechanical component, Ashby asserts, is specified by three groups of variables: the functional requirements F (need to carry loads, transmit heat, etc.), geometric specifications G and some combination of properties p so that maximized performance

$$P = f(F,G,p)$$

When these three groups are separable such that

$$P = f_1(F), f_2(G), f_3(p),$$

as they usually are, the optimum choice of material can then become independent of many of the design details and a merit index can be much more simply defined. Thus the steps in Fig 20 involve identifying primary function, writing down the equation for the objective (could be the energy stored per unit volume in a spring example) then identifying the constraints that limit the optimization by this last process in terms of a specific value of stiffness, say. The free variables can then be eliminated in leading to the index value (the objective function).

In the case of a beam, for a simply supported length L, under transverse load F, there might be a constraint on the stiffness S such that it must not deflect more than δ at mid-span. Here

$$S = F/\delta >/= CEI/L^3$$

where E is modulus and I section second moment of area, C being a constant depending on load distribution. For a square section beam, width b, I is $A^2/12$ where A is section area ,which can be reduced in order to lessen weight, but only until the stiffness constraint is met. Eliminating A in the objective function gives the mass

$$m \geq (C12S)^{1/2}(l^{5/2})(\rho/E^{1/2})$$

and the best materials for a light, stiff beam are those with large values of the material index

$$M = E^{1/2}/\rho$$

where ρ is density.

Because section shape is critical to bending stiffness a shape factor is defined as

$$\phi = 4\pi I/A^2$$

which is dimensionless and can take values as high as 100 for very efficient sections.

Cross-substitution of these equations can be carried out to eliminate the free variables and a

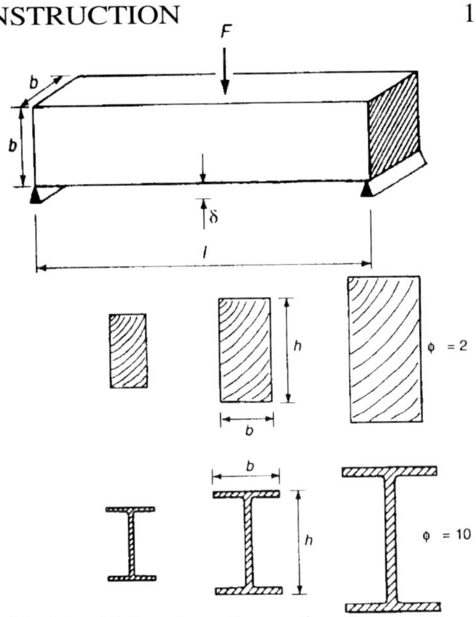

Fig 19: Identifying the primary function

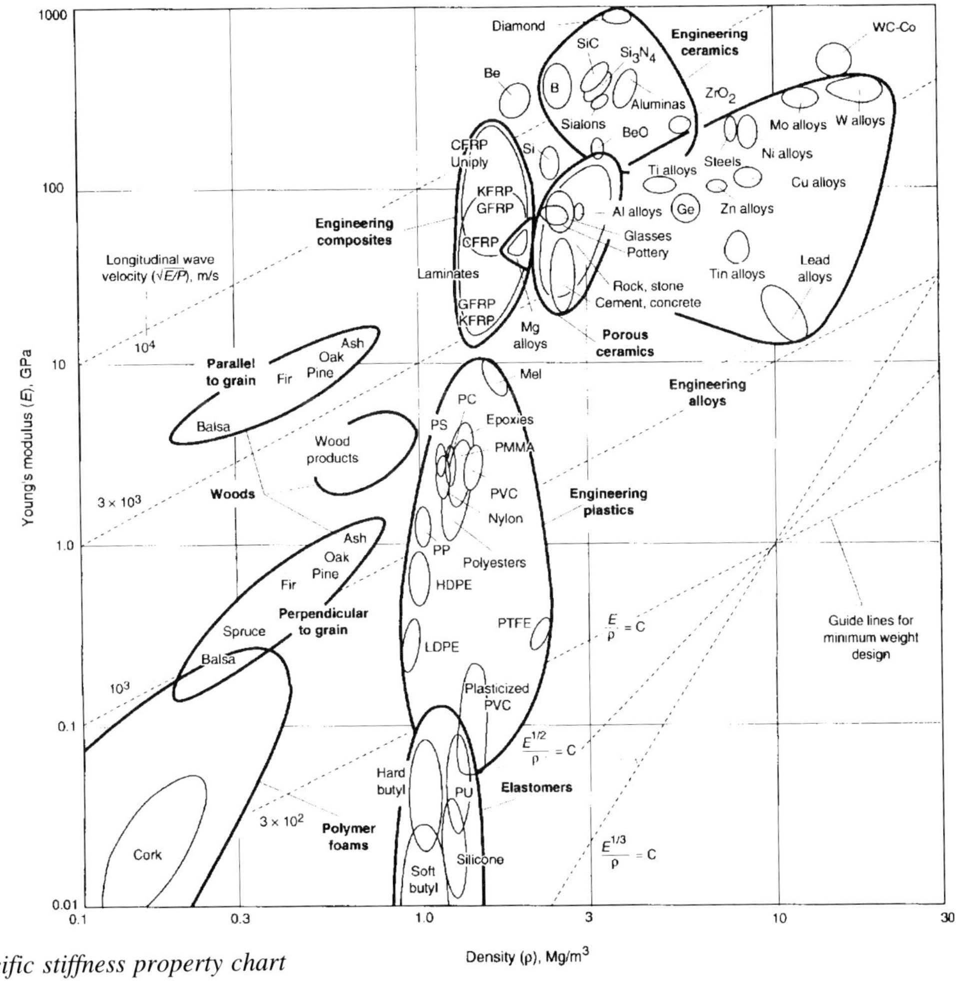

Fig 17: Specific stiffness property chart

modified performance index would include a shape-factor term.

The author also includes performance indices for panels, with in-plane compression allowing for buckling failure, among the many case studies in the book, as well as those for vibration limited designs.

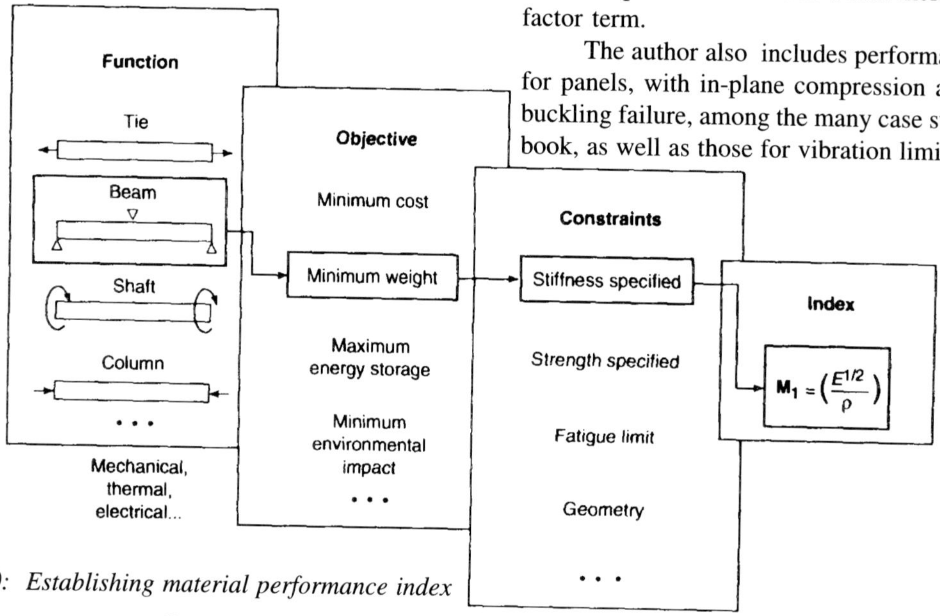

Fig 20: Establishing material performance index

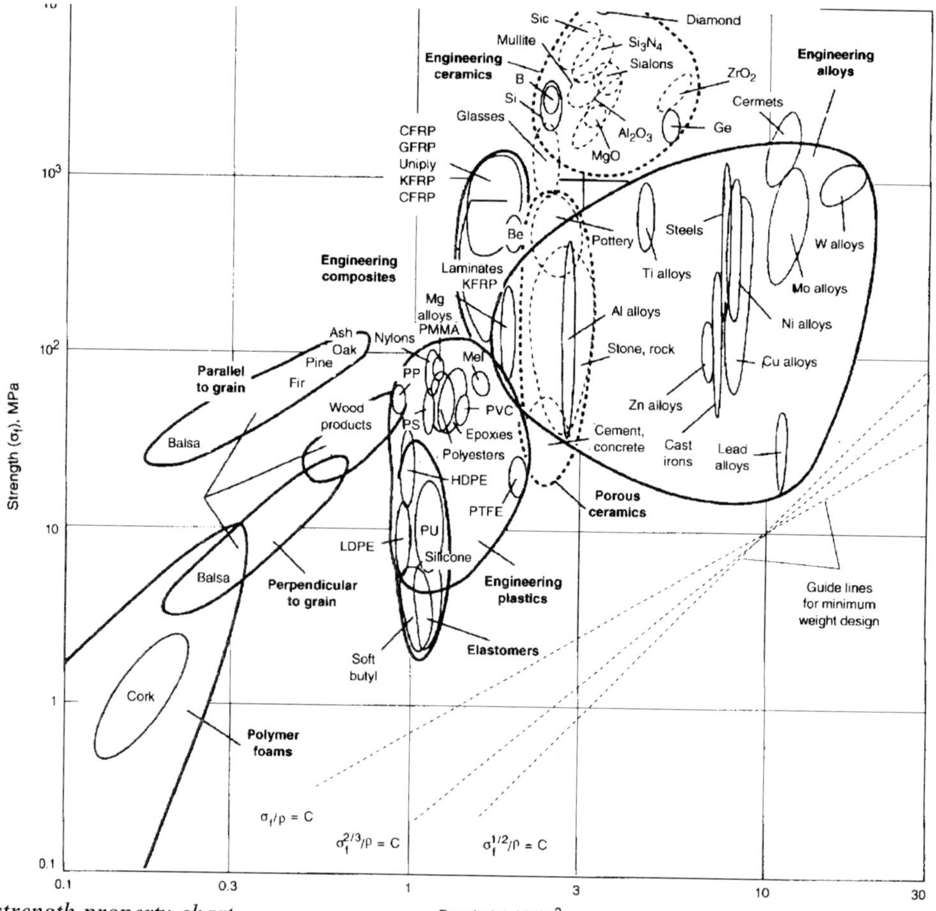

Fig 18: Specific strength property chart

Design for self-pierce riveting

More widespread use of aluminium alloys, and lightweight coated steels, in automotive body construction is leading to alternatives to the old certainty of spot-welding. Textron's Fastriv system is proving a reliable substitute, a leak-proof system with fully automated installation equipment, and claimed greater dynamic strength than spot-welding.

The Fastriv self-piercing fastening system, Fig 21, can join two or three metal sheets with combined thickness from 1.5 to 8.5 mm. In the placing sequence of Fig 22, a tubular rivet is driven by a hydraulically operated press tool into the materials to be joined; first the rivet pierces the top sheets then it flares radially into the lower sheet against the die, without breaking through, and forms a mechanical interlock, the assembly process being complete within 1–2 seconds.

A single Fastriv joint is said to be close to the shear strength of an equivalent spot-weld and stronger in peal strength, without the disadvantage of a heat-affected zone; the suppliers say at least 30% of spot-welding applications could be displaced by its use. There must, however, be access for tooling on both sides of the sheets and sheet thickness/hardness must fall within a given range.

Optimized Fastriv joint performance is obtained when the sheet material with the higher elongation is placed on the die side, when joining sheets of dissimilar material. For joining sheets of dissimilar thickness, the thinner sheet should be placed on the punch side and a large-head rivet should be used. Since joint elements of self-pierce fastening are not symmetrical on both sides, joint arrangements need special attention and suggestions are given in Fig 23.

Suggested automotive application areas are shown in Fig 24, the company recommending combination with adhesive bonding for load-bearing carrier beams. Steel panels can be joined to aluminium ones, where necessary, and the pre-painted panels can also be successfully joined by the process. A number of finishes are available on the rivets to ensure against corrosion. The standard finish used is electroplated bright zinc but Almac mechanically applied coating has higher corrosion resistance than zinc. Kalgard has an organic binder containing aluminium; Deltaseal has an organic microlayer top-coat with PTFE; finally, Dacromet has an organic binder containing both zinc and other protective solids.

Similar to that for spot-welding, installation equipment can be floor-mounted or portable, suspension-mounted, and have manual or robotic manipulation. Fasteners can be either tape-fed from reels or blow-fed from a vibrating bowl, through plastic tubing. Fig 25 shows proportions of the riveting module, from which an idea of joint access can be obtained. Various rivet head forms are available as well as the standard oval head, including flat counter-

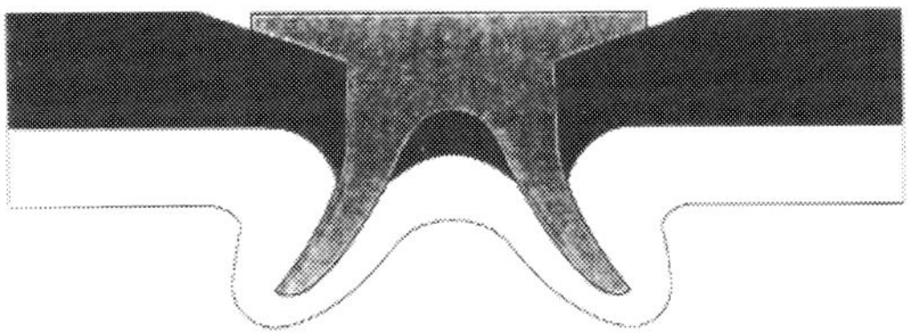

Fig 21: Fastriv configuration

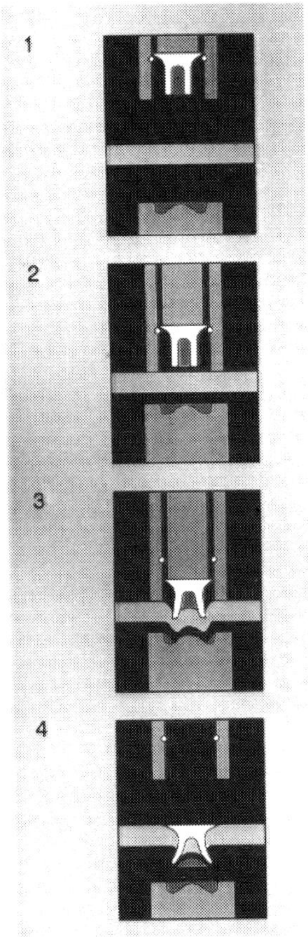

Fig 22: Installation sequence

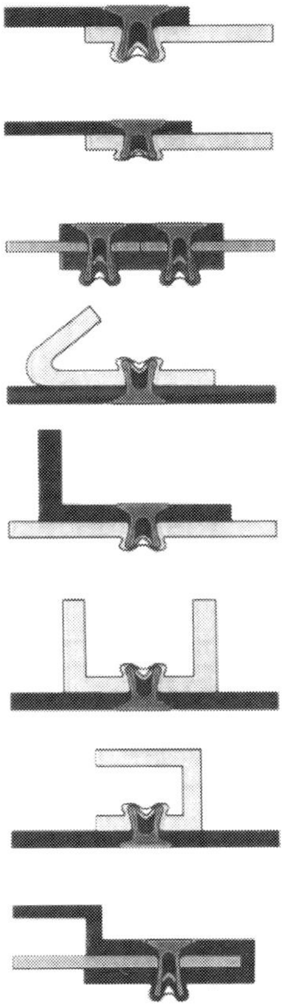

Fig23: Suggested rivet orientations

sunk and a 'tinman' head which is a flat head which stands proud for occasions when a visual feature is required to be made. Head diameters can range from 5.38 to 9.78 mm for rivet diameters from 3.10 to 4.78 mm. For the 5 mm (nominal) rivet size, fatigue performance is as shown in Fig 26.

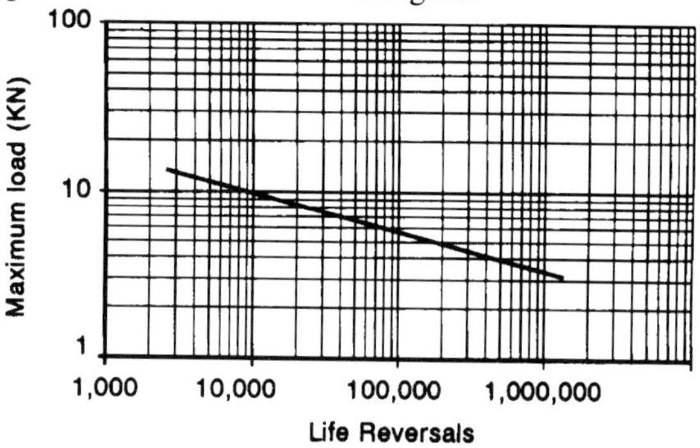

Source: Rover Group Ltd

Fig 24: Automotive body applications (below)
Fig 25: Riveting module sizes (right)
Fig 26: Fatigue performance (above)

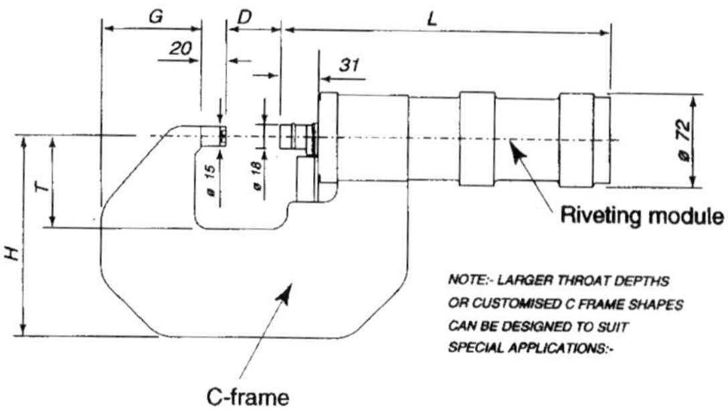

STANDARD RIVETING MODULE CONFIGURATION		
STANDARD DAYLIGHT SIZES "D" (mm)	RIVETING MODULE LENGTH "L" (mm)	APPROXIMATE WEIGHT (KG)
25	302	6.25
40	317	6.5
55	332	6.75

* PREFERED

STANDARD "C" FRAME SIZES (FOR 40mm DAYLIGHT)			
THROAT DEPTH "T" (mm)	ARM DEPTH "G" (mm)	FRAME DEPTH "H" (mm)	APPROXIMATE WEIGHT (KG)
50	50	120	5.5
100	71	190	8.5
150	81	280	15
200	96	360	20

NOTE:- FOR OTHER RIVETING MODULE SIZES PROPORTIONS WILL CHANGE.

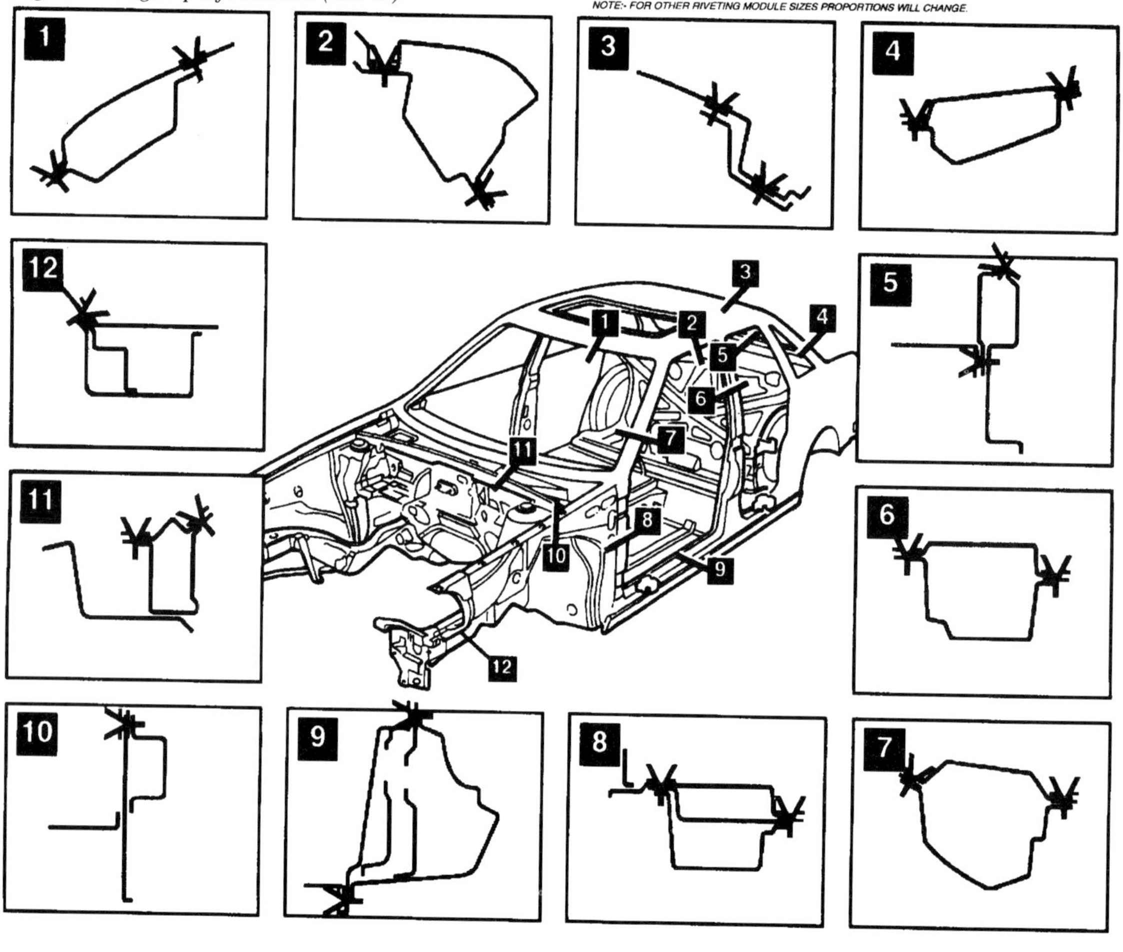

Chapter 2: Structure and safety

Vehicle structural design innovation continues to be driven by the impetus for improved passenger protection and light-weighting for lower fuel consumption and reduced CO_2 emission. With the perfection of controlled collapse front-ends the next stage in development is an analysis of footwell deformation so that controlled-collapse can be extended to that region. For further integrity of the occupant survival shell attention is also being paid to pillar/rail joint stiffness in reducing the likelihood of body shell collapse. Techniques of structural analysis are also being used for the prediction of noise level within the body shell. Attention is also finally turning from the protection of the vehicle occupant to that of the pedestrian road-user and realistic pedestrian models are now being produced for authentic laboratory crash analysis. In vehicle packaging, generally, is beginning to focus on the ageing and infirm section of the population and occupant restraint systems are being designed to be adaptive in their performance.

Understanding footwell deformation in vehicle impact

Recently published work by researchers at the US Automobile Safety Laboratory, and its collaborators[1], is suggesting a way of better understanding the collapse mechanism for a passenger-car footwell during frontal and front/side impacts. Hitherto, the authors maintain, the study of structural intrusion in this region has been based on toe-board mounted accelerometers set up to measure in the direction of the vehicles longitudinal axis. Since the instrument itself can easily be disoriented from its measurement axis, during the intrusion process, reliable dynamic deformation data has been difficult to forecast. In particular, a three-dimensional characterization of the collapse and buckling of the thin metal structures is difficult to obtain, except perhaps using X-ray speed-photography techniques for a qualitative appraisal. Obtaining orthogonal frames of reference is considered difficult with any conventional cine-photo technique.

Post-crash measurements can only provide intrusion data for the last instant of impact and no elastic effects occurring during or after the crash can be evaluated — nor can a progressive sequence of toe-pan translations and rotations be obtained. Hence the stimulus for these researchers to use a sled system for obtaining a toe-pan deformation pulse, akin to that obtained by sleds for vehicle deceleration as a whole. To design an intrusion system using a laboratory sled, work commences with obtaining a simple translational intrusion profile parallel to the ground, Fig 1. The intrusion simulator has been designed to allow rotational components and asymmetrical load-

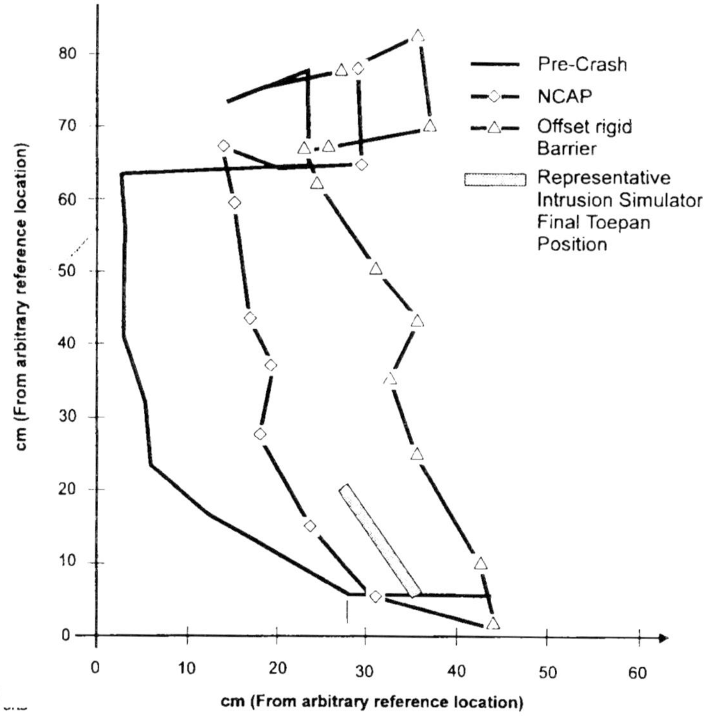

Fig 1: Post-crash profile of toe-pan intrusion

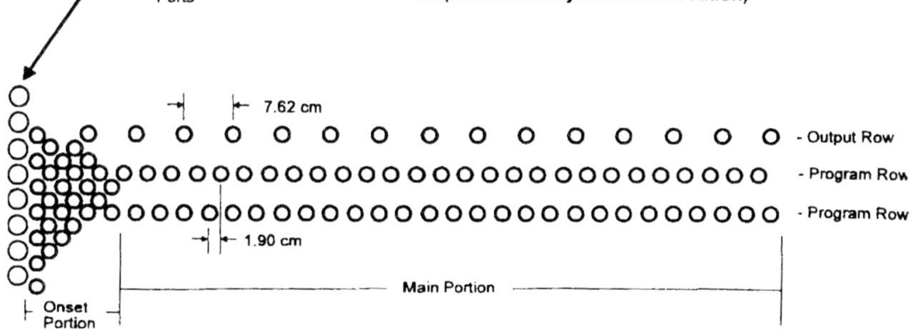

Fig 2: Hydraulic decelerator: projected view of programmed row of holes

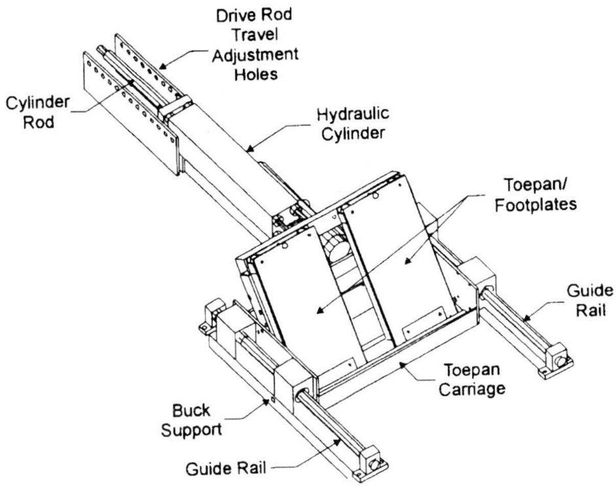

Fig 3: Intrusion simulation mechanism

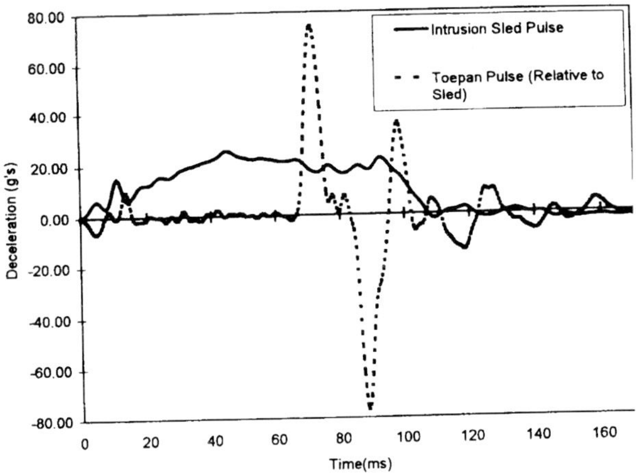

Fig 4: Sled deceleration vs toe-pan acceleration in a sled frame

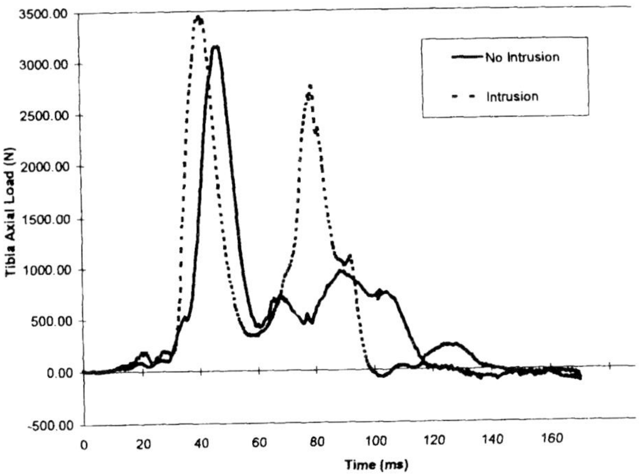

Fig 5: Left distal tibia load for intrusion and no-intrusion sled tests

ing effects, at a later stage.

The toe-pan produces a typical 20 kN force applied to both legs, about twice that of the axial tibial breaking strength. The panel is generally capable of 25 cm maximum displacement during a crash event and it has been found by accident-studies that 85.6% of below-knee injuries are sustained with less than 14 cm intrusion. The sled therefore moves the toe-panel in a progammed displacement time profile, with these maxima, by integrating the actuation of the panel components with the function of the sled's hydraulic decelerator. The sled and test buck are typically accelerated to a predetermined velocity and decelerated according tot he crash pulse. The decelerator is configured as a cylinder with an array of discrete orifices over its length, Fig 2 suggesting a flat-pattern view. At the start of an impact, hydraulic pressure is transmitted to both sides of the slave cylinder, holding the toe-pan stationary with respect to the buck. When the decelerator piston passes the control-side output port, pressure on that side drops and the toe-pan moves under the force of the hydraulic drive lines remaining in front of the sled piston. A constant force internal stop in the cylinders at the maximum mechanical stroke is used to determine maximum toe-pan travel according cylinder position. The intrusion pulse is controlled by variation of orifice sizes on the drive and control sides. Fig 3 shows the physical set-up, the simulator consisting of a toe-pan carriage with separate footrests for each foot. Toe-pan carriage travel is 25 cms and the hydraulic cylinder diameter is 6.4 cms. The rear mounting position is adjustable to limit toe-pan travel in increments of 2.5 cm. Controllable toe-pan intrusion parameters, during the impact event, include initiation, by location of a control line within the hydraulic decelerator; stroke, controlled by cylinder mount location and cylinder stop; and acceleration profile, depending on a combination of passive orifices in the drive lines forming an hydraulic accumulator so that dynamic effects of the sled pulse can be used to tailor the toe-pan pulse.

The researchers demonstrated the use of the intrusion simulator with sled tests carried out with zero to 7.9 cm intrusion, using the ALEX dummy developed by NHTSA. The test buck was configured as a mid-size car and based on a crash pulse measured by damped accelerometers. Sled velocity at initiation of the crash pulse was fixed at 58.2 km/hand full intrusion stroke is selected to be at 71 ms into the sled impact. In Fig 4 of a representative sled decel pulse and toe-pan intrusion pulse, intrusion decel relative to the test sled is approximately sinusoidal, with 80 g peak, or 100 g relative to the ground. Net effect of sled and toe-pan decels is movement of the toe-pan into the occupant compartment, providing a potentially realistic translation intrusion pulse most relevant for lower extremity injury. Left tibia axial compressions are shown in Fig 5 for both representative and no intrusion cases, the significant effect of toe-pan intrusion being demonstrated after about 70 ms. The second peak of the intrusion means tribial axial compression is more than doubled from the no-intrusion run.

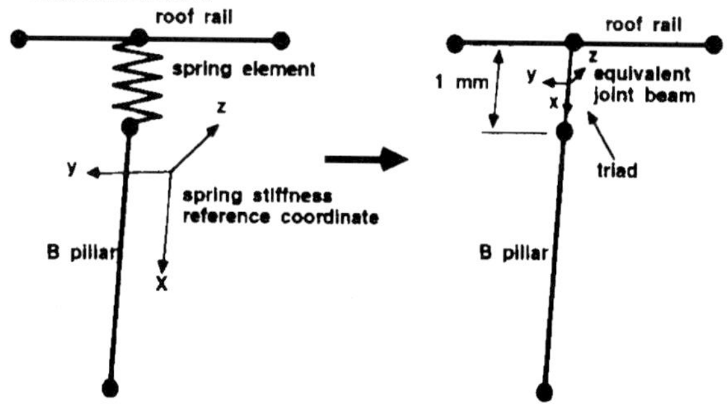

Fig 6: Traditional and proposed modelling technique

Pillar to rail joint stiffness

Low frequency vibration characteristics of a vehicle body structure are considerably influenced by the compliance of joints between the main pillars and rails, according to researchers at Kookmin University[2]. While these effects have been examined at the design stage by using torsional spring elements to simulate the joint stiffnesses, apart from the inaccuracy of the assumption it has been difficult to analyse the whole structure using the NASTRAN optimization routine because these mass-less spring elements cannot be represented. Because the rotational stiffness of the joint structure has a major influence, the authors have proposed a new modelling technique for the joint sub-structure involving an equivalent beam element. Fig 6 shows the traditional and proposed modelling techniques applied to a B-pillar to roof joint. The joint stiffness is computed using familiar finite element methods, Fig 7. Unit moments are applied at the tip of the pillar, with both ends of the roof-rail fixed. FE static analysis gives the deformed rotation angle from which rotational stiffness can be derived.

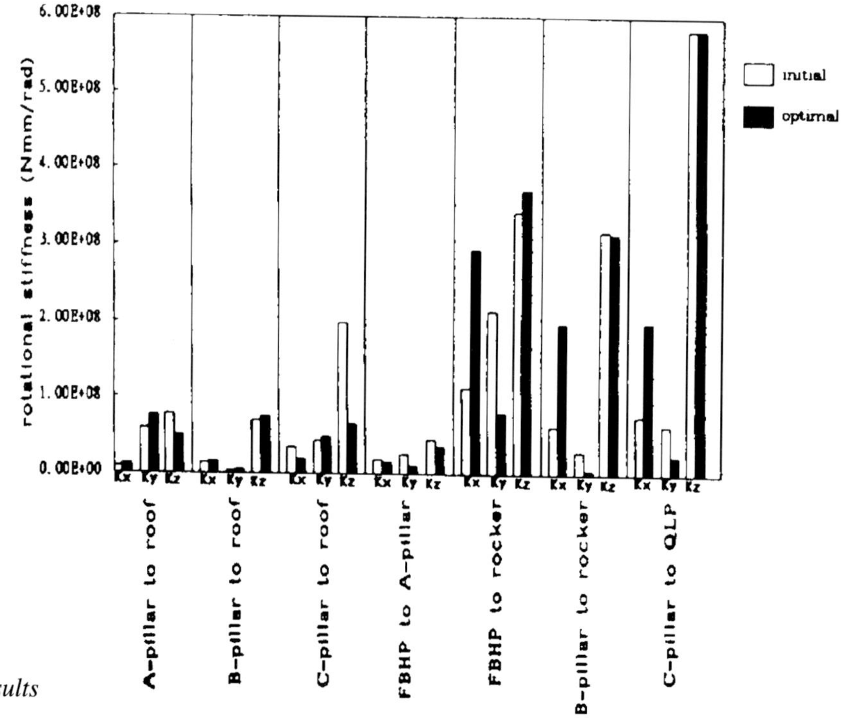

Fig 9: Joint stiffness results

In the subsequent MSC/NASTRAN optimization run, if the section moments of inertia are chosen as design variables, their optimal values can be obtained to meet the vehicle's frequency requirements. The table in Fig 8 shows a design change guideline to obtain optimum joint stiffness in this respect. The joint stiffnesses in the X and Y planes, plus the rotational stiffness about the Z-axis of the section centroid can be obtained. When these results are plotted out as a bar-graph as in Fig 9, the authors explain that it is relatively simple to which stiffnesses need increasing and which could perhaps be reduced. The example in the figure implies that the Y-direction rotational stiffness of the B-pillar joint needs to be increased while the other directional stiffnesses may be decreased. Detail study then showed the thickness and shape of the reinforcement panel needed to be altered in this case, establishing thickness and nodal co-ordinates of the panel as design parameters.

The reinforcement may be optimally positioned anywhere between the inner and outer panels of the joint, to satisfy the stiffness requirements, and by setting the objective function of the optimization run for minimizing mass, the constraints can be evaluated for achieving the required rotational stiffnesses, Fig 10.

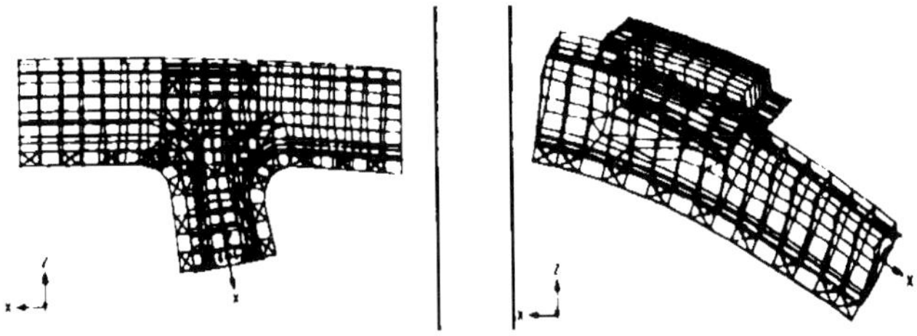

Fig 7: FE model of joint

unit : %

	I_Y	I_Z	J
A-pillar to roof	+25.6	-34.9	+39.9
B-pillar to roof	+30.0	+7.0	+7.0
C-pillar to roof	+16.9	-66.9	-46.0
FBHP to A-pillar	-63.6	-20.8	-16.1
FBHP to rocker	-0.2	+8.9	-3.1
B-pillar to rocker	+20.6	-1.1	-53.4
C-pillar to QLP	-63.8	+0.1	-0.6

Fig 8: Design change guideline

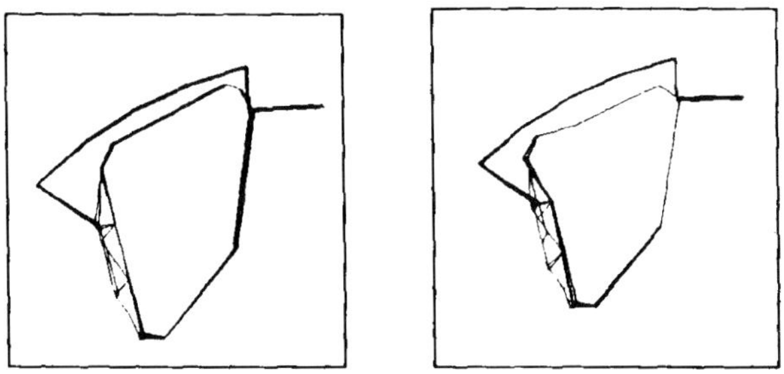

(a) initial **(b) optimal**

Stiffness	K_x	+7.48
	K_y	+ 34.15
	K_z	+ 7.18
design variables	thickness	+33.33
	shape	refer to Fig. 8

Fig 10: Joint configuration

Structure analysis for interior noise

A researcher at Rover[3] insists that refinement at his company is now designed-in at the concept phase of new products. In a recent paper he differentiated between component and system modelling for NVH and pointed out the difficulties of examining high frequency behaviour, above 70 Hz. At these frequency levels, the author explains that vehicle problems are dominated by flexible modes of the components direct-connected to the body and care is needed in interrogating information from the standpoints of non-linearities arising, for example, in rubber anchorages and structural complexity where crude beam models are typically 25-30% inaccurate.

Suppliers are now requested to carry out detailed modelling for analysis purposes with predictions of natural frequency for something like a subframe being to 10 % inaccuracy. Fig 11 compares a beam-element model with an accurate supplier model for a subframe and Fig 12 shows the driving point modal response of these two models. While it can be seen that the mass is predicted well in both model cases (amplitude at low frequency before resonance is proportional to the mass of the component), the bem element model only recognises two main resonances and misses the smaller ones between 300 and 350 Hz because these modes are out of the plane of excitation they are poorly excited.

Above 300 Hz, the author argues, FE modelling is unsuitable because the non-linear behaviour of most systems gives poor levels of prediction and experimental test methods have to be used. Currently, statistical energy analysis (SEA) techniques are being evaluated in conjunction with ISVR and a number of trim suppliers that eventually will allow accurate prediction of vehicle interior noise from 300 Hz upwards. Rigid lumped-mass representations of bodies are giving way to models which allow the effect of flexible modes. However, usually structural models for crashworthiness analysis are unsuitable as they do not incorporate closures and trim. Since the BMW acquisition of Rover the ability to add trim items to the models has been developed. Large trim areas such as head-linings are modelled as non-structural masses, additional damping effect being added to the elements representing the adjacent panel. For other items such as facia panels, which add stiffness to the structure, tuned spring/damper elements are added to the structural model.

The author suggests that a current 90,000 element model for a vehicle body will increase in terms of number of elements by 50% every time that a new model is introduced. He thinks model solution time will remain constant as machine speed is increasing at similar rate. Another development is studying effects of modifications to body structures by means of super-element whole-vehicle models which combine body and powertrain, Fig 13. The company also employs SYSNOISE software for acoustic modelling activities, specifically for problems with engine intake and exhaust systems. Vehicle interior acoustic modelling, Fig 14, has been used for some time to assess panel sensitivity and work is underway to couple the acoustic models to the full-vehicle models.

A case study in assessing the refinement ben-

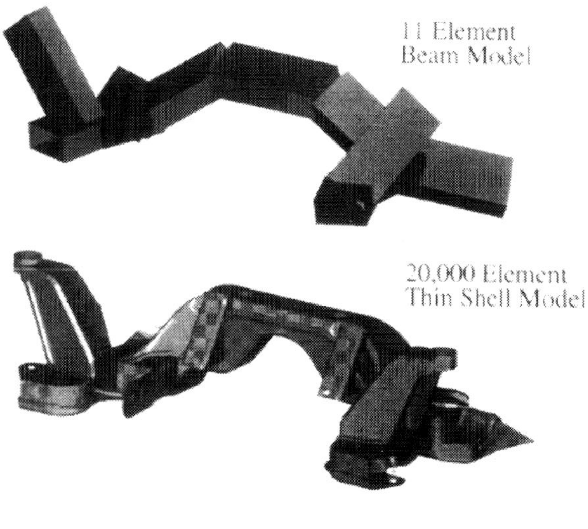

Fig 11: Beam model vs superelement

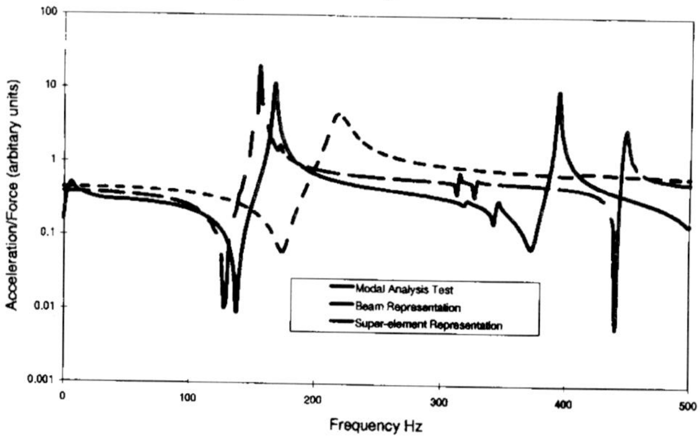

Fig 12: Correlation of FE models

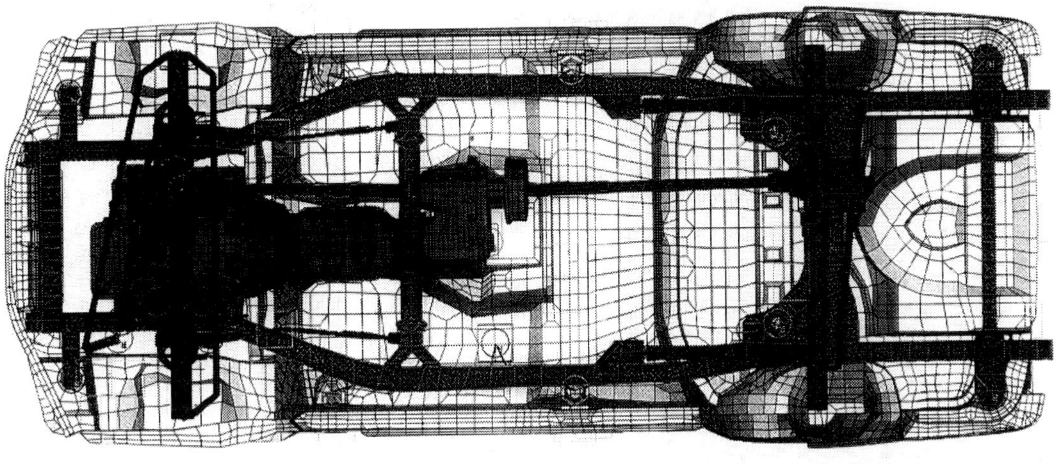

Fig 13: "Whole vehicle" model

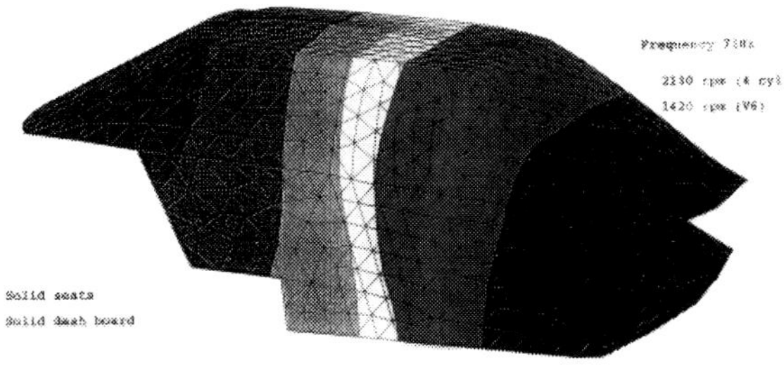

Fig 14: Acoustic model of vehicle cabin

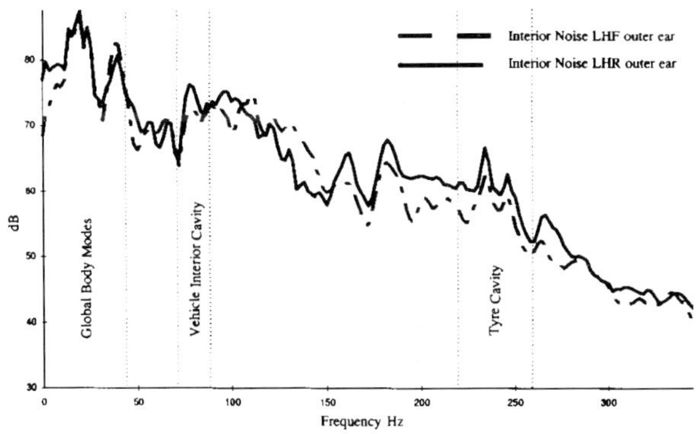

Fig 15: Road noise characteristics

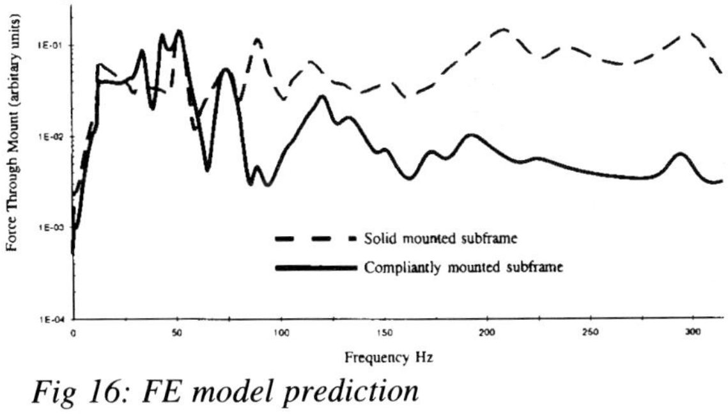

Fig 16: FE model prediction

efits of a compliantly mounted front subframe was given by the author. Fig 15 shows typical interior noise measurement for a vehicle driven at constant 50 kph along a surface of 20 mm cold-rolled chippings. At 25-45 Hz global modes of the vehicle body structure are evident; from 75-130 Hz excitation of the 1st and 2nd longitudinal cavity modes of the vehicle occurs and main noise source is tyre tread blocks impacting and releasing from the road surface; at 220-260 Hz strong excitation is generated at the wheel hub by an acoustic resonance of the air within the tyre, seen as two modes describing waves travelling forwards and backwards within the tyre.

The proposed subframe involved a front suspension system with body mounted MacPherson strut with an L-shaped lower arm mounted to the subframe at two locations; steering rack and anti-roll bar are similarly subframe mounted. The lower tie for the engine mounting is also attached to the subframe but the upper tie-bar is body-mounted. There is also a primary vibration route via the drive-shafts to the wheel hubs and back through the suspension system, the primary one for structure-borne high frequency gear noise.

When in-service recorded load inputs were applied to the model, results for the left hand front subframe mount are shown in Fig 16 as a level of force going into the body (not interior noise). Below 100 Hz there are additional resonances attributable to rigid body modes of the subframe on its mounts. Beyond 100 Hz there is marked improvement for the compliantly mounted set-up.

According to a researcher of Ove Arup and Partners, in his paper to the recent IMechE Vehicle Noise and Vibration conference[4], NVH refinement rivals crashworthiness as the major contemporary automotive design target. The author described an indirect boundary element method used to predict the noise sensitivity of a vehicle body. A modal model of the vehicle body is coupled to a boundary element model of the interior cavity, with unit forces, in the three transational degrees of freedom, applied at the engine mounts and damper towers. The method then depends on noise transfer functions being predicted between these force input locations and the occupants ear positions.

In normal finite element modelling, the author explains, predicting the body acoustic sensitivity starts with generating a finite element mesh, which might even simulate glazing, closures and trim (it could contain 200,000 elements for extracting some 800 modes); normal modes in the frequency range below 200 Hz are then extracted to obtain vectorial data, along the modal damping ratios, for making the modal model.

The airspace cavity is modelled using 10-20 modes, depending on cabin space geometry. A modal damping model is used which effectively 'smears' the damping over the entire model rather than just representing local area absorptions such as seat surfaces. Wavelengths of these acoustic waves are much greater than those of the body structure in this under 200 Hz range so elements up to 50-100 mm in length can be used for the model compared with element lengths for the structural model of 10-20 mm.

The two models are next coupled together and the loads applied to solve for airspace pressure. As the design progresses the models can be modified to account for the greater complexity brought about bey added features. At the early crude modelling stage focus is on the low-frequency noise sensitivity. The main drawback with FEM, however, is the time required for mesh generation.

With boundary element modelling (BEM), is suitable for acoustics applications where there are exterior or interior fields. The indirect BEM is proposed by the author in this case because the interior noise problem has several domains, cabin, seats and boot airspace. While BEM uses the same structural modal model as the FEM method described above, but cabin and boot airspaces are modelled using 2-D elements.

A typical structural model is in Fig 17 and a boundary element model in Fig 18. The latter can be created in a proprietary pre-processor and can be rapidly created from the structural mesh. A field point mesh, Fig 19, has also to be generated so occupant ear sound pressure level (SPL) functions can be calculated and the pressure field visualised. A surface impedance model is used to simulate seat and other trim absorption. The indirect BEM model works in the inverse of impedance, the admittance (velocity normal to the element divided by pressure).

The coupled indirect BEM requires a so-called surrogate structural model (SSM) and an acoustic model. The structural modal model is transferred on to the SSM and the search algorithm maps a vector component of each of the structural nodes on to the nodes of the BEM model. The modal model is considerably reduced by this process as the number of structural nodes is much greater than those of the BEM. For the example in the figures, a 2000 element acoustic model is involved; there are 531 structural modes, 91 frequencies and 24 load cases so considerable computer power is involved.

Results of the example analysis illustrated here, solved by supercomputer, gives the NTFs displayed

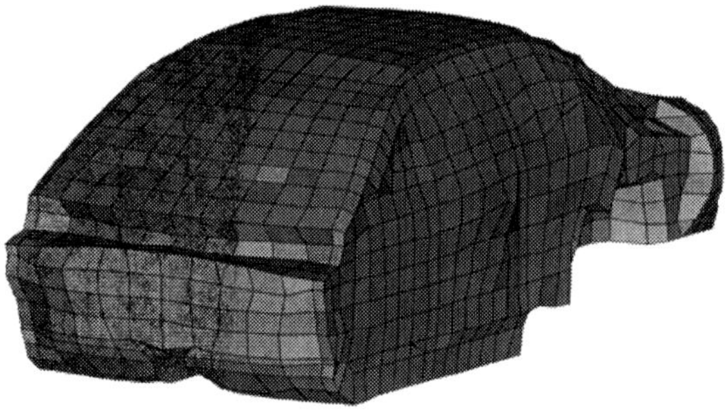

Fig 18: Cabin acoustic model

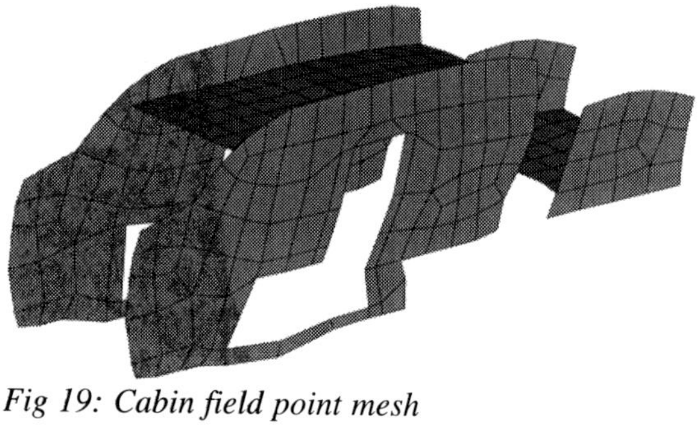

Fig 19: Cabin field point mesh

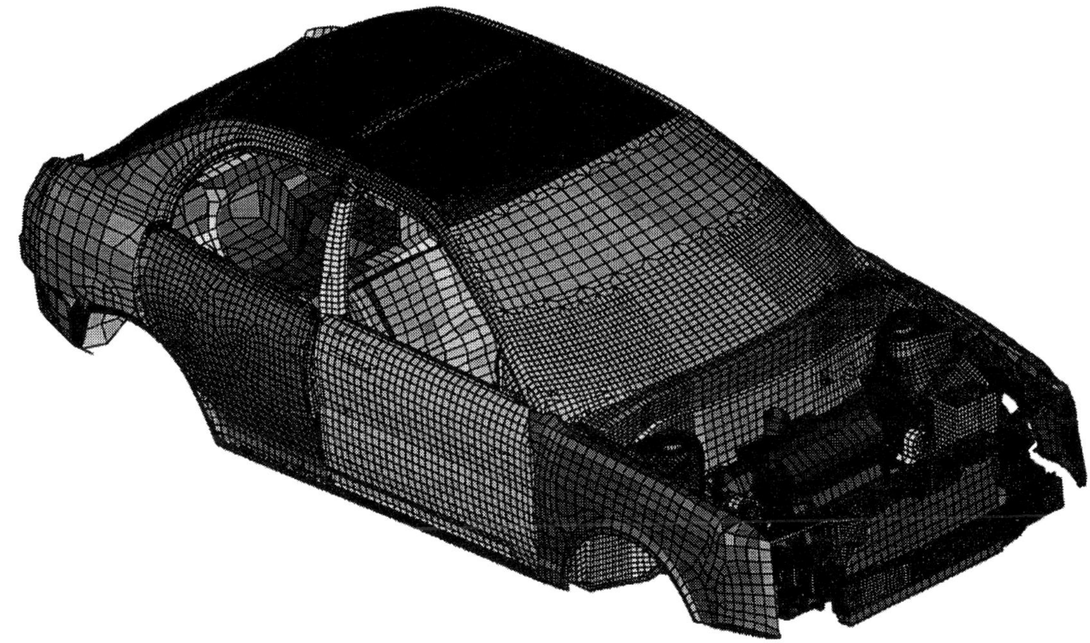

Fig 17: Vehicle body FEM

on a magnitude/phase plot, Fig 20, and cabin/boot airspace sound pressure distribution as in Fig 21. Where NTFs are above target, as at 144 Hz a noise boom occurs. Panel contribution and modal participation analysis is used to determine the controlling NTF factors at this frequency.

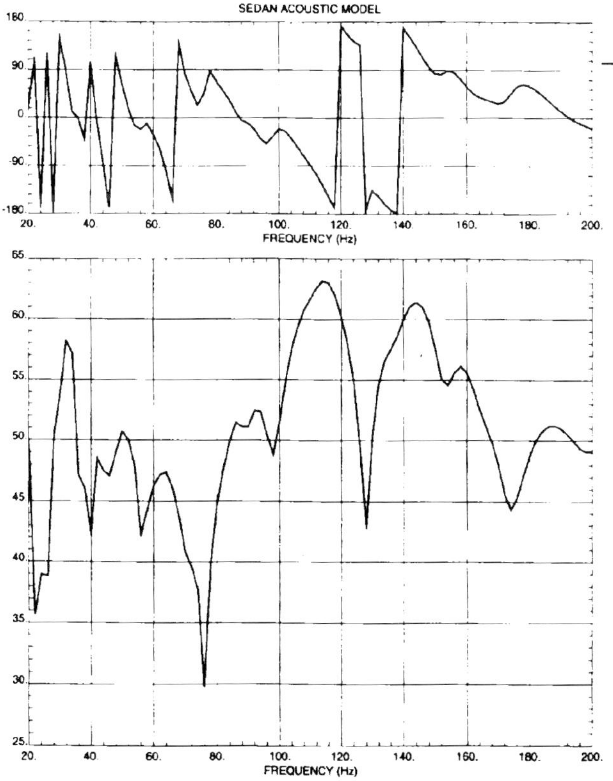

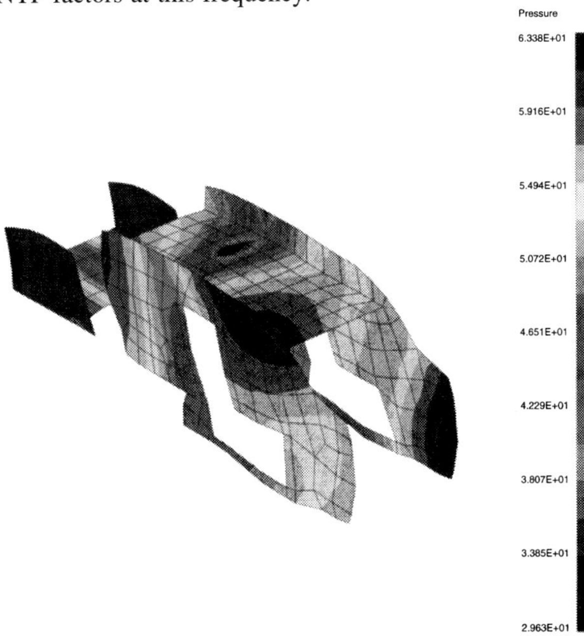

Fig 20 left*: Vehicle body NTF, phase pressure* top *and NTF magnification (dB/N) pressure* bottom
Fig 21 above*: Interior pressure distribution*

Preparing for statutory pedestrian protection

The next stage in structural design for safety, of trying to design-in protection for pedestrians subject to vehicle impacts, is being addressed with vigour by the UK Transport Research Laboratory[5] whose first priority is developing pedestrian impactor physical-models, to be a base for the development of standardised testing.

Since the historic work of Ralph Nader in the 1960s, culminating in his best-selling book 'Unsafe at any speed', little has been done for that most vulnerable of roadway cas ualties, the pedestrian. Even if motorists can be persuaded to respect the speed limits in intensively pedestrianised areas, or have his/her speed more forcibly reduced by traffic-calming measures, much can be done to vehicle front-end structures to increase the chance of pedestrian survival under impact.

According to TRL researchers a recent survey has shown more than one third of road deaths were of unprotected road users such as pedestrians and pedal cycles, some 60% of whom are struck by car front-ends. The laboratory produced an experimental safety car in 1985, with front-end modified using conventional materials, which demonstrated a considerable improvement in reduced injury level in 40 kph impacts, but this was largely ignored by UK legislators, despite DoT funding, in terms of ruling in measures to reduce the death/serious-injury toll of up to 100,000 per year.

Key impact areas, in order of occurrence after first strike, are lower-limb to bumper, femur/pelvis to bonnet leading-edge, and head to bonnet-top. Therefore, in response an EC directive, arising from research by the TNO-chaired Working Group, impactors now being developed represent knee-joint shear force and bending moment; femur bending moment and pelvis loads; child and adult head accelerations. Each impactor is propelled into the car and output from associated instrumentation establishes whether energy absorbing characteristics of the car are acceptable. TRL's responsibility is development of the upper leg-form to bonnet leading-edge test method and the laboratory now supply both leg-form and upper legform impactors to the car industry. They overcome the unreliability of testing using conventional dummies.

Initial impact with pedestrians is confirmed to be mostly side-on, in road-crossing situations; the bumper contacts typically below the knee and accelerates the lower leg against the inertia of the body

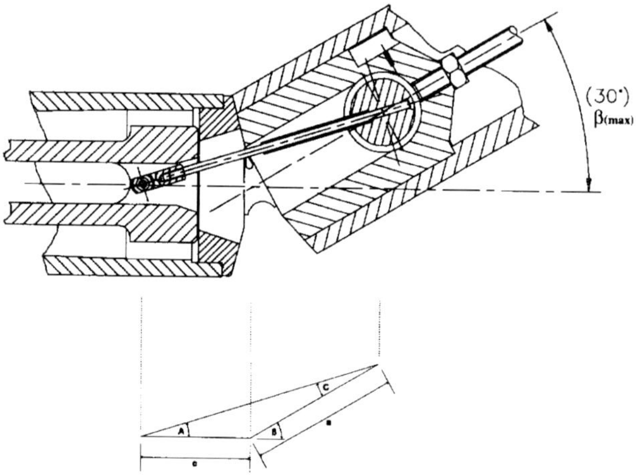

The knee bending angle (β) can be found from the following formula:

$$\beta = C + \arcsin\left(\frac{a}{c} \times \sin C\right)$$

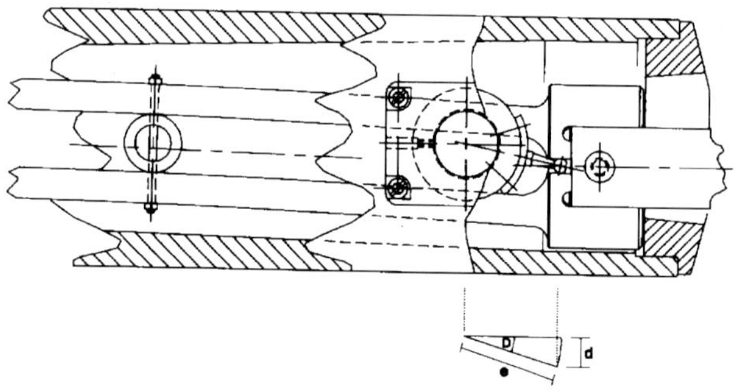

The shear displacement (d) can be found from the following formula:

$$d = e \times \sin D$$

Fig 22a: TRL knee joint at maximum bending displacement, above, and maximum shear displacement, below, showing trigonometric calculations for actual displacements

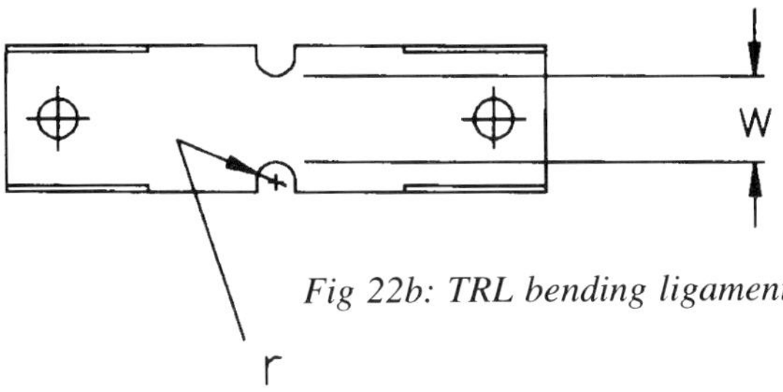

Fig 22b: TRL bending ligament

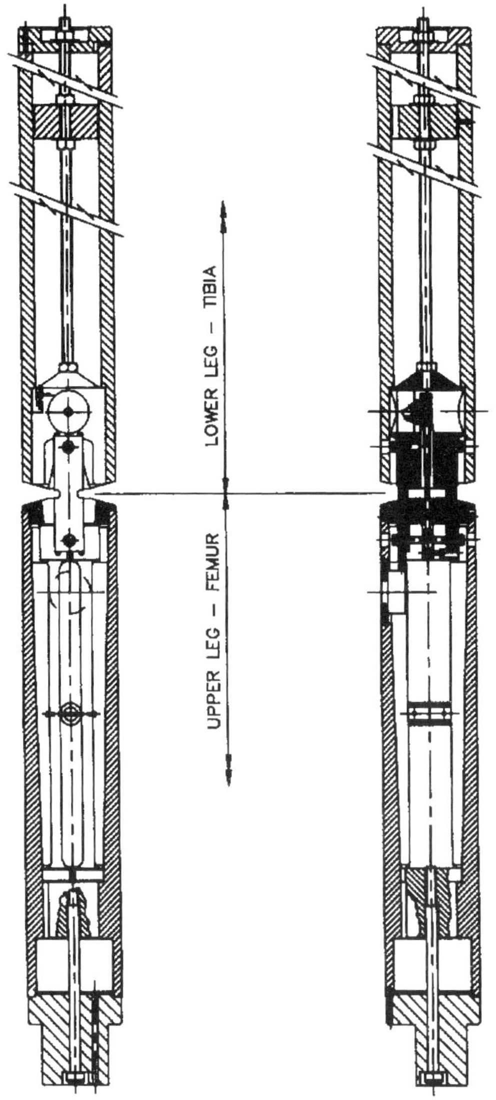

Fig 22c: TRL leg-form impactor

above and below the contact point, reaction forces causing bending and shearing with reactions between foot and road, hip-joint and upper body mass, adding to these forces. Resulting injuries are bending and local crushing of leg bones and/or knee ligament injury through bending and shearing.

The legform impactor thus consists of 'femur' and 'tibia' sections joined by a mechanical knee, with sideways loaded joint stiffness, the elements of the impactor having physical properties akin to a 50th percentile male. Studies with cadavers, and computer-simulations, have shown that leg inertia forces have major influence but foot-road reaction and upper body inertia only minor influence on these leg forces. The impactor is not 'fooled' by palliative efforts to soften only the bumper area which might reduce lower leg injury at the expense of knee joint damage. The TRL 'knee' comprises an elastic spring to produce shear stiffness and disposable deformable ligaments for the bending stiffness. The elastic spring is in the form of a parallelogram, a pair of bending ligaments joining the knee ends of each leg element; this particular configuration provides necessary shear stiffness whilst being unaffected by the bending moments required to deform the bending ligaments. Individual potentiometers separately determine bending and shear displacements, their angular outputs being converted trigonometrically, Fig 22.

The design of the upper legform impactor is geared to the realisation that while first contact is made between lower leg and bumper, next phase is for the pedestrian's legs to be swept away from beneath with contact then occurring between bonnet leading edge and upper leg, the first contact having been found to have an effect on the second one by modifying upper leg angle and impact velocity to an extent dependent on vehicle shape. Principally the upper leg impact causes bending of the femur and corresponding forces in the hip joint, the mass seen by the car bonnet being a combination of that of the upper leg and the inertia of body parts above and below it, its effective value again dependent on body shape. To determine the effect of shape, testing a wide range of body configurations has led to the recording of data in the form of look-up graphs for the use in the eventual assessment procedures.

Fig 23 shows the current upper legform impactor which comprises vertical front member representing the adult femur, supported top and bottom by load transducers on a vertical rear member, in turn mounted on the end of the guidance system through a torque-limiting joint. The front member has strain gauges to measure bending and is covered with a 50 mm foam layer to simulate flesh. Since the test procedure has been designed to make the impactor aim at a line defined by the car's shape, highest bending moment does not necessarily occur at the midpoint, if the cars peak structural stiffness occurs above or below it, so measurement is made at midpoint and 50 mm above and below. Provision is made to attach weights to the impactor so its mass can be adjusted to the effective mass of the upper leg on impact appropriate to each car shape, described by the look-up graphs. The torque-limiting joint is specified to protect the guidance system from damage on cars with poor impact protection, its minimum torque setting providing a rigid joint on cars meeting the desired protection criteria.

To meet the knee shear and lower leg acceleration acceptance criteria the car front must be able to absorb energy whilst to meet knee-bending criteria the car must make a well distributed contact along the length of the legform. While the second criterion might imply a requirement for a vertical front-end, in fact the TRL authors say that local deformation effectively "converts a streamlined shape into a more upright one". While the design of bumpers to meet mandatory low-speed impact legislation do provide some advantage to pedestrian protection, the researchers say it is those which have a substantial distance between the plastic bumper skins and the strong underlying structure that best meet the requirements of the legform test. Ideally a locally deformable lightweight outer bumper skin should be supported by an integrally stiff energy absorbing foam reacting against the strong underlying structure, the crush depth being at least 40 mm. A deep bumper/spoiler making a low contact with the pedestrian's leg at the same time as the bumper would reduce bending cross the knee.

In terms of upper legform impact, for a car with bonnet leading edge height of about 700 mm the depth of crush required would be 110 mm for a typical 100 mm bumper lead. At 900 mm bonnet leading edge height the crush depth requirement increases to 210 mm under the specified impact loading. The limit on bending moment also means that the contact force on the impactor must not be concentrated at one point, requiring generously contoured front ends. Usually the existing heavy outer bonnet reinforcing frame should be weakened and moved slightly backwards so as to allow easier deformation of the leading edge. The bonnet lock should also be move back and deformable clear plastic headlamp front-faces should be employed and a revised mounting system allow the lamps beneath to collapse inwards into space that is normally available rather than rigidly mounting them to a bulkhead.

In terms of the head injury criteria, a limit on acceleration is the key requirement with both child and adult head-forms having a minimum stopping distance of 70 mm which reduces according to the amount from which the bonnet inclines from the normal-to-surface impact situation. Some current bonnet skins are acceptable provided underlying localised reinforcements could be replaced by more evenly spread systems. Proper clearance is also required from suspension towers and engine surfaces, the authors point out.

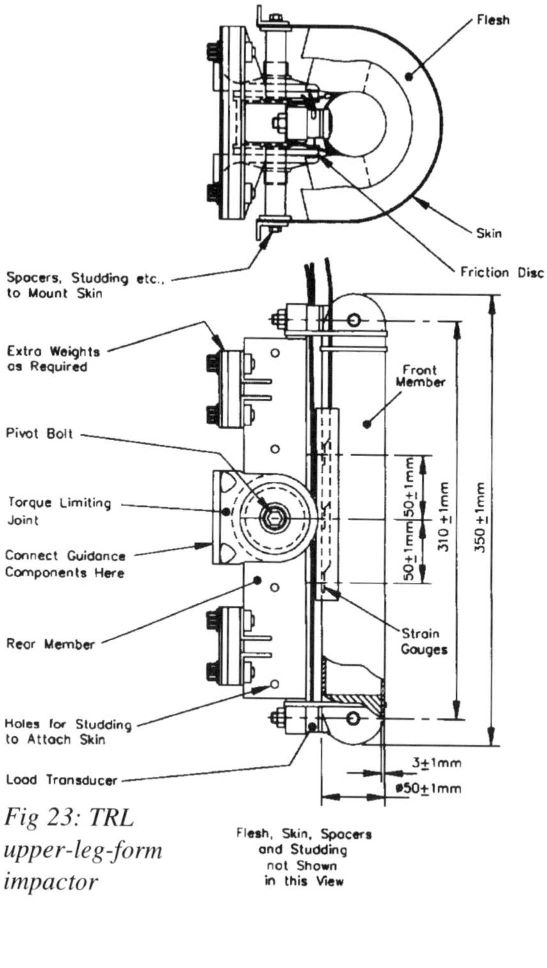

Fig 23: TRL upper-leg-form impactor

Design for the disabled

According to researchers of the Royal College of Art[6], a hard look at the realities of environmental stress, the ageing of populations and the crisis welfare systems are encountering, suggests that a new-concept "Mobility for all" vehicle may be a viable alternative to car ownership. At the same time, this shift from 'A Car for All' to 'Mobility for All' is quite in keeping with contemporary thinking about the shift from product to service and from the material to the virtual. Environmentalists talk of reducing the global impact of human activity by a factor of between five and 50 over the next 50 years, and a good point to start is transportation. The authors have been thinking about what this might mean and offer the following case studies of staff and student work as an insight into how this new mix of services and vehicle typologies might evolve.

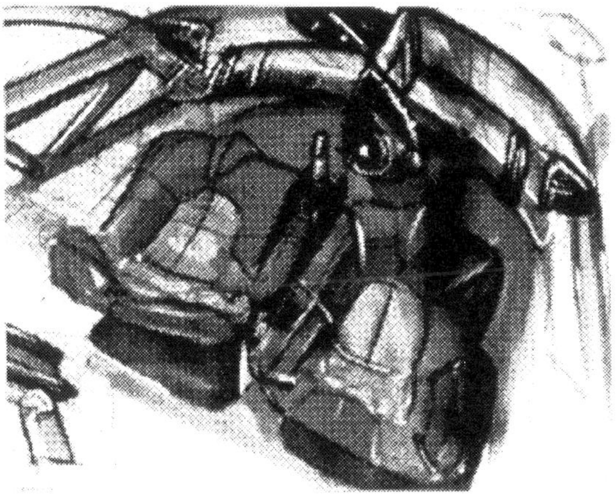

Fig 24: The "50+ coupe"

Fig 25: The "urban rickshaw"

They argue that for the generation now moving into retirement, personal transport is synonymous with freedom, dignity and security. This suggests that mobility, in the future, will mean having a greater range of transport solutions at our disposal — a more integrated transport policy that blurs the traditional boundaries between public and private services and vehicles. Patterns of car ownership will change as more vehicles are banished from town and city centres, increasingly pedestrianised to ease congestion, improve security and provide a better quality of life for all those who work, live and pursue leisure activities there. While reservations about public transport — mixing with other people, lack of comfort and security, etc — can be greatly improved by interior layout, better seating, lighting to aid security and privacy, and higher grade interior materials and accessories to promote a feeling of ambient luxury.

The increasing pressures to reduce materials and energy consumption in vehicle manufacture, and use, will lead to vehicles that last longer, can be adaptable throughout their working life and recycled or re-manufactured at the end of it. In simple terms cars will last longer and be more adaptable. For example, the traditional family car may encompass the needs of a growing family for its younger users, and act as a social space — and extension of the house — as they grow older. In America, older people are taking to the highways in their retirement to explore the vastness of their native country which they have been unable to enjoy during their working lives. Future trends indicate that vehicles and transport systems will have to respond to an increased emphasis on freedom, versatility, variety of use and social interaction, and do so by exploiting technical advances. Over the past twenty years personal security has become a prerequisite of everyday life: theft, vandalism and personal attacks are more prevalent and directed not specifically at the elderly or vulnerable. Consequently, security in and out of the vehicle has become a priority across the age ranges.

Where restrictions in the range of movements are encountered, ergonomics remains an area of significant interest to drivers of all ages, in particular as expectations of comfort, fatigue reduction and personalisation of the driving environment have risen substantially. This leads to a great deal of interest in increased flexibility, adaptability and user-friendliness in the mainstream of vehicle design.

The problem is that the major car manufacturers of the world continue to realise new designs around the slow and calculated evolution of a generic product. Experience tells them that if they get things wrong the costs can be enormous, like the profits if they get things right. So, advances in new directions are difficult to accommodate. Where age-friendly design can succeed is in offering a new direction for research and innovation, the results of which can be introduced to mainstream product ranges and targeted at the age sectors which are at present expanding and will continue to do so in future. Designed correctly, such improvements can appeal to a wider market sector by diminishing age differences and offering real benefits to younger purchasers—a 'car for all' that the marketing people can understand!

The 50+ private car

Vehicles targeted at particular user groups e.g. 'ladies' or 'oldies' cars, can be seen as patronising and be rejected by both target group and other potential purchasers, whereas off-road and MPV's have demonstrated the benefits of a large cabin area, easy entry and exit, high roofs and large door apertures which appeal to a majority of consumers and that can be particularly helpful for older people. The seat height allows the user to sit 'on' rather than 'in' the car, aiding parking, improving visibility, and making it easier to turn in the driving seat.

Although few older people are wealthy, a significant proportion are 'comfortably' provided for and some have considerable disposable income. Two examples from Royal College of Art students Jim Das and Geoff Gardiner, address this market sector by exploring a 'classic' solution in terms of styling, taking care to integrate many of the age-friendly elements identified above to appeal to drivers of all ages. Features include sliding doors, level floor, spacious cabin, increased roof height, lowered boot access etc., and the interior components are designed to be adaptable: e.g. the seats can be upgraded to give extra support or to pivot for ease of entry. The driver is assisted with automatic gear change and power steering, and is offered joystick type controls in one design, which give the driver more space and make the car usable for people who would presently require an adapted vehicle.

Instrumentation is by 'head up display', projected on the windscreen to aid driver concentration. Information is 'hierarchically driven', and only what is important is presented at any time so as not to overload the driver.

The format of a large coupe, Fig 24, reflects the lifestyle of a prosperous semi-retired couple, and the designs portrays qualities of sophistication and value. The cars features high shoulder lines to give a feeling of security and the front and rear sections have soft reactive surfaces that aid protection and reduce repair costs. The car would be manufactured from lightweight steels and re-cyclable plastics.

Taxis and people movers

The London Cab has been recently re-designed to offer more interior space and be wheelchair friendly. The London Cab is not only an icon of the capital, it also offers one of the most flexible transport systems

Fig 26: Short-haul taxi

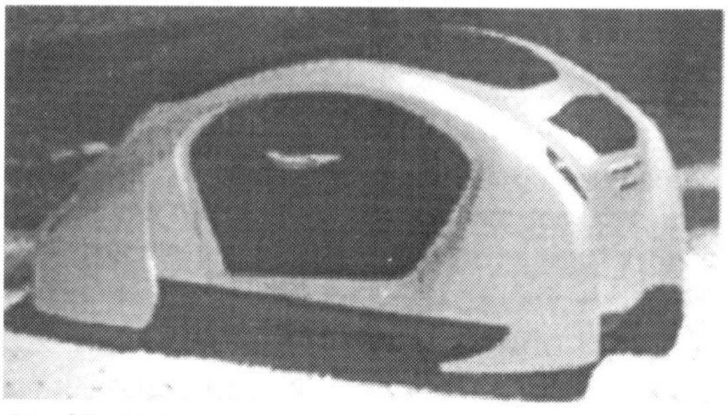

Fig 27: Driverless taxi

available. Ideally it should form the core of an integrated transport system within the capital, bridging the gap between people's homes and other transport services like the bus and underground. The problem is that it is relatively expensive and not easy to order for short trips, which means that for many older people it is not a viable travel option. A number of alternative scenarios have been investigated at the Royal College of Art, all of which seek to keep fares at an affordable level by reducing the capital and running costs of the taxi, use information technology to offer transport on 'demand' — with passengers booking by telephone or teletext — and extending its user profile to include disabled and older users.

The urban rickshaw

On London's congested streets a black cab is unlikely to travel more than 60 miles in a day, making an alternative power source (electricity, hydrogen, LPG) feasible. This would reduce space requirements and design limitations by eliminating engine and gearbox. Sotoris Kavros drew the inspiration for his lightweight taxi, Fig 25, from the Tuk-Tuk rickshaw of Indonesia, reworking the idea around an alternative power source. The two passengers sit behind the driver, and the Urban Rickshaw has been designed to be as narrow as possible, to aid travelling through city streets and parking, while offering good headroom and easy access through doors that fold back alongside the vehicle. The exterior is styled to look 'taxi-like' with soft bumpers to prevent pedestrian injury.

The greatly reduced capital cost of this vehicle compared with a new Metro cab costing £28,000, could cut fares in half and still offer the driver a good living.

Short-haul taxi

In designing this taxi, Fig 26, Peri Salvaag envisaged a future where large parts of thé city are closed to cars. The function of this vehicle is move people from place to place within the city centre and link tourist and leisure attractions with public transport. The taxi runs on electricity, with the batteries forming the floor pan of the vehicle. It would be used for short journeys, typically lasting less than 20 minutes at relatively slow speeds, and would be able to enter pedestrianised areas. Safety is therefore an important feature, along with accessibility. The passengers would order the cab electronically — in the foyer of a museum say — and be advised about when to expect it to arrive. The driver is there to provide security and assistance, and payment would be by credit card.

Short trips do not require the same level of passenger comfort as longer ones do and the vehicle exploits every inch of its tiny footprint with an interior design that features a centrally mounted driving seat with either two passengers perching behind on small half seats (bum rests), or a wheelchair passenger who would enter through a door and ramp at the rear. The exterior is designed to look strong, stable and dignified.

Driverless taxi

The most radical proposal for a city taxi that has been investigated is for a driverless vehicle that would work within pedestrian zones. The taxi, Fig 27, follows a guidance system in the road surface and is in constant communication with a control station that locates vehicles as demanded, manages their jour-

Fig 28: Kyoto tram (left and next page)

neys in the most efficient way and parks them in parking zones when not in use. During peak periods, the vehicles cruise the streets until 'called' by a passenger, just as in the case of the short-haul cab above. The individual vehicle then picks up the passenger, and the user also advises the drop-off point so that the control station can plan its next journey.

The cab is shared by up to four passengers to reduce costs and optimise use. It has large glass areas to give the passengers better visibility and a sense of security, while gull-wing doors aid entry and protect the interior from the weather. As there is no driver, the interior is designed so that passengers face one another around a central unit on which information about the city and route is projected.

Shopping ferry

This is a ferry service to be offered by a supermarket with the vehicle being commissioned or leased by the supermarket from the manufacturers. Out-of-town developments mean that more people rely on cars, to the detriment of older and disabled people. Background research for this vehicle demonstrated that most supermarket journeys are weekly and less than one trolley-load of goods per person is purchased. Regular transport to and from appropriate points, or following a route around housing areas could reduce road traffic and build customer loyalty.

In Mike Leadbetter's scenario, the bus would be owned and run by the supermarket and would be called or booked by the customer, who would be collected from a point near their home. The size of the bus is optimised to work within conventional city streets and carries ten passengers and their shopping to keep the journey and waiting times down. Luggage provision is under each seat, wheelchair access being by ramped entry aided by external wheels which allow the bus to kneel, and the bus is driven by hub motors in the wheel rims, saving valuable space. The driver will provide security and assistance as required.

Kyoto tram

Kyoto is an historic city and the challenge here was to design a tram which suits a very wide user profile, including older and disabled people, mothers with children, and tourists, as well as commuters. The internal layout is spacious with groups of seats replacing conventional benches, providing flexible seating options for single passengers, couples and people with small children as well as extending the clear floor area to aid wheelchair access.

To further aid access the tram has a low floor and large doors that open flush with the sides of the vehicle and the interior has built-in grab handles.

The tram, Fig 26, features large glazed areas that offer visibility to seated passengers and those in wheelchairs, as well as to standing passengers. Information is presented graphically to the passengers on the internal window surfaces with LCD technology, while the exterior of the tram has a low ride height to appear non-threatening and a soft, smooth front to minimise accidental injury. At night lighting effects similar to those used on aircraft would enhance the ambience, security, and safety of the vehicle.

28

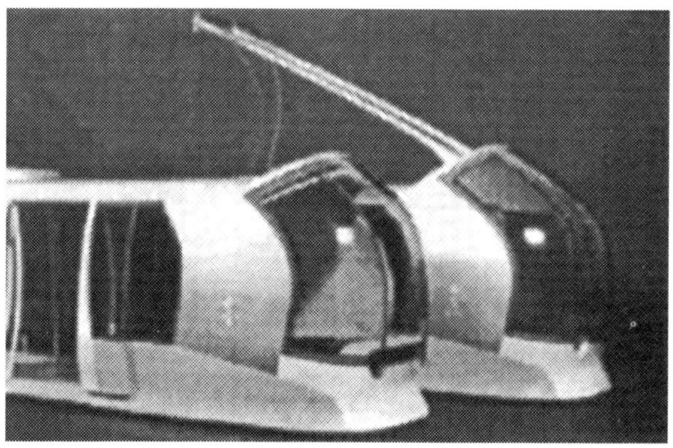

Adaptive restraint technologies

Airbag systems is considered by Delphi Automotive[7] to be the fastest growing sector of the automotive components market.

The limitations of current technology centre around the fact that most airbag systems today deploy with constant force, regardless of occupant position, weight, size or seatbelt usage. Although the majority of airbag deployments occur in impacts with significantly lower speeds than are used for regulatory testing, current system designs make no distinction in the level of protection for restrained or unrestrained occupants, crash severity or the type or position of occupant involved in the crash. The next generation of airbag system will take these factors into account, monitoring occupants' characteristics and the crash severity then tailoring airbag inflation to the situation, Fig 27.

Fundamental sensing and control elements for such systems include: characterisation of vehicle crash severity; detection and recognition of occupant size and location; sensing restraint system component configuration — seat belt usage, seat position and presence of a child seat; also adapting restraints to achieve maximum protection while minimising risk of injury. A Delphi team is developing sensors and restraints for Generation I ART systems targeted for late 1998 and Generation II systems targeted for 2001, Fig 28.

Crash severity sensing

Current airbag systems utilise a Single Point Sensing and Diagnostics Module (SDM), in the passenger compartment, to detect a crash through vehicle deceleration. Using a complex algorithm, the SDM determines the appropriate conditions and timing for deploying the airbag. Airbags are triggered at the lowest vehicle speed that might produce injury yet are designed to provide maximum restraint in the worst crash scenarios. Typically, airbags are deployed above 14-22 km/h Equivalent Barrier Speed (EBS) and require only 50-75 ms to deploy. Adaptive Restraints require an estimate of the potential occupant injury severity to make the best restraint activation decision. Although they are not necessarily the same, crash severity can be related to occupant injury severity potential. Total change in vehicle velocity during the crash, for example, can be used to estimate the maximum occupant injury potential. This information is needed early in a crash, to determine severity and appropriate deployment.

Due to the filtering effect of the vehicle structure during a crash event, the total delta velocity may not always be established early enough with a single point sensor. To support the crash severity algorithm in the SDM, an accelerometer-based sensor located in the crush-zone of the car could be used. This information would be sent to the SDM and used for situations where multi-level deployments are available. Delphi is also exploring Anticipatory Crash Sensing (ACS), which senses the speed at which the vehicle approaches exterior objects, to provide earlier crash severity estimates and significantly improve deployment decisions. A rollover sensing module is also being studied which can predict impending rollover and pitchover so as to trigger rollover safety devices, including seat belt reelers, inflatable curtains and side-airbags. Algorithm simulations suggest that it can predict rollover as much as 300 milliseconds in advance of those events, critical for the occurrence of occupant injury.

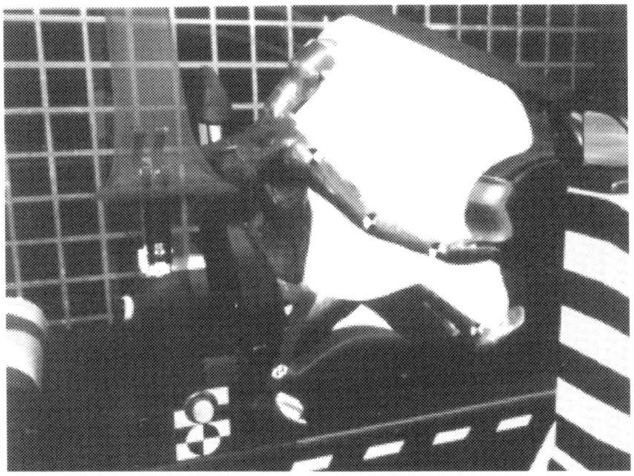

Fig 27: ASI passenger-side low-mount air-bag

Occupant sensing

The company is also developing technologies to measure occupant mass and position in order to enable airbag suppression for occupants below a specified seated weight. This technology could also be used to classify occupants into weight categories, thus improving the adaptation of restraint force for various impact situations. Transponders or 'tags' are also available for use with rearward facing child seats. They can thus be detected by a seat-based sensor and used to disable the airbag. Occupant sensing systems can work together with the SDM to control airbag deployment.

Delphi's Occupant Position and Recognition System (OPRS) recognises the passenger position as well as monitoring a specified zone around the airbag for occupant presence. The system determines whether a passenger is in the immediate proximity of the airbag at the time of a possible deployment and initiates appropriate action. Now under development it will initially detect empty seats, rearward facing infant seats and dangerously close occupants (passenger airbag only). Several sensing technologies are under consideration, including infrared, ultrasonic, thermal and capacitive. However it is likely that the final system will require a combination of these.

System integration

A complete system approach ensures that the limitations of current airbag systems will be remedied in the most efficient way. For example, the time at which the crash sensors have to make deployment decisions in order to properly protect the occupant is closely related to the characteristics of the airbag. This is also true for the desired classification of occupant weight and height based on the restraint system properties. Thus, sensors and airbags developed together will utilise the full capacity of both, whereas a 'system' formed by individually developed components may be less efficient and more complex.

Collecting, processing and utilising the occupant and crash situation data required for fully developed adaptive restraint systems is a challenge to be met by developing a Safety Bus and Distributed Restraint System Architecture, based on the 'plug-and-play' concept used in the personal computer industry. The new system allows the addition or deletion of sensors and deployment loops (actuators) without redesigning the system.

The Safety Bus concept uses two wires; each node derives its electrical power directly from the bus. The 'arbitration' bus architecture allows any device to initiate a transmission when the bus is in an idle state. The design accommodates two transmission rates: high speed (500 kHz) for rapid system parameter updates and low speed (10 kHz) for diagnostic/sensor information — seat occupancy, weight and seat belt status. Message throughput on the

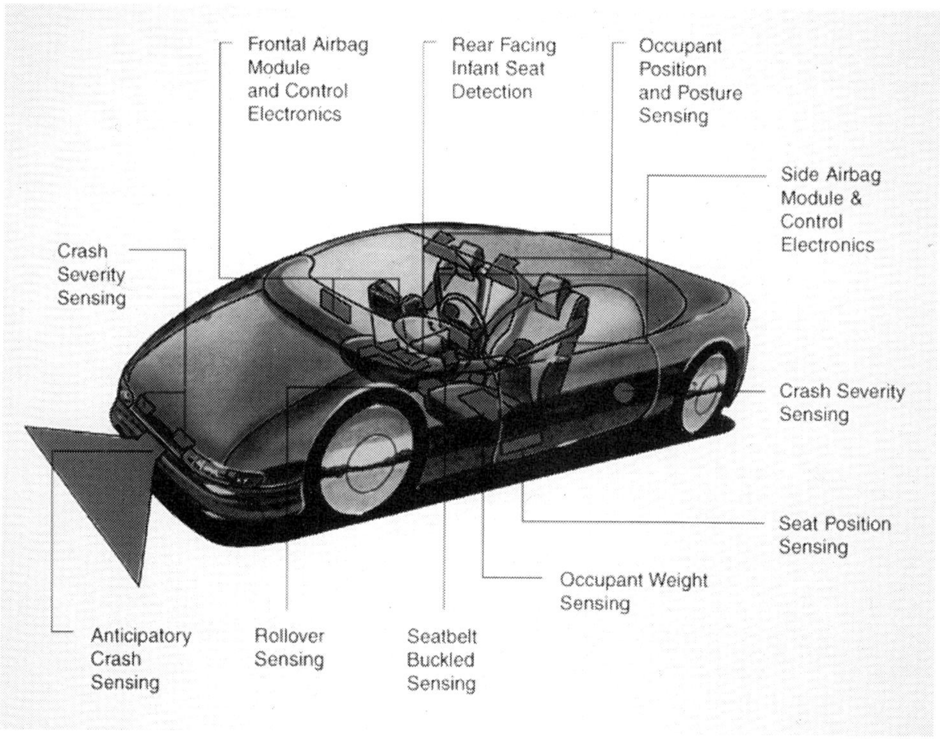

Fig 28: Adaptive restraint technologies

company's bus is eight times faster than existing CAN-bus protocols, it is claimed.

A complete system approach also ensures that many of the limitations of the current airbags will be effectively remedied. For example, the time at which the crash sensors make deployment decisions in order to properly protect the occupant is closely related to the airbag's characteristics. The desired classification of occupant weight and height based on the restraint system properties would work in a similar way. The company offers a complete line of highly integrated driver, passenger and side impact airbag systems and is an experienced manufacturer and integrator of steering wheels, dashboards, door modules, door trim and airbags leading to fully integrated modular cockpits. Products such as the Snap-In, Recessed and Invisa-ModTM Integral Steering Wheel and Driver Airbag Module consolidate component function in this way.

Attention to Leg Injuries

A conventional airbag system supplements the safety belt in that it offers additional protection to vulnerable areas such as the head, the neck and the spine. However, lower body injuries have been put in focus recently because the introduction and use of safety belts and airbags have greatly reduced the severity of most upper body injuries and also because many leg injuries have been found to have long-term consequences. Traditionally, a 'kneebolster' is used: a padded structure that faces the occupant's knees and applies restraint force to the occupant's lower body through the femurs. In order to work properly, the knee-bolster must be positioned fairly close to the

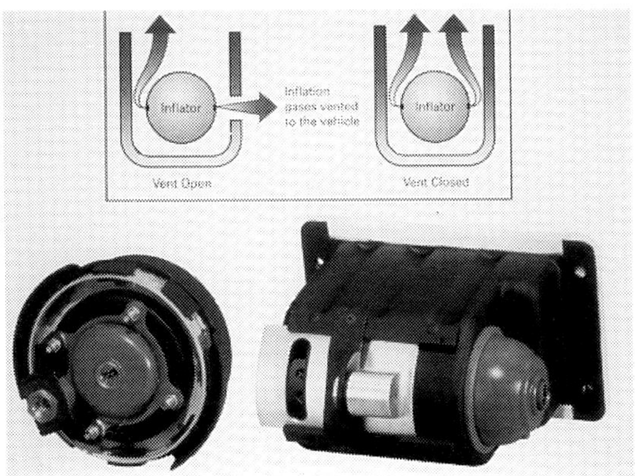

Fig 29: Pyrotechnic actuated venting

occupant, but that also means it takes up valuable space and reduces the leg room for the driver and passenger.

Hence, deployable knee-bolsters are now made. This makes it possible to move the dashboard forward in the vehicle, as the knee-bolster automatically positions itself in proper proximity to the occupants knees if a crash occurs. Recent development work has also shown that a properly designed, deployable knee-bolster has the potential to reduce lower leg injuries, keeping the legs in a more favourable position during a crash.

The company offers an unique, low-mount airbag for application on the passenger side. It incorporates a deployable knee-bolster, which is filled immediately when the bag deploys. An intricate chamber system then makes the gas fill different parts of the airbag at different times, thus eliminating the need for aggressive, immediate inflation of the whole system . The arrangement allows more freedom in the design of the passenger area, not only because it eliminates the need for a knee bolster, but also because it wraps around and covers the instrument panel completely, helping to protect the passenger from potentially harmful surfaces.

Controlling restraint deployment

The ability to control airbag deployment force is a critical aspect of any adaptive restraint system. Delphi's Pyrotechnic Actuated Venting (PAV) airbag module utilises a slide mechanism to direct inflator energy into the airbag, Fig 29. When combined with developing sensor technology, PAV will provide full or 'lower-power' performance based on sensor input on occupant and crash characteristics. The system — available for both driver and passenger applications — combines the advantages of variable inflation with the cost-effectiveness of a single output inflator.

Multiple inflators and pyrotechnically controlled venting help maximise occupant protection and minimise injury risk by adapting airbag inflation speed and stiffness to the occupant characteristics and crash situation. A new generation of algorithms for these technologies is being computer simulated and sled testing to evaluate the cost/benefit ratio. Airbag restraining stiffness can remain 'HIGH', as is the case with all current airbag systems, or can be modified to 'NONE' or 'LOW', based on various sensor inputs. Note that the 'HIGH' level airbag restraining force is only needed when the occupant is an unbelted adult and the crash severity is moderate to high.

While fully adaptive restraint systems are not yet available, Delphi already produces a unique airbag design to protect out-of-position passenger-side occupants from injury during airbag deployment. The patented Bias Deployment Flap airbag incorporates a box-shaped guide element, a 'flap' of material that diverts the deployment direction of the airbag away from an occupant located too close to the deployment surface. Through the Aegis joint venture, they are able to offer the D-60 driver airbag with pyrotechnic, non-azide inflator. The use of non-azide propellant means that it creates virtually no residue or toxic gas.

The Hybrid Inflator System, which was co-developed with OEA, provides a smoke-free gas to inflate the cushion; deployment is aided by a small pyrotechnic charge. Cushion inflation is precisely controlled for specified occupant energy management and a future adaptive inflation technology will provide dual-stage inflation of the airbag.

Safety seating developments

Other innovative restraint systems by the company include the Pro-tech self-aligning head restraint which has been specified for the latest Saab 9-5. It is designed to reduce whiplash injuries during low speed rear impacts, by lifting and moving forward and upward into the optimum position for occupant head and neck protection. The system is also available as part of the so-called Catcher's Mitt Seating from the company to be introduced this year, Fig 30, where it combines with a high-retention seat and the ABTS integrated seating and seat belt system in which the belt is attached to and moves with the seat. The seat absorbs energy during rear impact by allowing the pelvis and lower back to penetrate the seat squab in a controlled way, safely pocketing the occupant during the course of the impact. The Pro-tech head-restraint is activated by the energy of the impact, rearward motion of the occupant actuating a pressure plate and system of levers to move the restraint. ABTS involves the use of a high-strength seat structure and floor attachment to absorb belt anchorage loads. It particularly suits open car locations where no B-pillar anchorage is available, also removable seat situations or those where a large amount of adjustment is provided. A low-mass version is available weighing 25 kg. Bosch has announced that its Smart Airbag System is being implemented in three stages, the end of 1997 seeing start of production of the SMINC smart inflation concept to control the sequence of inflation according to the particular crash situation. Currently being introduced is the Automotive Occupancy Sensoring (AOS) system seen in Fig 31 which offers adult and child seat occupancy and out-of-position detection devices. In a latter phase a pre-crash sensor will be able to detect the relative crash speed before a collision, triggering individual protection mechanisms before impact.

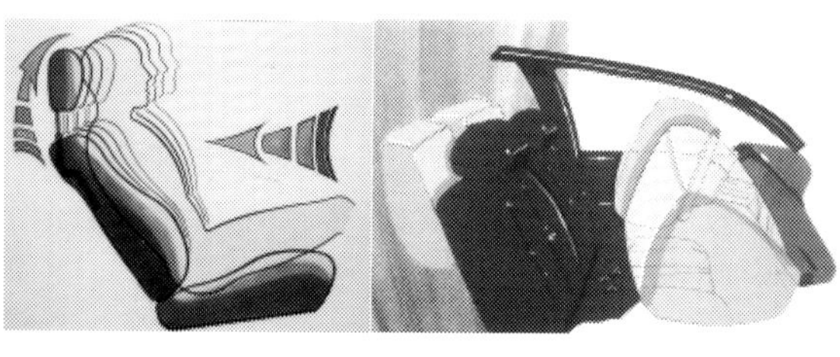

Fig 30: Catcher's Mitt seating

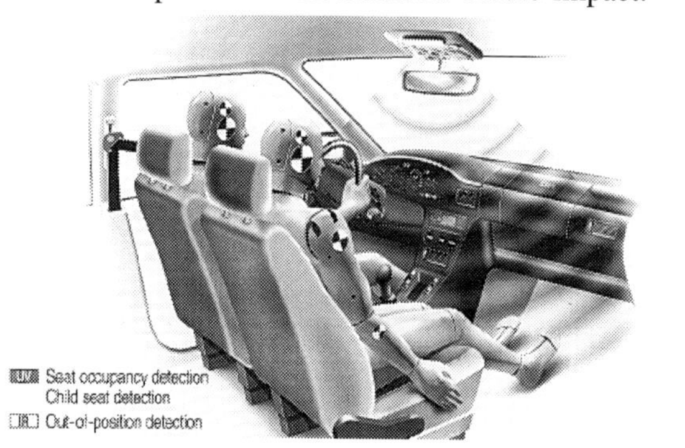

Fig 31: AOS system

Chapter 3: Powertrain/chassis systems

The challenge of natural gas engines increases as emissions legislation toughens for petrol and diesel engines. However, the further development of the small HSDI diesel, and the direct injection gasoline engine, is showing, thus far, that they are up to the challenge. The future could also lie in revised combustion cycles such as the constant-pressure one; there is considerable scope too in revised valve arrangements on otherwise conventional IC engines. In transmission design, clutch-to-clutch automatics vie with electronically controlled CVT, while in chassis systems, the mechanics of roll-over has come under the spotlight again, in vehicle handling analysis, as has the computer-simulation of suspension linkage motions.

Powertrains: the next stage?

Careful comparative studies are revealing some strong emissions advantages for natural gas engines, but LNG may be preferred to CNG for CVs. For lighter duty vehicles, lean burn natural gas competes with advanced diesel and gasoline concepts but the ultimate solution may be IC-engines operating in a restrictive speed band in conjunction with continuously variable transmission.

Important work on rating particulate emissions in terms of environmental damage has been carried out by MIRA's Powertrain & Emissions Technology team. Using the test set-up seen in Fig 1, a feel for the relative merits of gasoline, diesel and CNG engines is obtained in this respect. The research team under Simon Greenwood have shown that particulate nuclei start to grow as they leave the combustion chamber and pass down the exhaust tracts, as temperatures drop below 300C. Soot particles are found to agglomerate into chains with hydrocarbons and nitrates/sulphates condensing on to them. Some 90% of the particle size distribution is below one micron as measured by the Scanning Mobility Particle Sizer (SMPS)

Greenwood explains that particle sizing tests used a 20 cm diameter particulate dilution tunnel, with an exhaust dilution of about 20:1. The exhaust sources were various motor vehicles on a 50 kW emissions chassis dynamometer. The SMPS probe was a steel tube facing the aerosol flow connected to the SMPS inlet. Vehicle and tunnel were preconditioned by driving at 50 km/h and constant speeds were chosen as being representative of the constant modal conditions within the European emissions drive cycle. These speeds ranged from idling to 120 km/h. Two diesel vehicles were tested for this work. The vehicles had a normally aspirated IDI engine and a medium sized turbocharged IDI diesel with catalyst, Fig 2. All three gasoline-fuelled vehicles produce few particulates at idle and low power conditions. Particulates that were formed did not show any identifiable distribution. At higher power conditions there were large particulate populations, Fig 3.

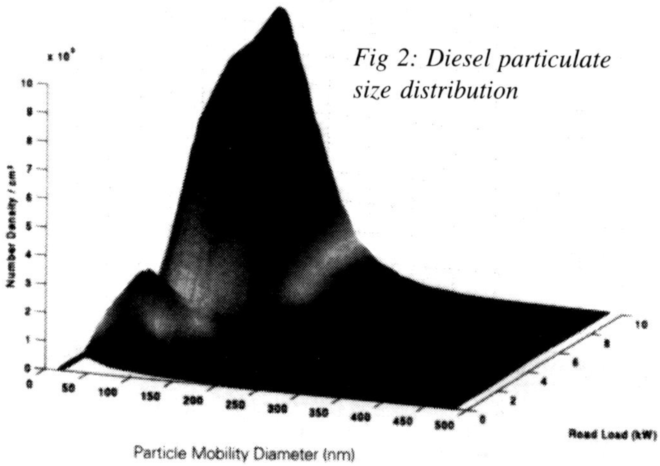

Fig 2: Diesel particulate size distribution

Of the fuels tested, 'Vehicle Petrol 1' produced the fewest particulates. A significant increase in particulate numbers started as low as 30 km/h and 1.0 kW. The peak position developed at around 34 nm. 'Petrol 2' produced distributions that were narrower and only developed above 80 km/h and 7.1 kW. The peak first formed at 25 nm, corresponding to the Nucleation Mode, before abruptly shifting to 70 nm at high powers. This shift occurred as the particulates started to aggregate and received condensate from excess fuel. 'Petrol 3' showed similar behaviour but at a higher particulate concentration, the higher power peak diameter being 40 nm rather than 70 nm.

The two CNG-fuelled vehicles produce few particulates at idle and low load, Fig 4. However unlike the gasoline vehicles/fuels described above, the particulates that are formed at low powers tend to form an identifiable and typically bi-modal nature. The first peak is at 60 nm, the second at 120 nm diameter. At higher powers a smaller diameter peak appeared and grew sharply to produce a very large but narrow particulate size distribution, with total particulate numbers similar to those seen in diesel and gasoline vehicles. In these tests, this vehicle peak occurred at 20–30 nm.

The diesel vehicles produced similarly shaped distributions but with the normally aspirated engine producing rather more particulates. The difference may be the result of the catalyst on the turbocharged car. The diesels were the only vehicles to produce significant quantities of particulate matter at idle. The measured loaded peak positions were in the range 70–90 nm for the normally aspirated diesel and 50–60 nm for the turbo diesel. The particulates at idle were smaller at 32 and 46 nm respectively. The particulate diameters pass through a maximum in the medium power range before decreasing again. This is the result of the two competing processes. At idle, there are fewer particulates produced as there is less fuel delivered to the engine, meaning that the number

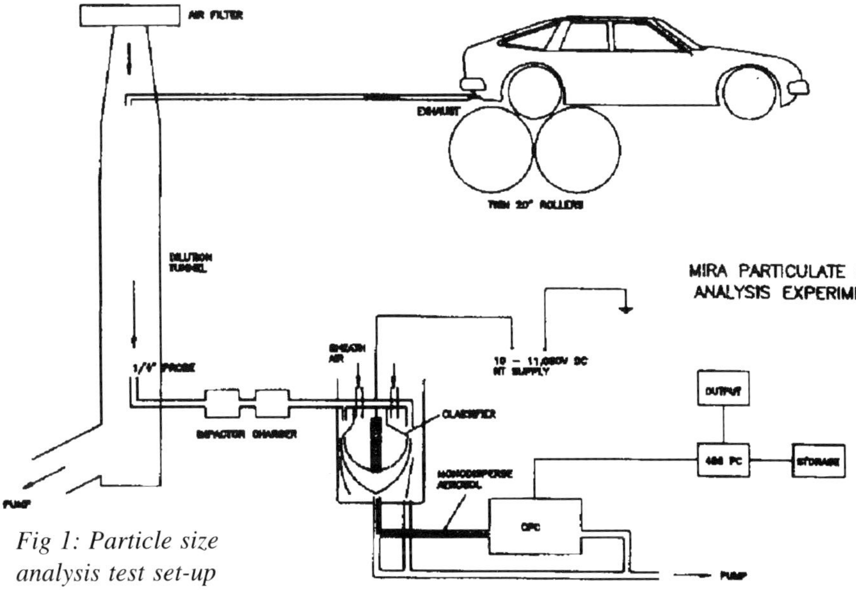

Fig 1: Particle size analysis test set-up

of particulate cores is lower than under load. With fewer cores available, the particulate chains cannot grow as long as under load. In addition, the concentration of excess hydrocarbons is low so less condensation occurs. As the fuelling increases, the mean chain length and condensation rate increase. However, at high powers, although the number of particulates is very large, the exhaust throughput is also rapid. This means that whilst agglomeration does occur, it has less time to proceed to completion.

In general the gasoline vehicles produced much lower particulate concentrations than the diesel at low and medium powers. At higher loads a peak forms in the range 20 to 30 nm, half that seen in the diesel experiments. The initial peak is formed at around 50 km/h and 3 kW for two of the three vehicles. The growth in the distribution is rapid at road loads over 10 kW. In two of the gasoline vehicles, this growth is paralleled with a shift to higher diameter, 40 nm and 70 nm respectively, clearly showing the bi-modal nature of the particulate size distributions as the shift is the change between the Nucleation mode and an Accumulation mode. It has been suggested that the engine oil also has a role in this behaviour, Greenwood points out. The particulate numbers at high power are similar to those seen in the diesel experiments.

Generally, the natural gas vehicles produced few particulates at low load. However there was a clear bi-modal distribution at this load. This indicates

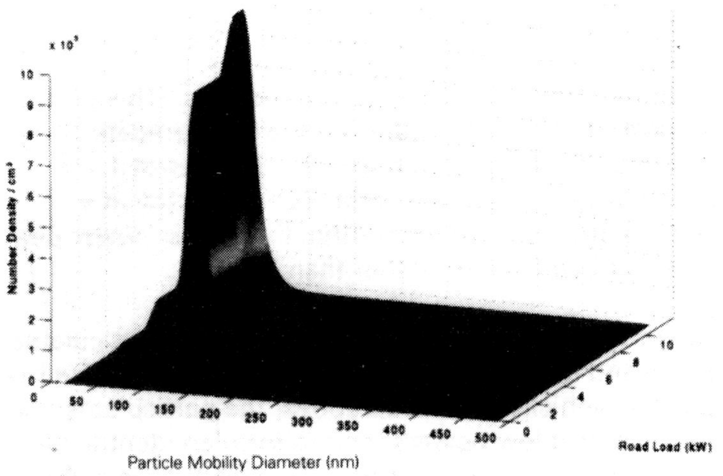

Fig 3: Gasoline particulate size distribution

that there is a mechanism working to produce these particulates unlike in the gasoline vehicles. At high loads, the number of particulates increased very sharply. The high load distributions are very narrow and have peak values an order of magnitude higher than for diesel and gasoline although at a smaller diameter. This allows the total number of particulates under the distributions to be quite significant even in comparison to the diesel and gasoline vehicles.

Greenwood concludes that all three fuel categories produce significant numbers of particulates at high engine loading but at low loads it is the diesel which is the significantly heavier polluter and particulate size, too, is significantly higher.

Truck and bus emissions

In a seminar held by Millbrook Proving Ground, the results of tests on truck emissions were reported which were based on what the GM facility calls 'real-world' emissions testing. This is an important distinction because current heavy-duty vehicle legislation is based on engine-dynamometer rather than whole-vehicle testing. Using Millbrook's VTEC facility complete vehicles are evaluated for emissions under simulated operating conditions approaching field service.

The Proving Ground's Andrew Eastlake demonstrated the importance of using equipment such as VTEC since on choosing a catalyst formulation for a London bus, to give best results on the legislative dynamometer tests, this actually showed a conversion efficiency of only 40% under the 'real-life' conditions of the VTEC tests. By examining tran-

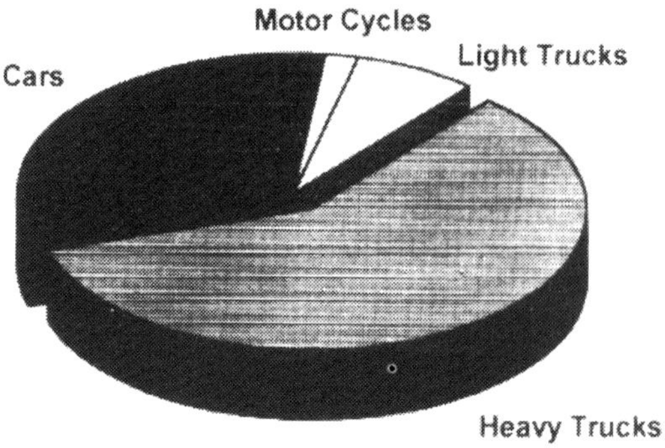

Fig 5: Road traffic NOx emissions, AD 2000 estimate

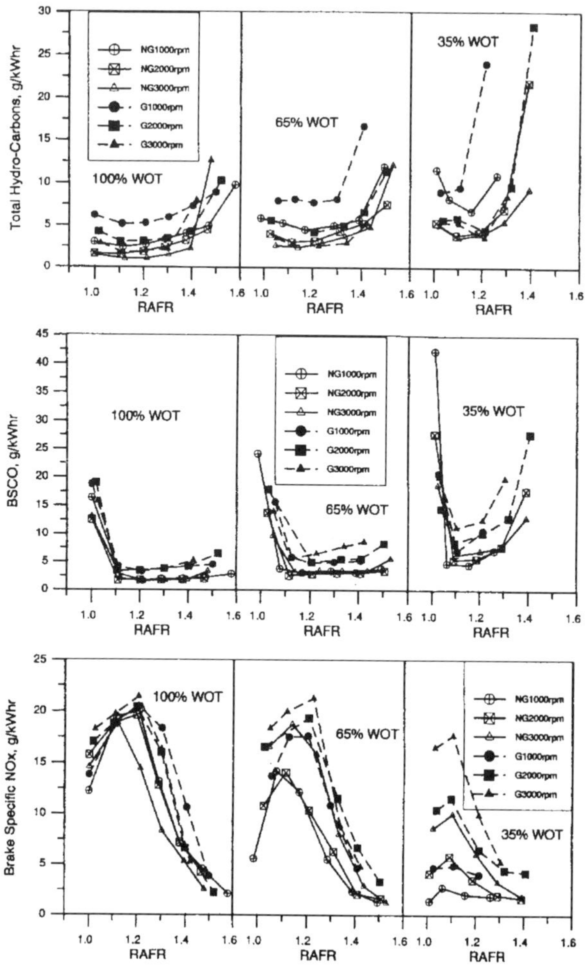

Fig 6: Brake specific HC (top) CO (middle) and NOx (bottom) emissions

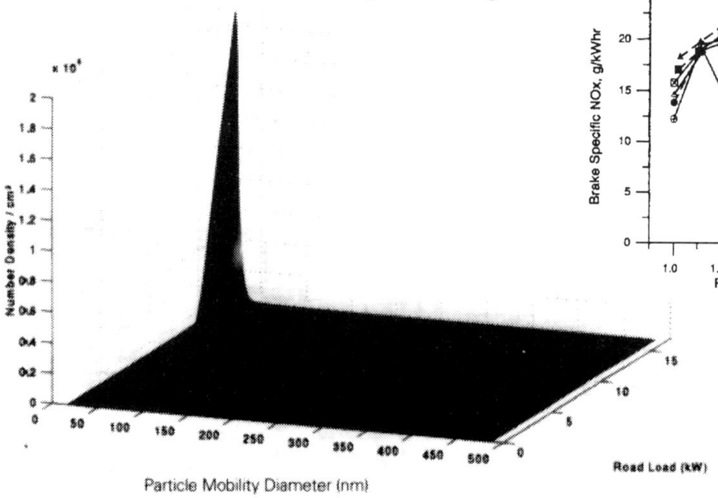

Fig 4: CNG particulate size distribution

sient- operation emission data Eastlake found considerable cooling-off of the catalyst over long periods, typified by long periods of idle experienced, say, in Oxford Street. On re-selecting a catalyst formulation to give best results on VTEC tests, 90–95% conversion levels were obtained.

The crucial importance of truck and bus testing was also demonstrated by the projected emissions levels for different vehicle categories, shown at the seminar by the pie chart of Fig 5, referring particularly to NOx emission in the year 2000. This is despite the far lower vehicle volumes in this category than those of passenger cars. In the case of buses, however, the engine emissions for passengers carried are much lower than those applying to passenger cars. With the European 13 Mode test for heavy-duty vehicle engines, testing is carried out at a series of steady speeds and loads, emissions being measured at each point. There is no measurement of transient acceleration effects nor cold-start emission effects. Because one type of engine is used in a variety of different heavy-duty vehicle types, the engine manufacturer only needs to design and calibrate one engine iteration. Once this is homologated it can be sold to many different vehicle builders who are likely to optimize the installed engine for fuel economy and driveability rather than reduced emission levels. It is therefore proposed to adopt a new transient cycle test for Europe in AD 2000. However, it is still likely to be based on engine, rather than vehicle, testing.

Fig 7: BOC DS LNG-powered ERF

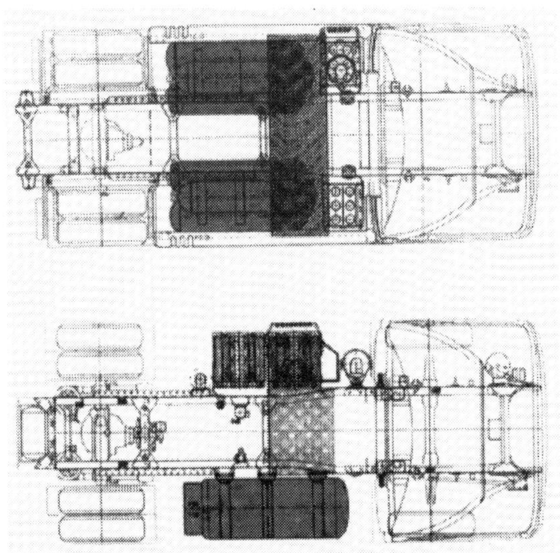

Fig 8: Gas storage requirement for equivalent range, CNG (top) against LNG (bottom).

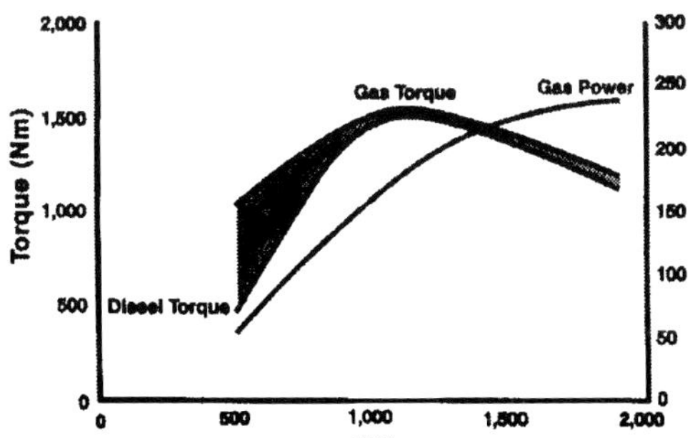

Fig 9: Torque characteristics: diesel vs gas

CNG's advantage over gasoline

Research at the University of British Columbia[1], carried out over a wide range of air/fuel ratios, has shown while generally power output is reduced by some 12% with CNG (compared with gasoline) due to the displacement of air by the gas, emission levels were substantially reduced. While both fuels exhibited nearly equal thermal efficiency, importantly natural gas showed increased efficiency at very lean mixture running, due to an extension of the lean limit of combustion over gasoline. The authors argue that by using a lean-burn strategy, with a carefully optimized SI CNG-engine, this might result in both light- and heavy-duty engines meeting current and proposed emissions regulations without the need for a catalytic converter.

Tests were carried out in the university's Alternative Fuels Laboratory in a Ricardo Hydra engine at three speeds and for one full-load plus two part-load conditions. Emissions results are shown in Fig 6 as a function of relative air/fuel ratio (RAFR) for gasoline (G) and natural gas (NG). At wide open throttle, HC emissions are some 50% lower with CNG while CO was 100% lower for all operating ranges. NOx levels have similar peak values, for both fuels, under full-load conditions but the advantage increases for CNG under part-load conditions.

Liquified natural gas (LNG)

The 12.2 litre Perkins Eagle LNG engine has demonstrated the effectiveness of natural gas in a real operating environment, used in ten ERF tractor units operating for BOC Distribution Services, Fig 7. The 320 bhp engine is developed from the Perkins 2000 Series of gas engines and Eagle TX range of diesels. The so-called TxSi unit is air-to-air charge cooled, fitted with a wastegated turbocharger and two-way oxidation catalyst. LNG has a fuel density over double that of CNG, allowing more energy to be stored in a smaller volume, Fig 8. LNG also has consistently high methane content, which allows reliable running in high compression ratio engines and gives more consistent emission control, say Varity-Perkins.

The IMPCO electronic air and fuel management system is a key feature of the engine. It simultaneously measures the mass of both air and gas, in a direct manner and with very fast response, so that the unit returns very low emission levels even under transient operating conditions. It meets AD 2000 legal requirements and, because LNG contains virtually no sulphur, long catalyst life is predicted. Noise emission is also very considerably less and at idle the level is down by 7 dBA against diesel. As seen in the torque characteristic of Fig 9, the engine can lug comfortably down almost to low idle speed.

Exploiting the low CO_2 emission of small HSDI diesels

Ricardo engineers[2] have shown the possibilities of using diesel propulsion for sub-B class cars which will emit less than 90 g/km CO_2 exhaust emission. Potential fuel consumption less than 3 litre/100 km is also mooted for a 1.2 litre 4-cylinder engine with 70 mm bore. Such a car would have weight of 750–800 kg and power rating of 45–55 kW. Fuel injection and maximum cylinder pressures (140–150 bar) are both considerably higher than existing HSDI diesels and common rail fuel distribution, with 17 mm dia injector, is suggested.

According to the authors, the current industry 'standard' 17 mm diameter fuel injectors are more suitable for truck engines than small passenger cars when packaging for diameter and length. Experience has proved that an injector mounted vertically and central in the bore between a 4-valve pattern typically provides the best overall performance for HSDI engines. However, with a relatively large injector and small bore, the space available for valves was naturally at a premium. The critical section was between the fuel injector and inlet valve centres, Fig 10. The maximum size of inlet valves was limited on one side by the clearance to the cylinder bore and on the other side by the requirement to achieve a coolant path between the inlet port and the injector, also between the valve springs and injector. The use of a separate injector sleeve, in preference to a cast boss, has allowed for the maximum possible inlet valve size whilst maintaining cooling and casting integrity. Together the inlet valves shown gave a total flow area of around 15% of the bore area which was shown by simulation to allow performance targets to be reached.

Fig 11 shows three possible porting arrangements: (a) shows two directed inlet ports and a tandem exhaust port. This achieved a swirl ratio of 2.5 Rs and was also unrestrictive, leading to high volumetric efficiency — but had the disadvantage that narrow water jacket sections under the ports would result in weak sand cores and potential casting problems. The small bore of 70 mm meant that it was not possible to separate, with coolant, the long inlet port wall from the exhaust port of the next cylinder,

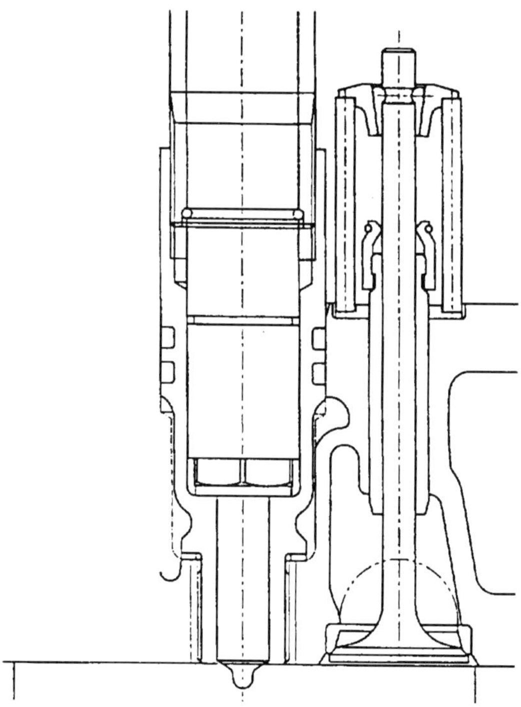

Fig 10: Injector/inlet-valve positions

introducing the possibility of inlet charge heating. (b) shows two helical ports, one long and one short. This arrangement had similar performance but the added advantage that if inlet port deactivation was considered, a better range and control of swirl was possible. (c) shows side by side helical inlet ports, found to be restrictive and also limited to a maximum swirl ratio of 2.1. Their structure does not improve cylinder head stiffness, unlike the tandem arrangements.

The packaging of the fuel injectors considered was dependent on their type, Fig 12. Two types of common rail injector were investigated: internal and external solenoid designs. Both B and C had external solenoid designs where the solenoid is attached to the injector either in-line or at 90 deg to the body. With this type the solenoid envelope requires a space of approximately 30 X 70 mm; A, however, has an integral solenoid which fits completely inside the 17 mm body of the injector. The figure shows that an injector with an integral solenoid offered improved packaging into the 300 cc/cylinder engine. This injector allowed for more freedom of valve-train design and achieved the smallest overall package size in the cylinder head area. Another advantage was that, due to its relatively short length and top fuel feed, it was possible to design a novel fuel rail which would also function as an injector clamp on a multi-cylinder engine.

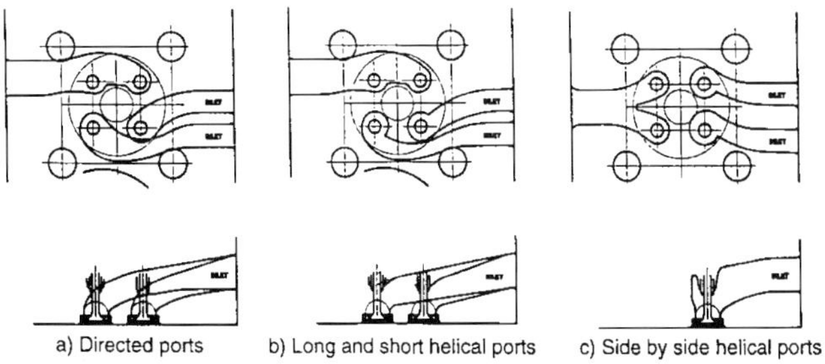

Fig 11: Alternative porting

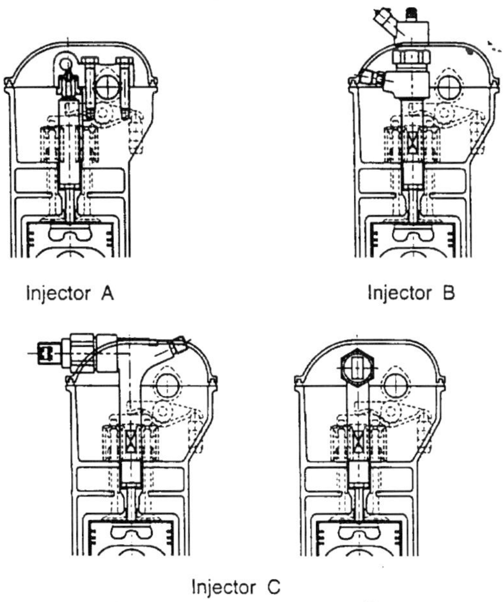

Fig 12: Alternative injector installations

The main constraints on the valve-train were that the four valves had to be close to vertical and the injector had to package in the centre of the bore. Fig 13 shows the alternative valve-train layouts that were considered most feasible. DOHC direct-attack was the simplest and potentially least expensive valve-train, but required a 9 deg valve included angle to allow for the central injector. This meant that dead volume in the head increased, which dropped the 'K factor' below the minimum desirable level of 70%. A DOHC valve-train was feasible, however, when using finger followers which allowed positioning of the camshafts away from the centre of the engine and provided adequate clearance to the fuel injector. Although this design would be more expensive and less stiff than the direct-attack design the vertical valve inclinations allowed the least possible dead volumes and hence the lowest predicted engine emissions. Both alternative SOHC valve-train layouts used bridge pieces to connect the inlet and exhaust valve pairs.

The SOHC-rocker design had a low camshaft and hence potentially the smallest overall package height when also considering the camshaft pulley. However, this design would be more expensive than the SOHC-finger design because of the extra rocker shaft machining required. OHV-pushrod/rocker was considered feasible for a low cost design but the performance of the engine would probably be limited by the high inertia and low stiffness of the valve-train components, making it impossible to produce the accelerations required to achieve the desired valve performances. Electro-hydraulic operation of the valves was investigated briefly, and potentially the freedom to have variable control over valve timing, lift, and duration would naturally allow for better optimization of the gas exchange process including internal EGR.

The main structure could be conventional, using separate cylinder block and head, or a monoblock, Fig 14. Aluminium was considered a good compromise between cost and performance, a possible engine structure being shown *left*, the aluminium head and block being through bolted from underneath, which allows extra packaging space for valve-train and fuel equipment in the cylinder head. The pistons are shown running in the parent bore aluminium which would be surface treated to achieve acceptable wear resistance. A monoblock engine structure is shown *right*. The head and block are integral, and the structure is completed by a bolted bedplate. The more uniform loading and cooling of the upper cylinder would reduce bore distortion and hence reduce oil consumption and oil sourced emissions. Satisfactory manufacture of a monoblock presents some difficulties. Firstly, the cylinder bore cannot be through-honed due to the closure of the cylinder by the gas face, which means honing would stop short of the gas face. A piston with a deep top land would then be required, penalizing emissions due to the increased crevice volume. This honing problem could be solved by inserting a fully honed dry liner into the bore, the only penalty being a small increase in engine length and weight; also machining the valve seat insert bores into the gas face has to be done through the bore. However, using a dry liner means that extra clearance is available before the liner is inserted. With this arrangement it was possible to achieve acceptable tool clearance with valve sizes large enough to achieve the required performance. In order to achieve good fuel consumption, parasitic losses to engine ancillaries must be minimized. Mechanical drives to these components means they are driven at speeds in a fixed ratio to engine speed rather than at their most efficient operating conditions. An alternative to a conventional alternator is to use a flywheel starter/generator which could be used to crank the engine at high speed for starting and then generate the required electrical output once the engine is running. High speed cranking would help to reduce start- up emissions and electrical drive input could improve idle stability and reduce emissions during critical phases of the drive cycle. This system would be more commercially feasible if the HSDI engine were used in a hybrid diesel and electric vehicle. With vehicle demands for greater electrical outputs the demands on alternators has increased. Alternator manufacturers are developing water cooled alternators which have higher outputs and are less noisy than their air cooled predecessors, the authors point out.

A novel approach to packaging the ancillaries for a 1.2 litre HSDI engine is shown in Figure 15. The air conditioning compressor is mounted at the front of the engine with its clutch being driven directly from the crankshaft. A belt from the crankshaft

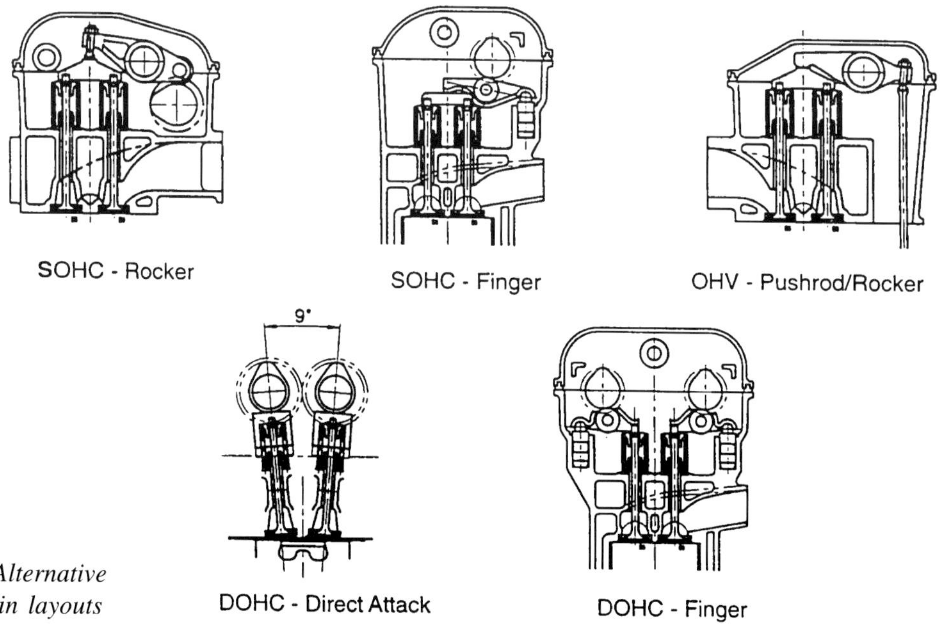

Fig 13: Alternative valve-train layouts

pulley would drive a water cooled alternator, also front-mounted. Electrical power steering would be used. This arrangement reduces the width of the engine and in a transverse installation would mean either increased frontal crash protection or the possibility to reduce bonnet length. Two types of transmission configuration were considered, conventional transaxle and an integrated engine and transmission structure. A transverse mounted engine with front wheel drive was assumed in both cases. A conventional bolt-on transaxle transmission provides the opportunity to easily adopt a range of transmissions for the vehicle. These would probably include manual, automatic, CVT and AWD. It is also known current technology. An integrated engine and transmission is shown in Fig 16. This arrangement has the transmission components housed in the cylinder block with the final drive on the split line between the block and the bedplate. This powertrain would be stiffer and hence quieter than a conventional design, and would also be lighter. Overall powertrain length is reduced, which allows the engine to be mounted more centrally and for the drive shafts to have equal lengths, reducing torque steer and part-count. The disadvantage of this arrangement is that it would be more difficult to have vehicles with a range of transmissions when compared to a conventional bolt-on transaxle transmission.

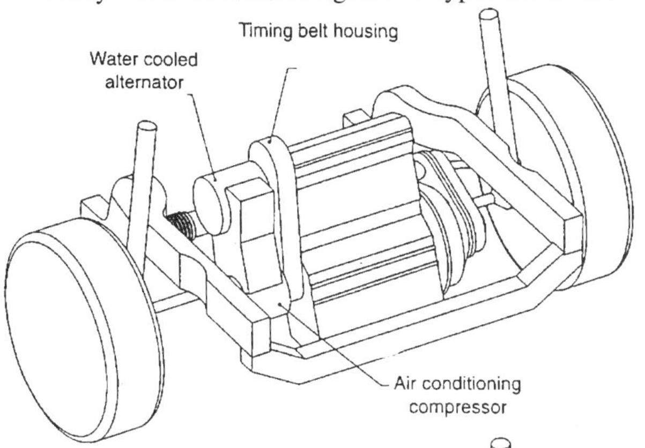

Fig 15: Ancillary packaging

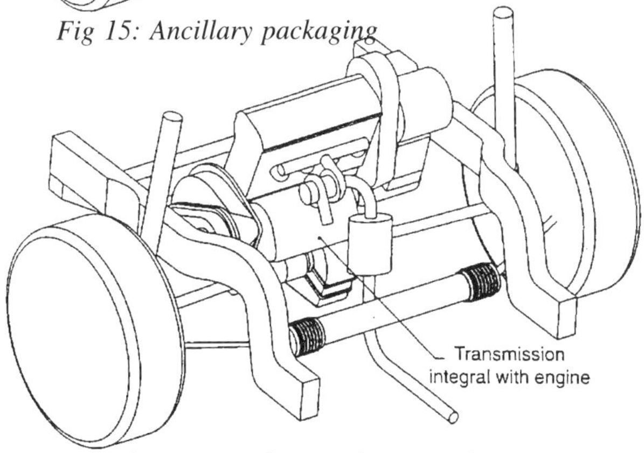

Fig 16: Integrated engine/transmission

Direct injection gasoline

Ricardo engineers are also taking direct-injection gasoline engines closer to production realization with work on the optimization of the combustion system[3]. In-cylinder gas sampling has been used to obtain local air-fuel ratio measurements in the development of a suitable piston bowl design. The measurements are used to validate CFD codes valuable to the design process. Stable combustion at 50–60:1 A/F ratios has enabled unthrottled operation down to

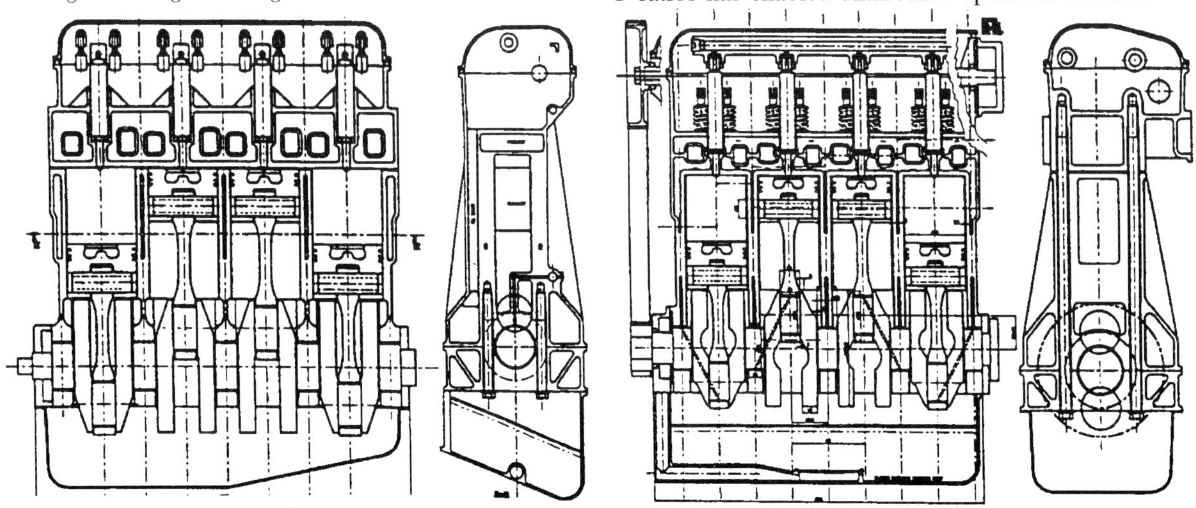
Fig 14: Through-bolt (left) and monoblock (right) configurations

zero load.

Relatively low cost electronically controlled common-rail fuel injection systems are spurring a renewed interest in stratified-charge combustion, say the authors, using the technique of early injection at full load and late injection at part-load. Reduced cold-start enrichment and improved fuelling control in transient operation are also in sight, with the eventual goal of achieving the theoretically possible fuel economy of a diesel coupled with specific power of a gasoline engine. Success will depend on efforts now being made to control NOx emissions in the lean exhaust.

The optimum-shaped piston bowl of the combustion chamber is one which can produce an ignitable mixture at the spark-plug at a crank angle for phased combustion to occur, producing best trade-off between thermodynamic efficiency and HC/NOx emissions over a wide speed/load range. Ricardo have modelled the process in the four steps of Fig 17: (a) applying Bernoulli's theorem to the fluid flow in the injector nozzle to calculate spray characteristics; (b) comparing spray penetration with piston position to predict time when impingement on the piston is reached; (c) fuel flow along the piston until it detaches as a vapour cloud; (d) assistance by air motion in bringing the vapour to the plug in time for ignition; Fig 18 shows a dynamic visualization of the flow-field using CFD.

Fig 19 shows an example result of the total calculation over the four steps, its use being in the quick screening of combustion chamber design options. Here, four different start-of-ignition timings have been used in calculating the fuel spray trajectories prior to piston impingement. After flow along the piston and departure as a spray-cloud, it is assumed the tumble motion, measured at BDC by CFD, is compressed into the bowl as the piston rises and an angular momentum calculation is used to estimate air velocity at the edge of the piston bowl and hence the transportation time for the cloud to reach the plug. In-cylinder sampling tests were carried out using flat-piston crown, square and spherical bowl shapes, start-of-injection being varied between 310 and 330 deg and injection duration held at 20 deg, equivalent to a load of 2.5 bar BMEP. Resulting A/F ratio at the plug is seen in Fig 20.

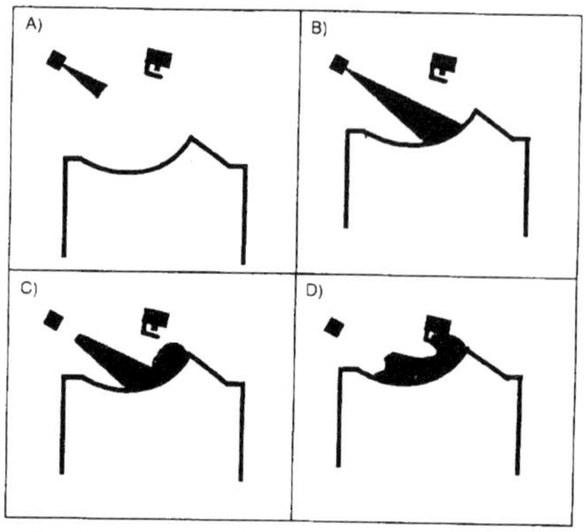

Fig 17: Fuel stratification and transport mechanism

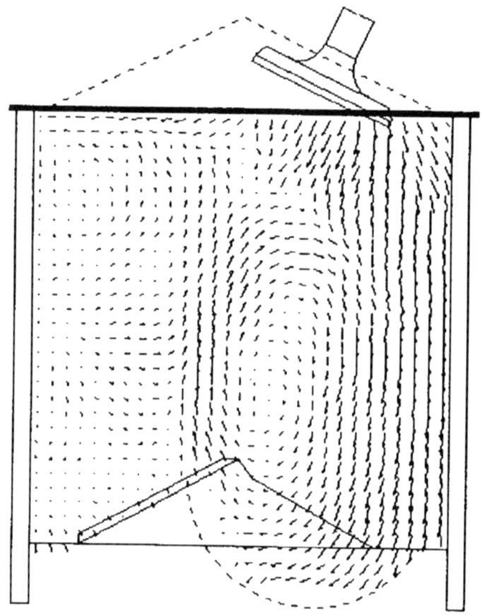

Fig 18: Dynamic visualization of in-cylinder flow field

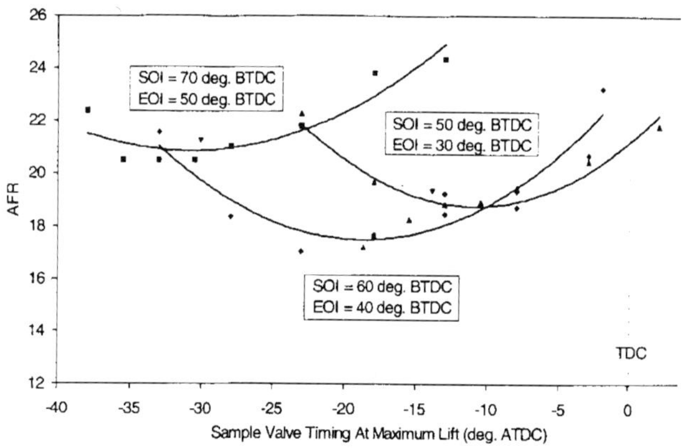

Fig 19: Fuel spray transport calculation result

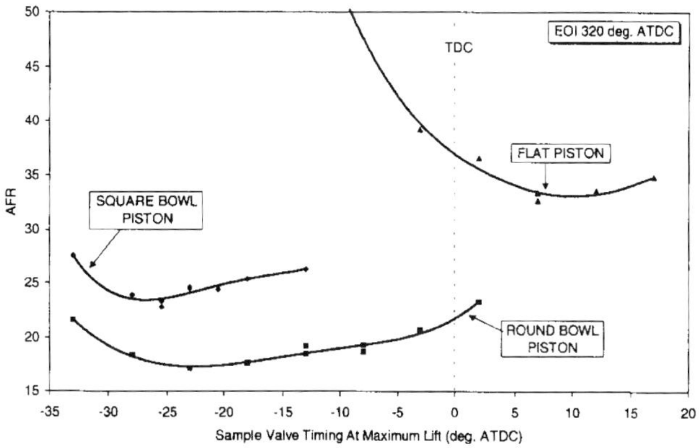

Fig 20: Effect of injection timing on A/F ratio at the plug

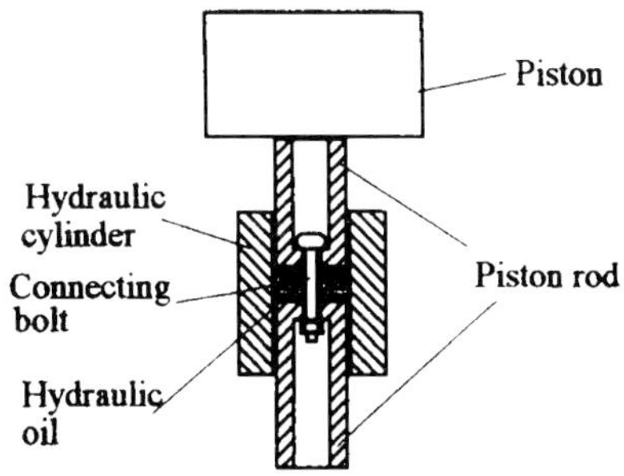

Fig 21: Hydraulically-sprung connecting rod

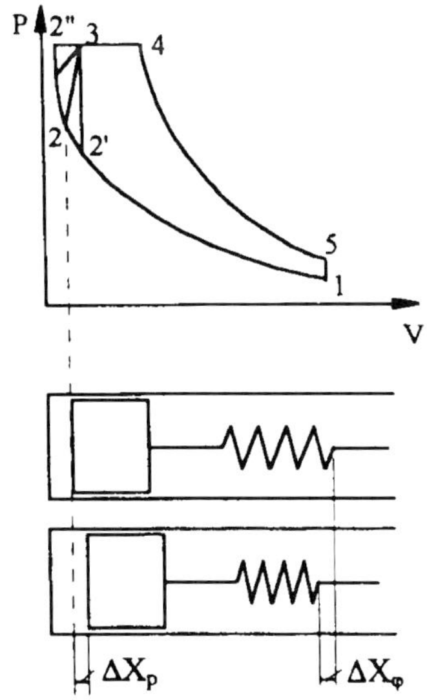

Fig 22: Ideal cycle for Deocp engine

Constant-pressure cycle: the future for diesels?

Researchers at Newcastle University and the Dalian Maritime University in China[4] have hinted that the limitations on diesel development, imposed by severe exhaust pollution control regulations, may be solved by unlocking the cycle efficiency of the constant-volume cycle.

A variable engine cylinder volume controlled by cylinder pressure has been achieved in a diesel engine modified by a hydraulic cylinder, Fig 21, inserted into the connecting rod. The diesel engine with oil-cushioned piston gives the unit its name of Deocp. It is characterized by high cycle efficiency, improved low-load performance and good starting ability. The theoretical combustion cycle for the engine is seen in Fig 22 (points 123451) overlaid on the conventional cycle (12'3451).

The effective stiffness K (with corresponding constant pressure P_k) of the hydraulic cylinder can be altered to change the characteristics of the engine, the table of Fig 23 indicating that engine compression ratio ε and efficiency η increase as the engine load p_l decreases. A solid connecting rod is given by $K = 2.26 \times 10^9$ and here it is seen that efficiency does not change with load.

In tests of a physical engine, account had to be taken of the fact that piston movement on the Deocp is not exactly dependent on crank angle, owing to the hydraulic spring, hence the measurement technique shown in Fig 24 was used by the authors. Tests were carried out at 100, 90, 75, 50 and 25% full load. Results in Fig 25 (the PV diagram) show the 9% increase in thermal efficiency at 25% full load and the 1.6% gain at full load.

The authors attribute the gain in thermal efficiency to the piston of the Deocp reaching its highest position, at the end of the exhaust stroke, because of its inertial force. Improved scavenging results from small clearance volume, and high static compression ratio, especially at low loads. Swirl is also created at this point by the rapid increase in cylinder pressure.

The authors have calculated that the combustion process finishes about 7 degrees of crank angle earlier than that of a comparable conventional diesel. In examination of the indicator diagram, in crank-angle form (Fig 26), it can be deduced that the cylinder pressure of the conventional engine (CE) is higher than the Deocp at the beginning of the compression stroke because the CE has a relatively large amount of remnant waste gas. After about 55 degrees of crank angle BTDC the Deocp's compression pressure increases faster because of its high compression ratio. During the expansion stroke, as cylinder pressure falls, the compression energy stored inside the hydraulic spring is released to the engine cylinder and the expansion line of the Deocp is resultingly higher. The diagram also shows significant compression of the hydraulic spring occurs at about 4 degrees BTDC, where combustion starts, with maximum compression taking place 10 degrees ATDC.

SFC results are shown in Fig 27 and the detailed test results in Fig 28.

K (N/m)	P_k (bar)	P_1 (bar)	ε	ρ_2	λ_2	η
1.49×10^7	49.05	2.455	12.52	1.235	1.569	60.27
		1.962	13.22	1.228	1.608	61.20
		1.471	14.17	1.218	1.653	62.37
		0.981	15.46	1.197	1.719	63.86
2.26×10^9	39.24	2.455	11.67	1.010	1.732	61.57
		1.962	11.67	1.008	1.895	61.57
		1.471	11.67	1.006	2.116	61.64
		0.981	11.67	1.004	2.548	61.70

Fig 23: Ideal cycle efficiency

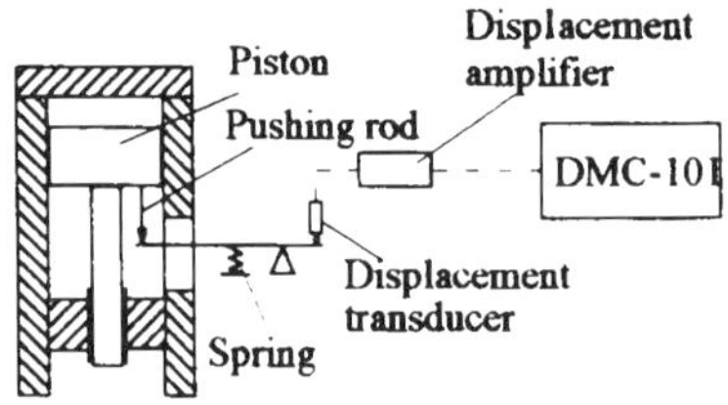

Fig 24: Piston displacement measurement

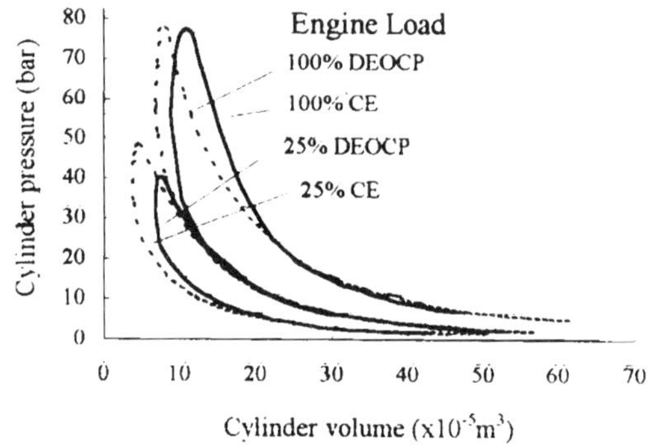

Fig 25: Cylinder pressure vs volume

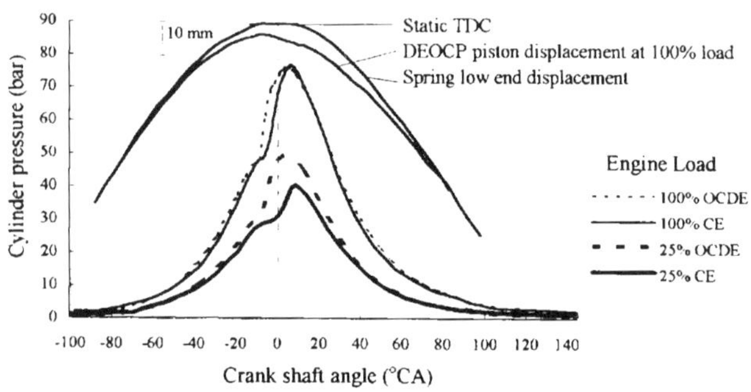

Fig 26: Cylinder pressure vs crank-angle

	Diesel engine with an oil cushioned piston					Comparison engine				
Load (%)	100	90	75	50	25	100	90	75	50	25
ε	11.72		13.28	14.95	15.93	13.29	13.29	13.29	13.29	13.29
p_1 (bar)	1.76	1.65	1.47	1.23	1.05	1.76	1.65	1.47	1.23	1.05
n (r/min)	1010	965	905	799	636	1010	965	907	794	630
N_i (kW)	6.29	5.01	4.59	3.34	1.97	6.20	5.61	4.63	3.14	1.59
η_i (%)	39.17	35.82	37.39	36.26	32.91	38.56	40.19	37.46	35.28	30.19
g_i (g/kW h)	215.22	235.38	225.47	232.50	256.21	218.63	209.79	225.05	238.94	279.23
p_3 (bar)	79.59	76.75	69.53	58.66	50.47	80.14	70.88	63.99	51.43	41.37
$\lambda(p_3/p_1)$	45.07	46.29	47.25	47.46	47.63	45.38	42.75	43.48	41.61	39.05
N_e (kW)	5.23	4.68	3.86	2.62	1.13	5.42	4.96	4.06	2.73	1.33
η_e (%)	32.59	33.64	31.42	28.47	21.83	33.70	35.55	32.91	30.70	25.18
g_e (g/kW h)	258.67	250.60	268.28	296.09	386.30	250.15	237.13	256.13	274.64	334.78
t_1 (°C)	16.5	17.0	17.5	17.5	18	11.5	11.5	11.5	11.5	11.0
t_e (°C)	250	242	244	236	200	250	242	249	242	224
Remark	$\varepsilon_0 = 18.47$, $K = 1.68 \times 10^7$ N/m, $P_k = 38$ bar									

Fig 28: Full test-results

Valve arrangements for enhanced engine efficiency

Three recent contributions to the literature of engine design and development show how major changes in engine efficiency can be effected by different valve control systems and configurations

Intake valve disablement

According to Rover researchers[5] the technique of disabling airflow through one inlet valve (Rover Asymmetric Combustion Enhancement, RACE) is a way of generating stronger axial and barrel swirl to improve engine combustion efficiency. In conjunction with EGR it improved efficiency from 3 to 7% and with lean burn from 7 to 11%. Because high levels of EGR or excess air normally lead to poor ignitability, slower combustion, instable combustion and unburnt HC emissions, a method has to be found of increasing the tolerance of charge dilution. Development work was therefore carried out on a 10.5:1 compression ratio 4-valve per cylinder engine, fitted with air-assist fuel injection.

Previous work had shown that increasing barrel swirl alone was not enough to achieve charge dilution tolerance and hence the decision to increase both axial and barrel swirl with an arrangement as seen in Fig 29 used for high activity, homogeneous charge control. Prior to disabling the second inlet valve, an inducer in the form of a port mask was fabricated to give barrel swirl ratio of 2.4 with both ports flowing. Flow testing of the cylinder head the gave the results of Fig 30.

Air/fuel ratio loops at a 2000 rpm/2-bar BMEP test condition were obtained for the RACE engine, with and without the inducer, and compared with the base engine as shown in Fig 31. A very good lean burn limit and large fuel economy benefit resulted, with the low-swirl engine working slightly better than the high swirl one. Generally, high activity of the RACE engine speeds up the early burn and stabilises lean combustion. Lean burn NOx limit emissions were also lower than the base engine although emissions were higher at stoichiometric due to the faster burn of the RACE system. Higher EGR tolerance and greater fuel consumption better than the base engine were also achieved.

Hot-wire anemometry measurements were used to show that high levels of turbulence were present while maintaining low bulk velocities in the combustion chamber, believed by the authors to be the key to igniting a dilute mixture.

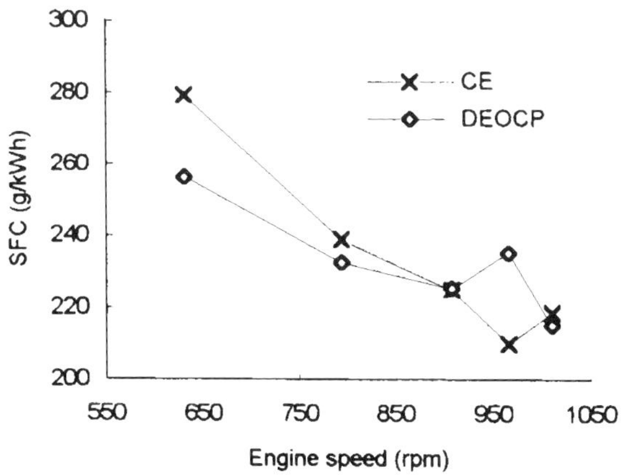

Fig 27: Specific fuel consumption

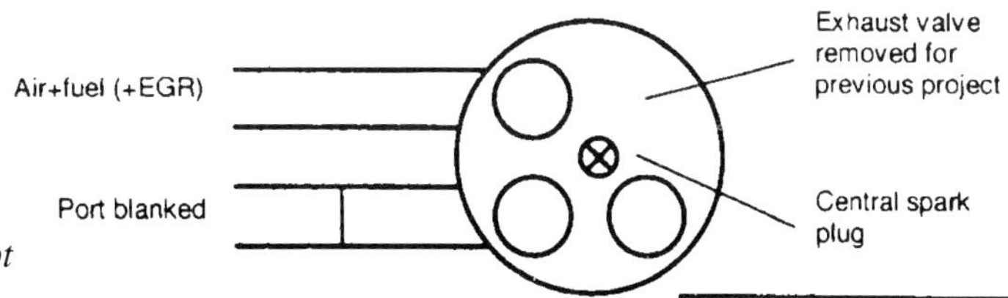

Fig 29: High activity, homogeneous charge concept

Fig 30: Results of flow tests

	Barrel swirl ratio	Axial swirl ratio
Base head	1.2	0.0
Base head with port blanked ("RACE, low swirl")	2.2	2.6
Port blanked + 2.4 inducer ("RACE, high swirl")	3.2	4.5

Late intake valve closing

Researchers at Sheffield Hallam University[6] have shown fuel savings as much as 7% can be obtained by the elimination of most of the pumping work energy loss inherent in late intake valve closing (LIVC) load control. They have also shown that, combined with a variable compression ratio device, LIVC can achieve up to 20% savings and a two-state LIVC control mechanism is under evaluation.

Energy loss in pumping gas past the throttle, to the point where no useful output is generated at idle, is considered by the authors to be an important contributor to engine inefficiency. The LIVC concept achieves engine power modulation without the need for a throttle valve, trapped charge mass being reduced by keeping the valve open during a portion of the compression stroke, allowing excess induced charge to be returned to the intake manifold. Fig 32 compares the throttled and LIVC engines by their PV diagrams, shaded areas depicting pumping losses. While conventional variable valve-timing mechanisms can alter valve phasing by some 30%, the authors consider that for pure LIVC operation alterations in excess of 100% are necessary.

Earlier workers reportedly noted a slight loss in indicated thermal efficiency with LIVC, mostly caused by a reduction in effective compression ratio. Some workers used a secondary camshaft above the rockers, on a push-rod engine, to achieve LIVC of up to 90 degrees of crank angle. Other workers reportedly tackled the loss of thermal efficiency by adopting the Otto-Atkinson combustion cycle, restoring the compression ratio at light loads by a variable compression device for reducing clearance volume.

The Sheffield Hallam approach of allowing a two-state LIVC control involves a relatively simple mechanism compared with that used by earlier workers, some of the load control being carried out with a conventional throttle. Seen in Fig 33, the arrangement consists of an intake camshaft with a profile to cause late closing of the intake valve at all times while for full-load operation a reed valve in the intake manifold prevents the charge being rejected from the cylinder. At low loads the reed valve is bypassed to allow the charge to return to the manifold. The authors point out that the reed valve must be positioned as close as possible to the inlet valve while leaving sufficient space for the bypass. Future development is seen as incorporating the system within an Otto-Atkinson cycle engine

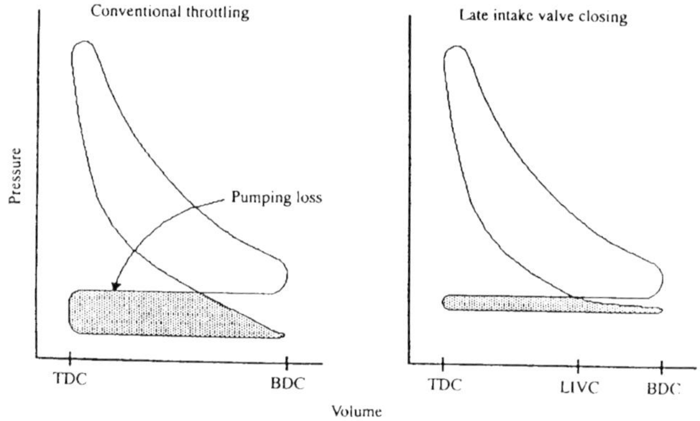

Fig 32: Part load modulation by late intake valve closing

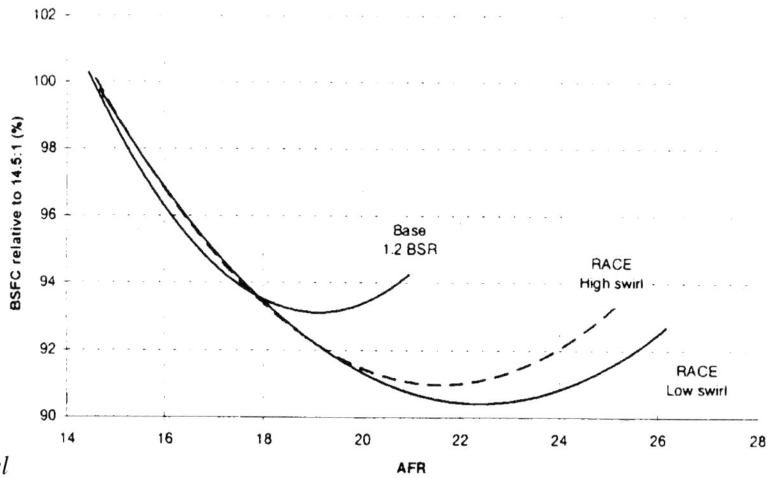

Fig 31: Effect of AFR on fuel

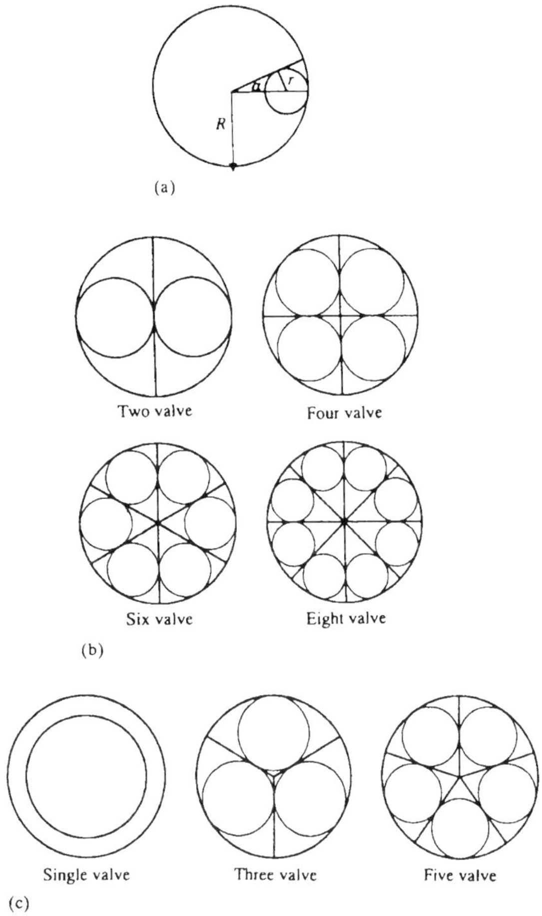

Fig 34: Generic areas for (a) single, and Fig 35:(b) even number and (c) odd number of valves

Multiple valve arrangements

With the trend to increasing efficiency by the use of multi-valve combustion chambers, researchers at Queen Mary and Westfield College, London University[7], have developed a simple method of making preliminary assessments of such schemes, based on calculation of valve effective flow areas for chambers with eight or less valves. Seven valve heads are seen as an upper limit, with 5–7 valve systems meriting investigation in view of 4-valve heads being shown to have a 37% increase in valve area over the traditional 2-valve systems.

Effective area is calculated from the steady flow equations applied to a constant pressure model for obtaining steady mass flowrate. The authors quote expressions for effective area in both choked and unchoked modes of flow. They obtain a non-dimensional effective area by dividing the above values by the generic area of the valve seat as $\pi d^2/4$. The resulting values are compared in Fig 34/35 for different poppet valves as a function of valve lift.

Maximum ideal area, generic area, of a single valve (Fig 34) is given by the authors as

$$[R/(1 + \operatorname{cosec} \alpha)]^2 \pi$$

and for N valves the half angle α is π/N and the generic area ratio

$$4n_j(r/R)^2$$

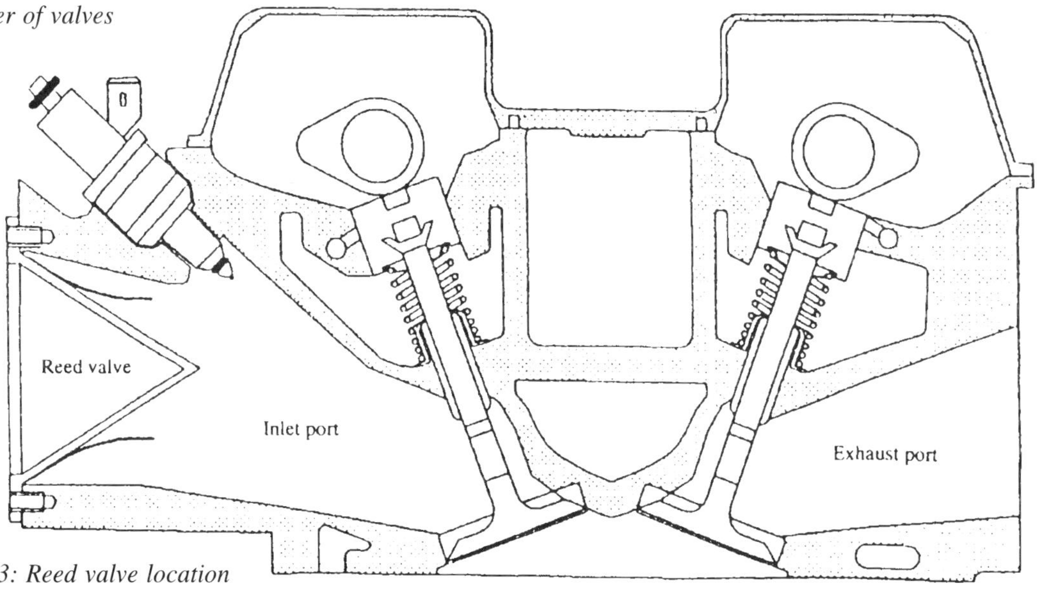

Fig 33: Reed valve location

where $j = i$ or e, the ratio of valve and cylinder radii and area ratios being shown in the table of Fig 36; the generic flow area ratio of the different valve arrangements is seen in Fig 37. These two figures describe the authors' key ideas, the last figure having curves intended only to show curves, just the points having physical meaning.

The analysis above is based on inlet and exhaust valves of identical head diameter and Fig 38 has been provided to show an extension of the analysis for unequal size heads. This three valve case has the same exhaust area as that for the two valve head while the inlet valves have a 58% combined larger area. The method could be used to optimize valve areas, possibly, the authors suggest, by slightly increasing exhaust while reducing inlet valve head sizes, to give significant overall gain.

A special case for 5-valve heads is made (1) with two inlet valves the same size as in the four-valve head, giving 37% increase in area over the two-valve case and (2) with three exhaust valves the same size as those for the six-valve case, giving a 33% area increase. These points are shown in Fig 39.

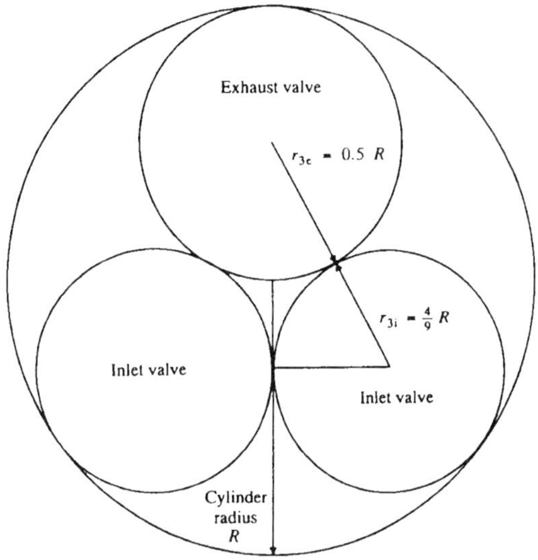

Fig 38: Ideal 3-valve scheme

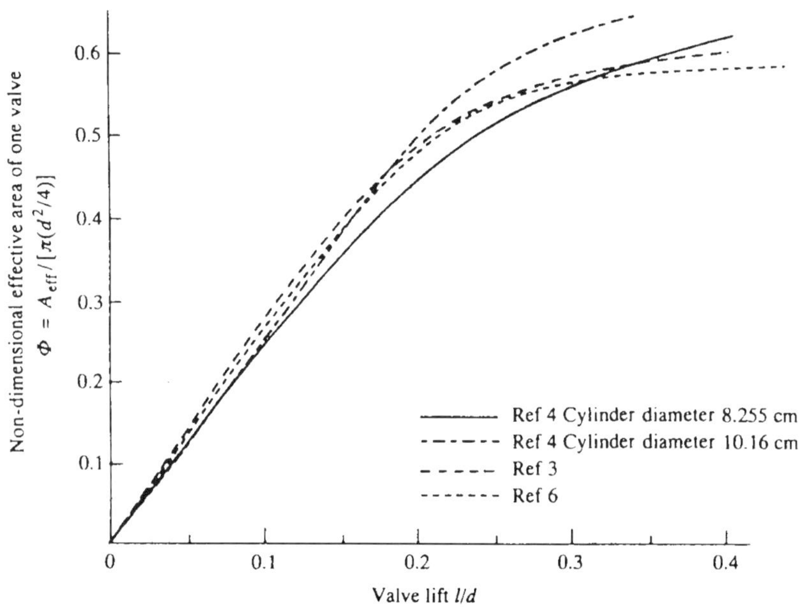

Fig 39: Poppet valve effective area comparison

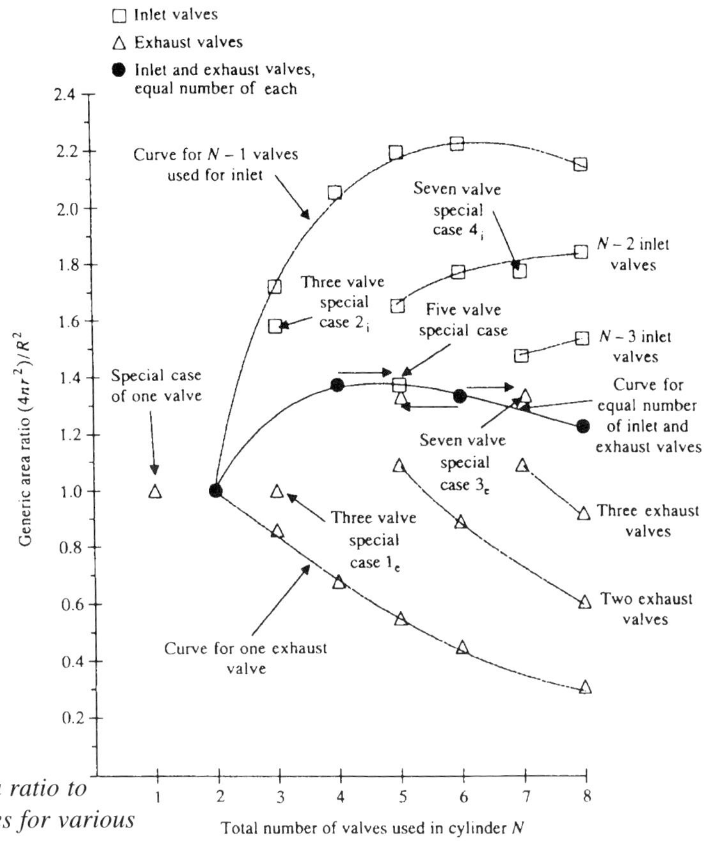

Fig 37: Generic area ratio to total number of valves for various multi-valve schemes

Total number of valves used N	Half angle subtended by one valve α	Radius ratio $\frac{r}{R}$	Cases considered: Number of inlet valves n_i	Cases considered: Number of exhaust valves n_e	Area ratio for number of inlet valves Ω_i	Area ratio for number of exhaust valves Ω_e
1	180°	Practical value say 0.5	—	1	—	1
2	90°	$\frac{1}{2} = 0.5$	1	1	1	1
3	60°	0.4641	2	1	1.7231	0.8616
4	45°	$\frac{1}{1+\sqrt{2}} = 0.4142$	3	1	2.0592	0.6864
			2	2	1.3728	1.3728
5	36°	0.3702	4	1	2.1927	0.5482
			3	2	1.6445	1.0963
6	30°	$\frac{1}{3} = 0.3333$	5	1	2.2222	0.4444
			4	2	1.7778	0.8888
			3	3	1.3333	1.3333
7	25.7143°	0.3026	4	3	1.4656	1.0992
8	22.5°	0.2768	7	1	2.1448	0.3064
			6	2	1.8384	0.6128
			5	3	1.5320	0.9192
			4	4	1.2256	1.2256
7 Special case	—	$\frac{1}{3} = 0.3333$	4	3	1.7778	1.3333

Fig 36: Poppet valve half angles, radius ratios, area ratios and number of valves considered

Trends in transmission design

The design of the 4-speed all-clutch-to-clutch electronically-controlled automatic transmission developed by Honda for the 1998 Accord was described recently by company engineers[8]. Newly developed technology guarantees improved shift quality and fail-safe at high speed without dependency on one-way clutches as in the more conventional system it replaced. Already, by 1996, the company had developed an all-clutch-to-clutch 3-shaft design of short overall length. Old and new gear train schematics are shown in Fig 40.

In the conventional system seven engaging elements involve five multiple disk clutches, a 1–2 one-way clutch and a servomechanism. The countershaft-mounted one-way clutch transmits driving torque when accelerating in 1st gear but it overruns without torque transmission when decelerating in this gear or when driving in other gears. In 1st gear the holding clutch engages to apply engine braking for deceleration. The one-way clutch thus performs dual functions of improving up-shift quality from 1st to 2nd and preventing 1st gear drive when a high speed malfunction occurs.

In the new system there are five engaging elements comprising four clutches and a servomechanism. The 1st clutch transmits torque, while accelerating in 1st, to the 1st range and applies engine braking for deceleration. Cross-section of the new transmission is shown in Fig 41 and its specification in the table of Fig 42. The elimination of the two parts meant that two separate approaches to maintaining the performance criteria had to be taken.

When the conventional transmission shifts up from 1st to 2nd, engine torque output first is transmitted to the main shaft via the torque converter, then on to the idle gear, secondary shaft, 1st clutch, 1st gear, one-way clutch, countershaft and finally to the drive shaft. The computer starts to engage the 2nd clutch when 1st to 2nd gear requirement is indicated, and when 2nd-gear torque has increased to equal mainshaft torque, the transmission changes to the inertia phase and 2nd gear takes over all power transmission. At this point the one-way clutch overruns so that 1st gear does not contribute to power transmission and the shift quality is governed by the drive shaft torque as the product of mainshaft torque T_m and ratios i of 2nd and final drive F gears.

When shifting up from 1st to 2nd gear with the new transmission, operation is the same as for the conventional one up to the point when the computer starts to engage the 2nd clutch. Then, when second gear torque has increased to the sum of mainshaft and 1st-gear torques, the transmission changes to the inertia phase. As the transmission does not have a one-way clutch, the 1st gear contributes to power transmission and the drive shaft torque is given by:

$$T_m i_2 i_F - T_1 (i_1 - i_2) i_F$$

The minimum drive shaft torque value, when

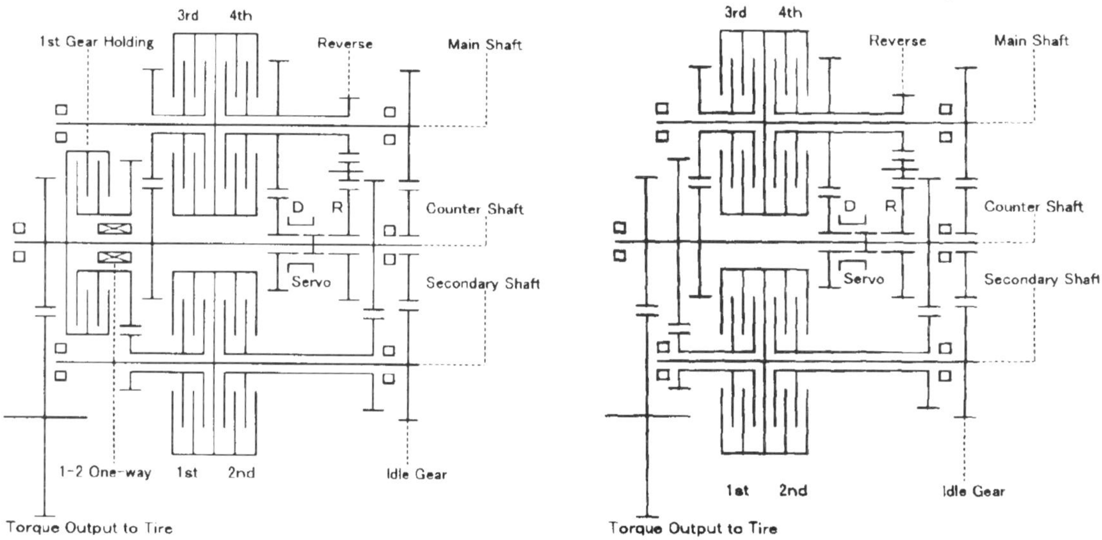

Fig 40: Gear train schematics for conventional (left) and new Honda (right) automatic transmissions

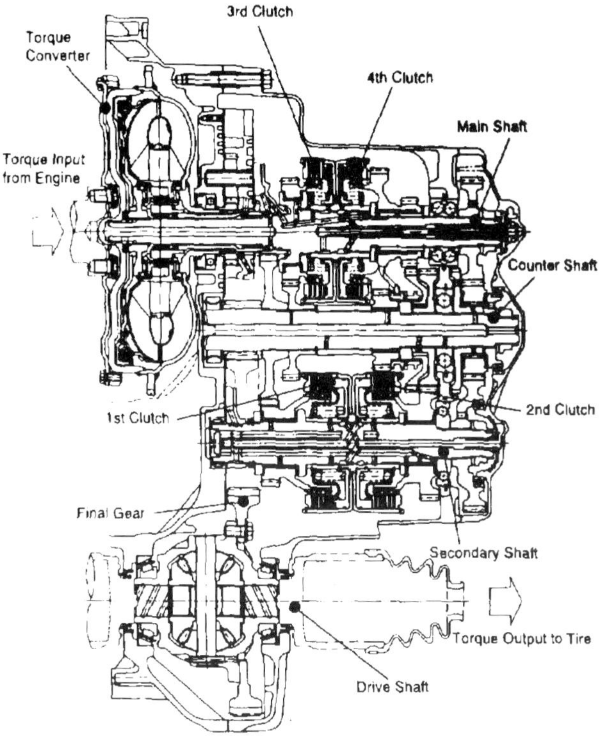

Fig 41: Honda automatic transmission

the clutch-to-clutch transmission changes from the torque phase to the inertia phase, is reduced by $T_l(i_l - i_2)\, i_F$. At this time the driver senses that the torque has dropped and the shift quality has deteriorated. Therefore, in order to prevent this deterioration in shift quality, it was necessary to establish a method for accurately controlling the 1st and 2nd clutch hydraulics at highly precise timing so that $T_l = O$ when $T_m = T_2$. Dealing with the occurrence of high speed malfunction in the conventional transmission involved a fuel cut operating as engine speeded above its set limit and overunning of the one-way clutch. With the new transmission a centrifugal hydraulic cancellation device eliminated centrifugal pressure build-up by the rotating 1st and 2nd clutches; the control system is reworked and two linear solenoid added which switch clutch hydraulics at the appropriate timings. There is also a new shift-logic hydraulic circuit employing a fail-safe system, involving three non-linear switching solenoids, to prevent downshifting into 1st with an engine overspeed.

CVT for 2-litre engined vehicles

Nissan engineers[9] have succeeded in commercializing continuously variable transmission for the upper-middle class of car, and improving accelerative performance from rest over existing CVT systems by the incorporation of a torque converter which has also resulted in better creep capability. The electronic

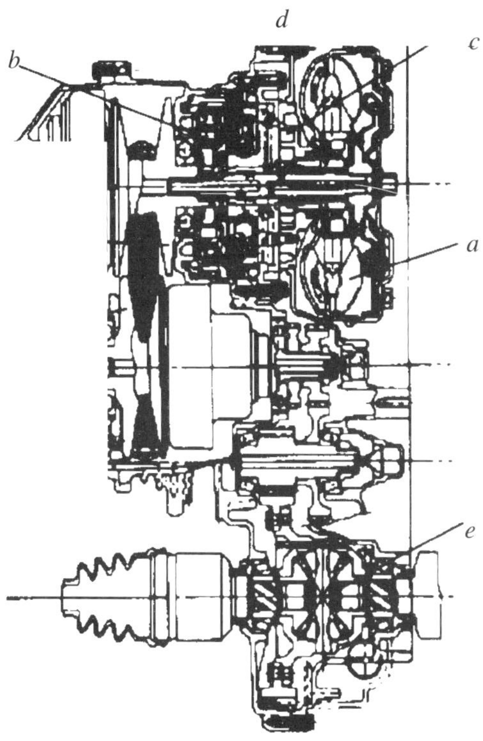

Fig 43: Nissan CVT for 2-litre cars

Engine	Model	F23A
	Version	2.3l SOHC VTEC PGM-FI
Torque Converter	Type	3-element, 2-phase with Lock-up Clutch
	Size	232mm
Gear Train Type		Constant-mesh Type with Three Parallel Shafts
Gear Ratio	1st	2.528
	2nd	1.427
	3rd	0.931
	4th	0.620
	Reverse	1.863
Reduction Ratio		4.466
Used Oil	Type	Honda ATF
	Total Amount	6.2l
Repair Weight (wet)		83.5kg
Overall Length		390mm

Fig 42: Honda transmission specification

control system has achieved improved fuel economy by attaining fine-tuned driveability and lock-up operation at low speeds; also a manual shift mode has been successfully incorporated.

A cross-section through the new unit is seen in Fig 43 indicating the principal elements of (a) torque converter, (b) forward/reverse actuator, (c) ratio-change mechanism, (d) electro-hydraulic circuits and e. reduction gears. Specification and configuration are in Figs 44 and 45. The structural layout is similar to a conventional automatic transmission, only the change-speed mechanism being adapted to belt and pulley layout of a CVT. Overall length, including torque converter, is thus much the same as a conventional automatic gearbox for a 2-litre car.

Less booming noise than that of a conventional automatic's geartrain is inherent in the new unit, however, because inertia of the belt and pulleys is larger. The lock-up clutch has been given facing material with twice the durability of that on a conventional automatic to give lock-up speed of around 20 kph compared with the normal 50 kph, providing the main fuel economy benefit. The ratio-change mechanism uses a newly developed 30 mm wide Van Doorne belt, with nine layer steel band and 400 friction elements. The additional width over the normal 24 mm gives the unit its higher torque capacity. Fig 46 shows the enhanced performance over a system used for a 1.3 litre Nissan car.

The variable-ratio pulleys are arranged such that a tandem piston design on the primary pulley gives two-stage operation to generate the necessary force required in overdrive operation. On the secondary pulley, because of its particular high speed operation, a cancelling chamber is incorporated into the hydraulic cylinder; it prevents any loss in performance due to centrifugal effects on the hydraulic fluid.

High-torque manual transmission

Engineers at New Venture Gear[10] described the development of the company's NVT-750 gearbox, de-

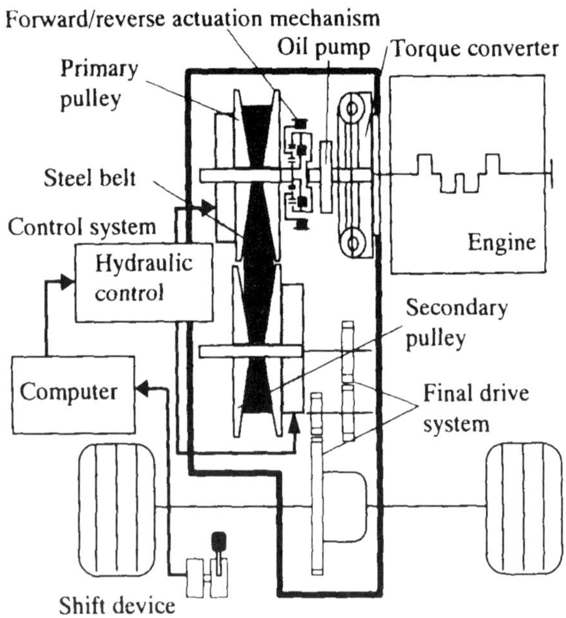

Fig 45: Nissan CVT configuration

Applicable engine		Displacement: 2.0L Max. power: 190ps/7000rpm
Ratio change range		2.326 - 0.434
Basic construction	Launch element	Torque converter with low-speed lock-up clutch
	Forward/reverse actuation mechanism	Planetary gearset + wet multiplate clutch
	Ratio change mechanism	30-mm-wide steel belt + pulleys
	Oil pump type	Hypotrochoid gear pump
	Final gear ratio	5.473
	Fluid type used	Nissan Genuine ATF (special-purpose CVT fluid)

Fig 44: Nissan CVT specification

Nissan belt-drive CVT	1.3-L class		2.0-L class
Belt working radius at Low ratio (Primary pulley)	27.1 mm	<	33.9 mm
Pulley force at Low ratio (Primary pulley)	28.9 kN	<	37.6 kN
Friction coefficient between elements and pulleys μ	0.09	=	0.09
Pulley angle	11°	=	11°
F x R	783.2 kNmm	<	1274.6 kNmm
F x R ratio	1	<	1.63

Fig 46: Torque capacity potential of belt-drive CVTs

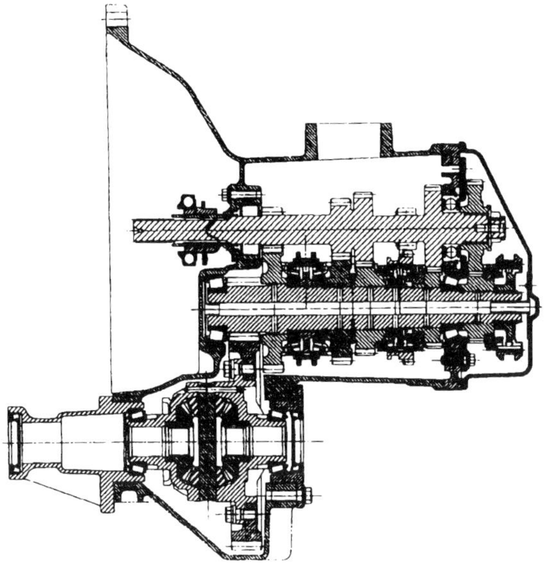

Fig 47: New Venture Gear NVT-750 gearbox

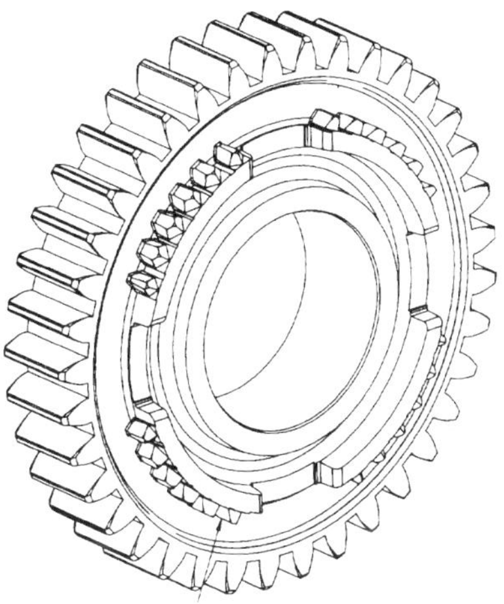

Fig 48: Net form clutch teeth

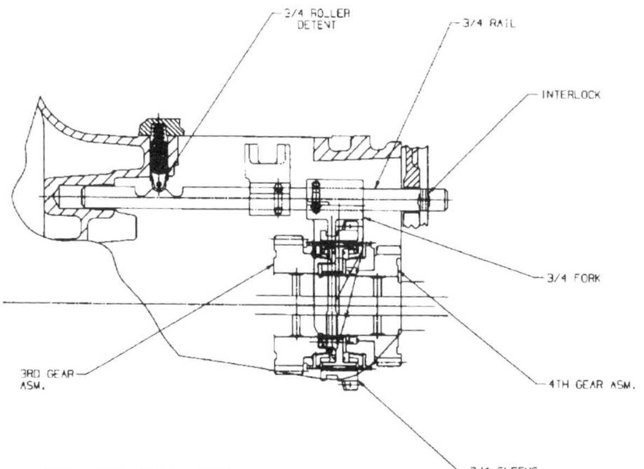

Fig 49: 3/4 shift system

signed as a high torque density transaxle suitable for heavy duty FWD operation. The development objectives were the achievement of greater durability than the 650 unit while achieving world-class gear noise reduction targets. The 5-speed unit has torque capacity of 290 Nm, synchronization on all forward gears, 13.08 maximum reduction and a 200 mm distance between input and output, Fig 47.

Greater gear durability was obtained chiefly with a 15–20% increase in face-width but this was obtained without any change in the length of the gearbox over the previous model. This was achieved by developing net-forming clutch teeth tucked inside the gear width and reducing shaper cut tool clearance by going from one to two-piece welded construction, Fig 48. Gears were designed with total overlaps less than three times the sum of helical and profile; some 10–15% compressive stress reduction resulted.

The bearing installation was redesigned so as to increase effective shaft rigidity, the input shaft now being supported on a roller-ball combination designed to negate the effect of differential thermal expansion between aluminium alloy and steel without loss of support. The front intermediate shaft taper-roller bearing is now fitted with a high temperature polymer cage which allows more rollers to be included within a given space envelope, despite the narrower width constraint placed by the increased gear face-width.The taper bearing cup is first pressed into the case, during assembly, the bearing then seated in the cup and, finally, the whole intermediate shaft assembly pressed into the taper bearing cone bore without brinelling the rollers. Easy gauging of the bearings is thus permitted with a simplified assembly process.

The dual-cone synchonizers have organic friction material, the 3/4th and 5th synchonizers being arranged with a common blocking ring, giving a claimed 400% increase in energy absorption over the conventional brass ring set-up. The reverse idler is engaged between the input shaft and an external spur gear on the 3/4 synchronizer sleeve, torque transmitted resulting in lateral motion of the sleeve during drive and coast reverse engagements. This motion is limited to 1–1.52 mm by using the 3/4 fork as a thrust surface, Fig 49; clutch teeth are located away from the sleeve by 1.27 mm to compensate for the remaining motion left in the sleeve, preventing premature engagement with the blocker ring.

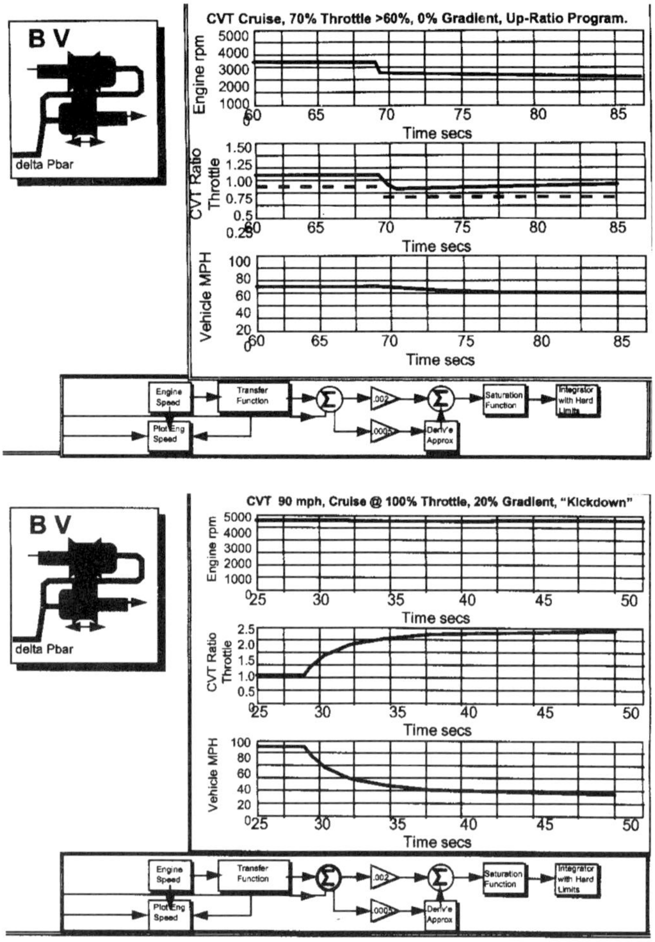

Fig 50: Base (above) and high level (below) strategies

Control strategies for CVT

A timely study of the control criteria for continuously variable transmission has been made by researchers at Ricardo-FFD[11] who have provided a most lucid explanation of engine/transmission matching. They remind us that too little overlap in the torque curves for different ratios of a stepped transmission results in gaps in the shift sequence with engine speed, and hence power, having to drop too far between shifts — thus breaking the gear-change rhythm. Connecting the peak power points in a group of typical 2-D graphs of this sort would show a 100% efficient transmission characteristic working with an engine held at full-power continuously. When plotted in 3-D form, physical 3-D steps can be seen between the torque curves and these become a smooth surface with CVT. The surface has depth in the sense that part-throttle performance characteristics also exist and navigating through the resulting solid volume is the job of a CVT control system.

The FFD PDSL simulator allows control systems to be tried out on a sophisticated computer model of the powertrain. Control strategies analogous to those in-built into human responses for manual gearbox control can be tested, for example. Fig 50 compares PDSL interpretation of a base strategy 'if in doubt move up the ratio range' with a higher level one of 'kick-down' for possible overtaking, say. Similar strategies are also provided by the authors to demonstrate the operating requirements of racing cars and delivery vehicles, as extreme exam-

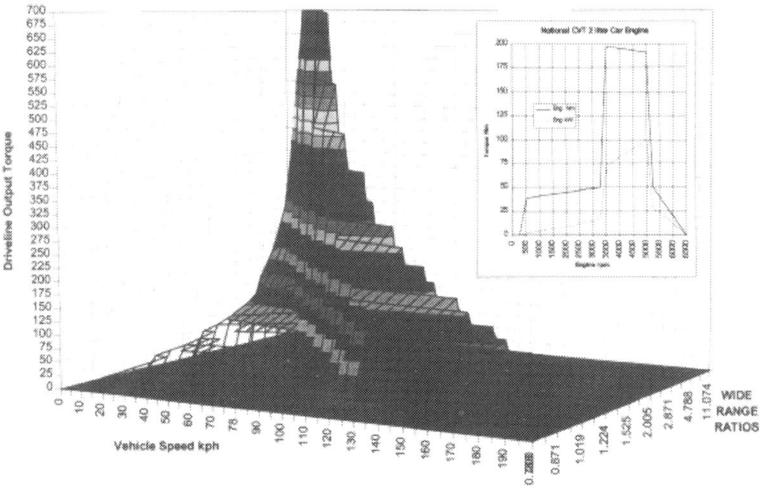

Fig 51: Proposed CVT engine torque curve matching to wide range CVT

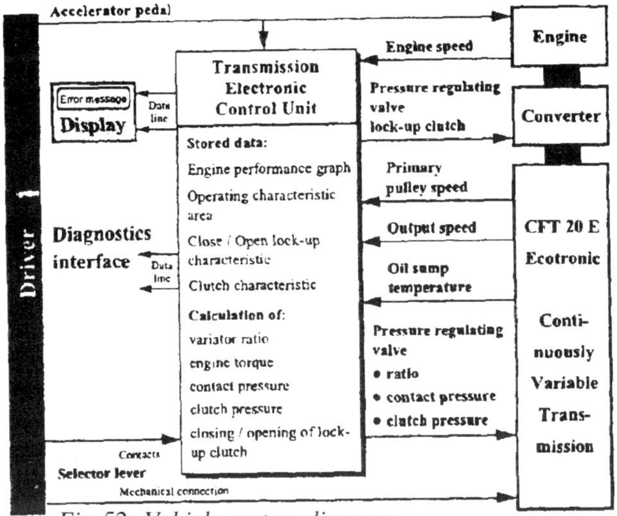

Fig 52: Vehicle system diagram

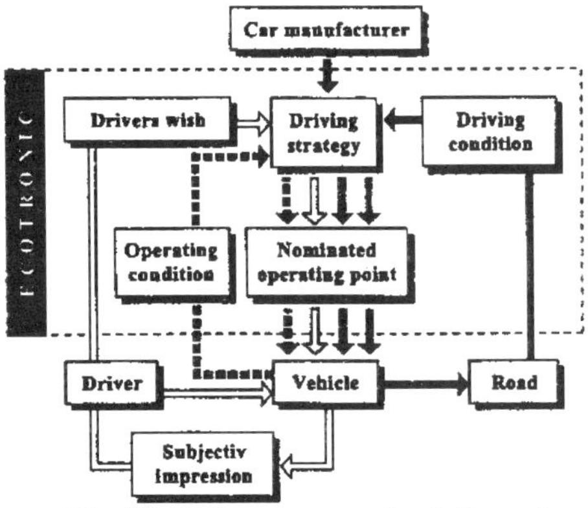

Fig 53: Drive strategy for influencing factors

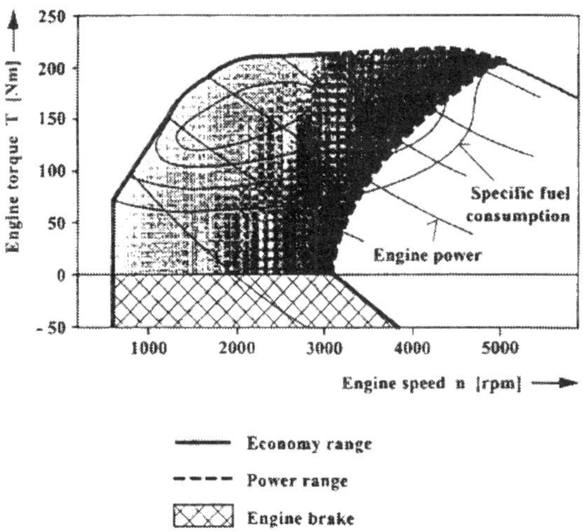

Fig 54: Adaptation ranges of drive strategy

ples, the first wanting to hold maximum power through the accelerating sequence while the second is wanting a narrowing of the performance corridor to preclude cargo-damage. FFD believe a CVT strategy should be broadened from merely providing an infinite number of ratio selections to sophisticated engine/transmission interaction in a way that the driver sees no limiting characteristics of the transmission. The opportunity to map the control domain exists rather than moving from one rigid strategy to another, giving the driver hints and feedbacks as to the right way to use the transmission. Matching conventional engines to conventional transmissions requires, in the case of heavy haulage, for example, engine torque back-up for reducing the need to change gear. However, this flexibility in power delivery is paid for because the engine is compromised between two duties, providing maximum power at maximum speed and maximum torque at usually a much slower speed. With a wide-range CVT there is no need for such flexibility and engine operating speed range can potentially be considerably narrowed down. This scenario has been modelled in PDSL, Fig 51.

In the case of the German maker ZF's CFT20 CVT, flow of signals between driver, engine and transmission is seen in Fig 52. Fuzzy logic control is involved and the ECU is adaptable to different driver strategies. Achieving optimum drive characteristics was done by arranging for automatic specification of the engine operating point, accounting for the special features of each driveline. With the extensive range of possible system modes available an adaptive strategy is necessary and this led to the use of fuzzy logic in specifying the operating point, Fig 53 showing the system structure. The operating point is generally fixed by the need to optimize fuel economy and is only adapted to higher engine speeds (more power) by driver demand or operational task such as gradient negotiation. Converting performance demand into increase in operating point involves the evaluation of many parameters; for example, a distinction being made between setting off in an urban environment and overtaking out-of-town. Performance level evaluation takes the form of an adaptation factor infinitely variable between minimum fuel consumption and maximum road performance. A large adaptation area is also covered by the range of engine characteristics, plus accounting for all driveability requirements, Fig 54.

The mechanics of roll-over

As economy vehicles for urban operation become shorter, and therefore taller so that volume capacity and visibility are not compromised, the problem of roll-over intensifies. The appearance of new software from American Technical Publishers for allowing calculation in the fundamentals of vehicle dynamics gives an opportunity to better predict likely vehicle performance in this respect at the design stage.

According to Gillespie[12], who originated the software, the initial simplified quasi-static behaviour of the vehicle in cornering can be studied with the elementary model of Fig 55. The vehicle is assumed to be in a steady-state turn without any roll acceleration and the tyre forces shown represent the sum of front and rears. Super-elevated (banked) highways help to resist the overturning effect and the symbol j for the road-banking angle is used in calculation as positive when tilted towards the cornering vehicle. Lateral acceleration in gravities can thus be expressed as:

$$\{(t/2) + \varphi/h - (F_{zi}/Mg)t\}/h$$

and the limiting value when load on the inner wheel becomes zero is:

$$\{(t/2) + fh\}/h$$

thus on level roads only the CG height and track are necessary for a first estimate of resistance to roll-over. Though this term gives conservative values for propensity to roll-over it is useful for comparison purposes and the author gives the table of Fig 56 to illustrate this. Since the peak tyre-to-ground friction co-efficient is quoted as 0.8, the table suggests that the lighter vehicles are safe from roll-over as they would slide out of the curve, which of course has been disproved in practice.

In fact, by neglecting the compliances of the tyres and suspensions, roll-over threshold is overestimated as the lateral offset of the CG is unaccounted for. Thus by considering the more refined model of Fig 57, incorporating the hypothetical roll-centre, lateral acceleration, related to roll rate R_ϕ, becomes:

$$(t/2h)\{1/[1 + R_\phi(1 - H_r/h)]\}$$

the curly bracketed term showing the factoring roll-over threshold, usually around 0.95 the author explains, but quite a lot higher for a low-slung sports car. Solid axles, with high roll-centres, reduce the effect of lateral CG shift compared with many independent suspension systems with roll-centres at near ground level. Lateral deflection of the tyres adds a further 5% reduction of the threshold, it is explained.

However, for an accurate assessment complete modelling of the vehicle suspension is required accounting for such factors as lateral shift of roll-centre and lateral movement of the line-of-action of tyre vertical force, due to cornering forces and deflections arising from changes in the overturning moment under combined cornering and cambering effects. Such modelling would result in quasi-static roll response, indicated by Fig 58. This shows the linearity of the response breaks when one of the inner

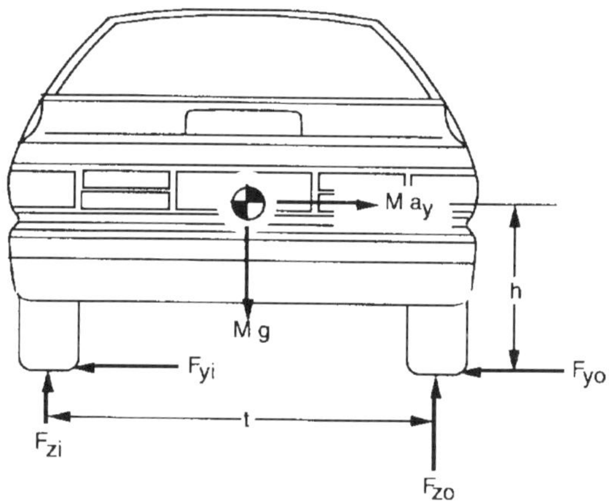

Fig 55: Forces inducing roll-over

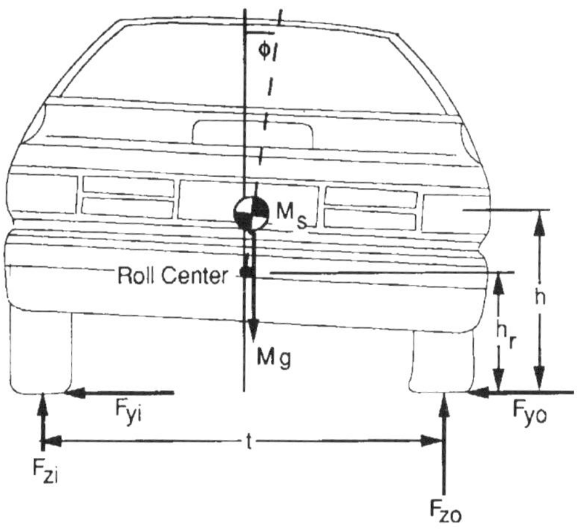

Fig 56: Roll reactions on suspended vehicle

Vehicle Type	CG Height	Tread	Rollover Threshold
Sports car	18-20 inches	50-60 inches	1.2-1.7 g
Compact car	20-23	50-60	1.1-1.5
Luxury car	20-24	60-65	1.2-1.6
Pickup truck	30-35	65-70	0.9-1.1
Passenger van	30-40	65-70	0.8-1.1
Medium truck	45-55	65-75	0.6-0.8
Heavy truck	60-85	70-72	0.4-0.6

Fig 57: Threshold comparisons

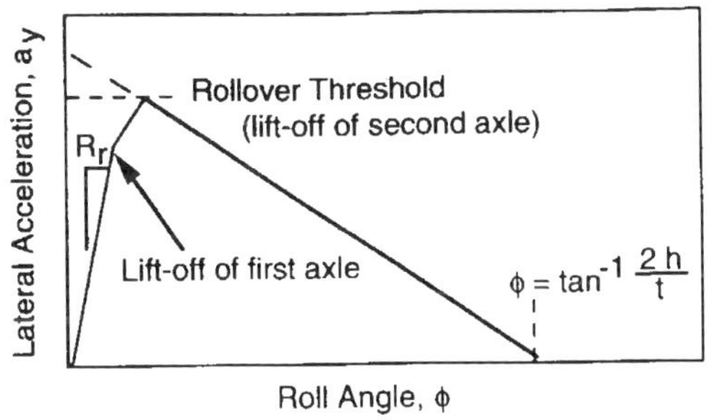

Fig 58: Equilibrium lateral acceleration

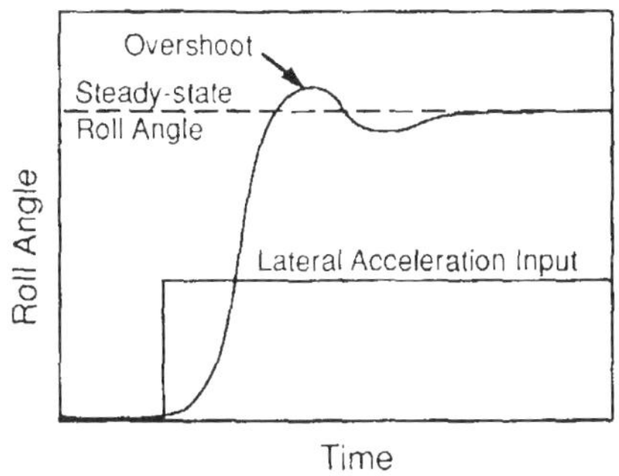

Fig 59: Roll response to step input

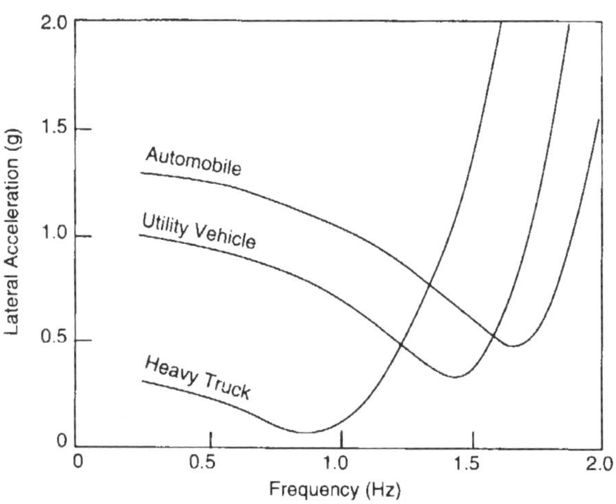

Fig 60: Threshold as a function of sinusoidal steer

wheels lifts off. The plot suggests that highest roll-over threshold will be achieved by maintaining sprung mass roll rate at the highest level by using high roll stiffness suspension and designing so that simultaneous lift-off of front and rear inner wheels occurs at the cornering limit.

Transient roll-over

To examine vehicle response to rapid changes in lateral acceleration a transient model is required. The simplest one would be that of Fig 57 but with a roll moment-of-inertia term added. This is useful for responses to step inputs such as changes in road surface friction or release of brakes after wheel-lock, the response of the system being similar to a single degree of freedom system, Fig 59. The fact that roll-angle can overshoot, depending on the degree of roll damping, the author explains, means that wheel lift-off may occur at lower levels of lateral acceleration input in transient manoeuvres, hence a lower roll-over threshold.

In the case of sinusoidal lateral acceleration input, as approximated by the slalom 'moose' test, roll-over threshold response is shown in Fig 60 as a function of frequency for different classes of vehicle, the values approaching the steady-state cornering case at zero frequency. Heavy trucks are seen as being particularly sensitive to this effect since their roll resonant frequency is often less than 1 Hz and lane-change or certain roundabout manoeuvres of around 0.5 Hz could result in roll-over. Cars having lower CG positions usually have roll resonances at about 1.5 Hz.

The most complete and accurate picture is provided by models which simulate both yaw and roll response. Using such a model to examine sinusoidal steer inputs reveals the important additional effect of front and rear tyre force phasing lag caused by the rear tyre not building up side force until a slip angle builds up. Fig 61 illustrates the effect. Here the front to rear lateral force build up lag is 0.2 seconds for the 1 Hz steer input case. The lateral acceleration for the whole vehicle is diminished from 0.8 to 0.5 g and to the driver there is a perception of lack of responsiveness in transient manoeuvres. Four wheel steer cars eliminate this lag and improve responsiveness; however, the author points out, it could contribute to roll-over propensity.

In tractor-trailer vehicle combinations the phase

lag is particularly pronounced and the response in Fig 62 is typical. Here a 2 second sinusoidal steer input excites both a rearward amplification of the yaw response and roll-resonance of the full trailer, such that it experiences much larger lateral acceleration than the tractor, creating a dangerous whip-lash effect. The effect is prevented by using a tractor-trailer coupling which transfers roll-couples from tractor to trailer.

Software example

Gillespie used the quasi-static model to demonstrate the effect of highway super-elevation on vehicle occupants. The objective here is to find what angle would make the occupants experience 0.1 g lateral acceleration in a 40 mph steady-state turn of 500 ft radius in a vehicle of 60 inches track, 20 inches CG height and weighing 2200 pounds.

Fig 63 shows the data input spread sheet with these values entered. On solving, the sheet generates the first three views of Fig 64, the third of which is a vector diagram showing the relationship between the vectors *ac* and *g* (relative to horizontal and vertical) and *ay* and *az* (relative to the vertical axis). The point common to the upper left corners of both rectangles represents the CG location while the line from the CG to G is centrifugal acceleration ay and the vertical line from CG to the other G is gravity acceleration *g*. On the other rectangle the line slanting right and upwards from CG to point V is lateral acceleration experienced by the occupants ay and the one slanting right downwards to the other V is the vertical acceleration they experience.

Neutral speed is that at which the occupants experience no lateral acceleration relative to the vehicle and the program allows back-solving for this case by changing φ from an output to an input variable, changing the input instruction for ay and blanking the V input field. This produces the sheet of Fig 65 and the fourth view of Fig 64, a vector diagram in which G = global axes (*ac* and *g*); V = vehicle axes (*ay* and *az*).

The neutral speed for this curve is 29 mph and the equilibrium lateral acceleration plots are the same for this solution because the radius, super-elevation, track and CG height are unchanged. However, the vector plot is different because centrifugal acceleration ac is smaller and the lateral acceleration *ay* is now zero.

St	Input	Name	Output	Unit	Comment
					Fundamentals of Vehicle Dynamics
					by Thomas D. Gillespie
					Chapter 9: Rollover
					Quasi-Static Rollover of Rigid Vehicle
					Eq. 9-1, p. 311, and Fig. 9.2, p. 313
	40	V		mph	Velocity
	500	R		ft	Radius of turn
		ac	.21394745	g's	Lateral acceleration (horizontal)
	60	t		in	Track width (tread)
	20	h		in	Height of CG
	2200	W		lb	Weight of vehicle = M*g
		g	1	g's	Acceleration of gravity (default: 1 g)
		M	2200	lbm	Mass of vehicle
	.1	ay		g's	Acceleration along vehicle y-axis
		az	1.0177296	g's	Acceleration along vehicle z-axis
		Fzi	1046.1692	lb	Normal force on inside tire
		Fzo	1192.8359	lb	Normal force on outside tire
		phi	6.4644744	deg	Road cross-slope angle
					Rollover thresholds:
		Vroll	120.5635	mph	Velocity
		acroll	1.9436535	g's	Lateral acceleration (horizontal)
		ayroll	1.8187084	g's	Acceleration along vehicle y-axis
		azroll	1.2124722	g's	Acceleration along vehicle z-axis

Fig 63: First input spreadsheet

St	Input	Name	Output	Unit	Comment
					Fundamentals of Vehicle Dynamics
					by Thomas D. Gillespie
					Chapter 9: Rollover
					Quasi-Static Rollover of Rigid Vehicle
					Eq. 9-1, p. 311, and Fig. 9.2, p. 313
		V	29.109577	mph	Velocity
	500	R		ft	Radius of turn
		ac	.11330756	g's	Lateral acceleration (horizontal)
	60	t		in	Track width (tread)
	20	h		in	Height of CG
	2200	W		lb	Weight of vehicle = M*g
		g	1	g's	Acceleration of gravity (default: 1 g)
		M	2200	lbm	Mass of vehicle
	0	ay		g's	Acceleration along vehicle y-axis
		az	1.0063988	g's	Acceleration along vehicle z-axis
		Fzi	1107.0387	lb	Normal force on inside tire
		Fzo	1107.0387	lb	Normal force on outside tire
	6.4644744	phi		deg	Road cross-slope angle
					Rollover thresholds:
		Vroll	120.5635	mph	Velocity
		acroll	1.9436535	g's	Lateral acceleration (horizontal)
		ayroll	1.8187084	g's	Acceleration along vehicle y-axis
		azroll	1.2124722	g's	Acceleration along vehicle z-axis

Fig 65: Second input spreadsheets

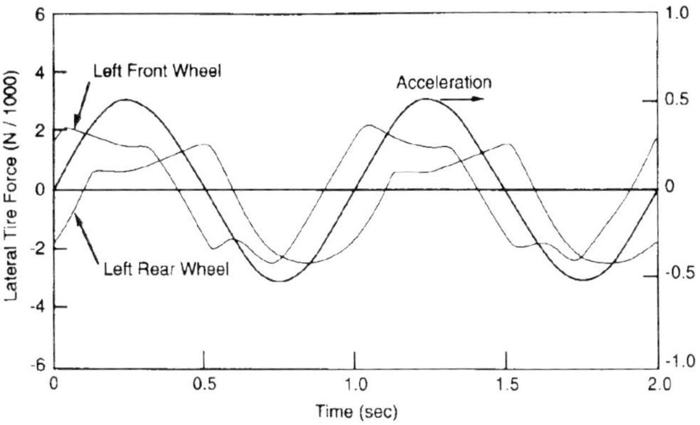

Fig 61: Phasing of tyre forces in sinusoidal steer

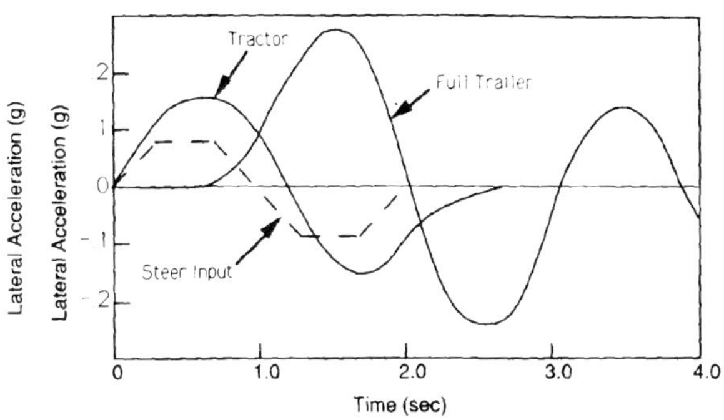

Fig 62: Lateral acceleration of tractor/trailer

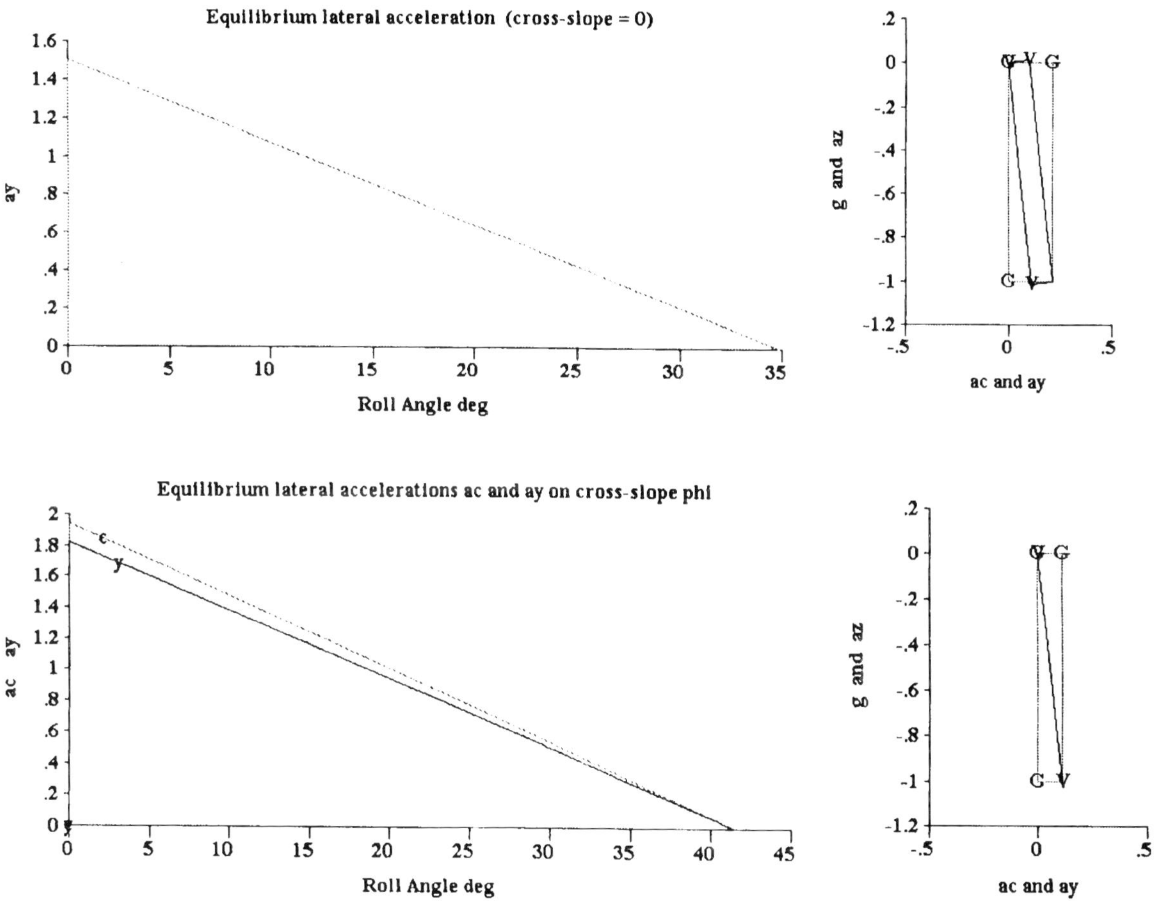

Fig 64 : Graphical output

Suspension and steering linkage analysis

For light vehicles, advances in modelling techniques are making the analysis of handling behaviour a much more realistic process than was possible with classical quasi-static techniques. The same is true of analysing suspension behaviour with respect to road damage by heavy vehicles.

The change from active suspension systems in Formula One race-cars brought about by the 1994 revised regulations focused a new interest in the perfection of passive systems. According to Brescia University researchers[13], double-wishbone configuration, with either tie-bar or compression-strut brake force reaction, remains the favourite means of controlling wheel-to-chassis relative movements important for maintaining roll-centre location and stability of the tyre-ground contact patch.

Continuing conflict existed, the authors maintained, between low roll-centre to reduce lateral scrub at full-bump spring deflection, as well as large weight transfer and roll angles in cornering, against, with high roll-centre, reduced weight transfer but increased lateral scrub and the jacking effect. Key objective in design was to restrict the movement of the roll axis to a minimum to ensure constant levels of weight transfer in different operating conditions. F1 practice is to adopt long parallel wishbone arms in order to minimiZe camber sensitivity to change in suspension ride height.

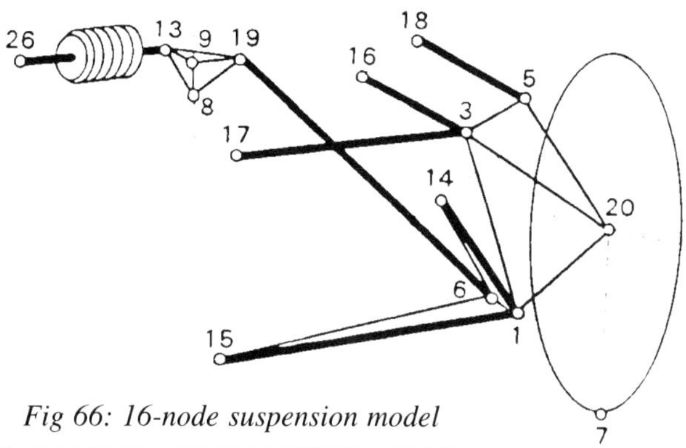

Fig 66: 16-node suspension model

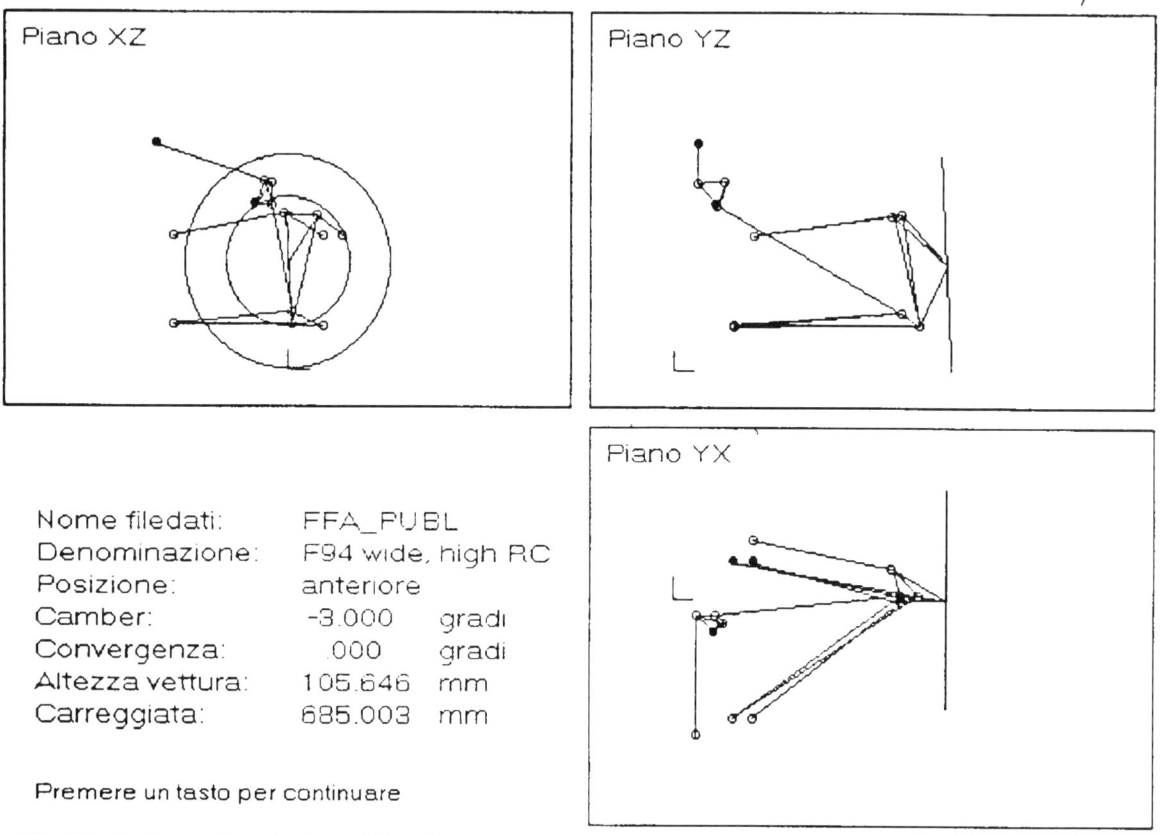

Fig 67: Orthogonal projection of Fig 66

The authors have developed MMGB design software for race-car suspension, in conjunction with Dallara Snc, for a complete 3-D analysis. Fig 66 shows a prototype configuration involving a 16-node kinematic system, the wheel, hub-carrier and spring rocker lever being assumed to be infinitely stiff. The wishbones are each represented by two pin-ended members connected at the outer node. The spring-damper unit is modelled as a variable length beam while the spring-reaction and steering arms are considered to be rigid. The program permits the linkage to be seen in orthogonal projection as seen in Fig 67. Data input files handle spring rate and road loads and the program is based on algorithms for kinematic solution and force equilibrium.

The kinematic module computes geometry variation during wheel travel and full bump rebound situations can be examined with the chassis effectively held still. Main suspension property outputs are seen in the graphs of Fig 68. A graphic animation is also available to simulate the motion of the suspension for observation by the designer. The force module computes link loads, rocker-pivot loads and spring force for given wheel displacement and external loading. There is also a compliance module which effectively measures the interaction between the first two modules but taking into account link elasticity.

Additional modules have been developed since the program was first launched. These allow study ofwheel rate and sprung mass frequency variation as the vehicle progresses along its track, arising from rising-rate or preload effects, also the effect of reduc-

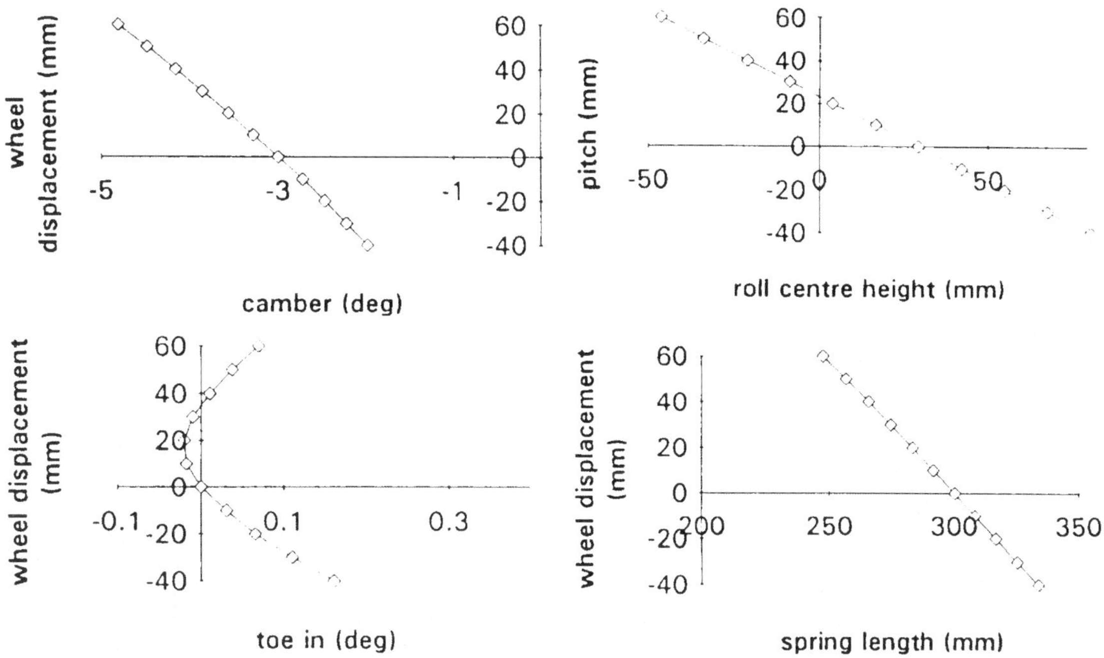

Fig 68: Main suspension geometry outputs from analysis

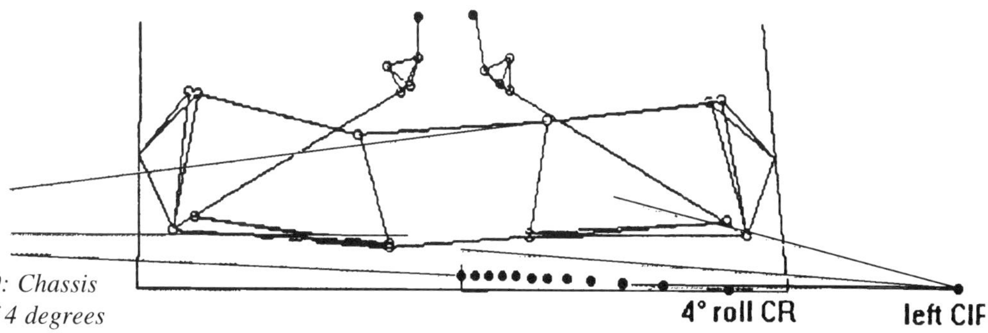

Fig 69: Chassis roll of 4 degrees

ing fuel weight and aerodynamic downforce inputs. Anti-dive and anti-squat configurations can also be examined. A roll-centre module can examine not just vertical but also lateral motions of the roll-axis. The very considerable lateral movement for a nominally ground-level roll-centre suspension design is seen in Fig 69 for a chassis roll of 4 degrees.

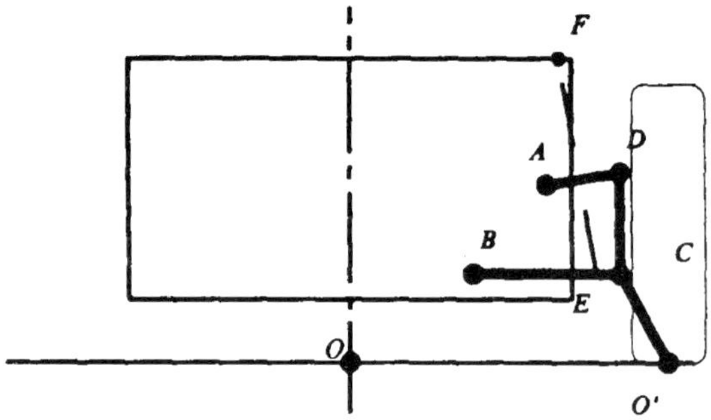

The authors also referred to the work on suspension compliance of John Ellis whose book on vehicle handling dynamics was published in the same year[14]. Ellis recommends a vectorial approach to suspension analysis which is not dependent on location of roll-centre and which can be applied at any vehicle attitude. For two-dimensional analysis, in Fig 70, the origin O is fixed in the vehicle body plane and an outboard point O' is at the wheel-ground contact. His analysis considers the velocities of these points in relation to the body and wheel velocities, relating them by the rigid link which effectively exists between the inboard and outboard points.

Referring to the co-ordinates of these points, shown in the figure, velocity of the inboard point is

$$V_a = (-pz)j + (W + py)k$$

That of the outboard point is composed of the wheel velocity V' and a component p' due to rate of camber rotation of the wheel in

$$V_d = (v' - p'z')j + p'k$$

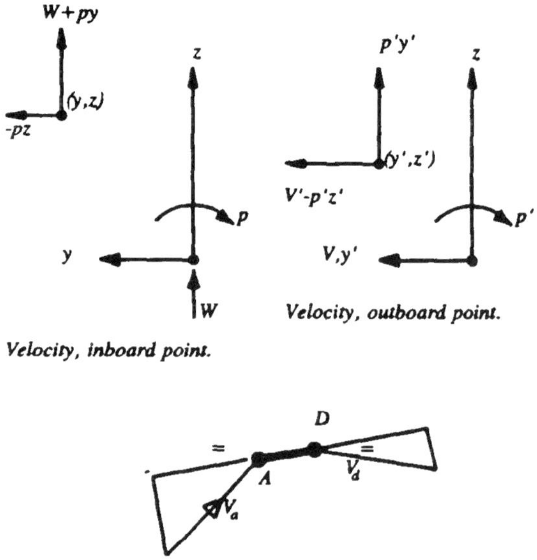

Fig 70: Two-dimensional analysis

the two velocities acting at the opposite ends of link AD. Since the relative velocities of these points along the line of AD is zero, the scalar products of these velocities with the unit vector of the link must equal each other, Ellis explains. By representing the link as unit vector

$$\underline{ad} = mj + nk$$

where m and n define the spatial position of the link defined by co-ordinate geometry. Velocities of body and wheel/road interface then become related by the expression:

$$(-pz_a)m + (W + py_a)n = (V' - p'z_d')m - p'y_d'n$$

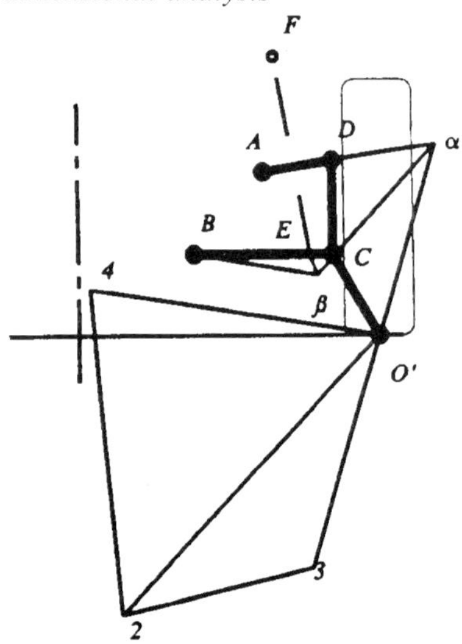

Fig 71: Compliance representation

By repeating this procedure for each link connecting body and wheel, a set of equations is obtained which can be solved simultaneously by matrix methods to give useful design relationships such as the rate of contact patch lateral velocity with bounce velocity of the suspension; tyre scrub due to roll; and camber change due to bounce and roll.

Ellis also carries out a three-dimensional suspension analysis using these techniques and finds the interrelationship with the steering system, as well as modelling the effects of tyre stiffness. Compliance effects are examined for the flexible mountings of the suspension. He defines the change in suspension characteristics due to external forces as the compliance and describes its quantification using the two-dimensional representation of Fig 71.

Forces on the inboard bearings are found by static equilibrium equations, the stub axle O'CD being under the action of normal and lateral forces at the road/tyre interface, which can be expressed algebraically for this purpose. The line of action of the force within the suspension link AD intersects the force vector at α. The vector O'3 is known, so also are the directions of the force vectors for joints D and C. The lower suspension link is under the action of spring force along FE and the forces at B and C. Directions of the reaction C and the spring force are known and intersect at β, so the direction of the reaction at B can be obtained and the vector diagram drawn. By inputting the stiffness of the joints 1 and 4, the deflections of these joints can be obtained and new positions of the inboard nodes of the suspension linkage obtained and the change in handling behaviour calculated. A similar analysis for a 3D suspension model is given in the book.

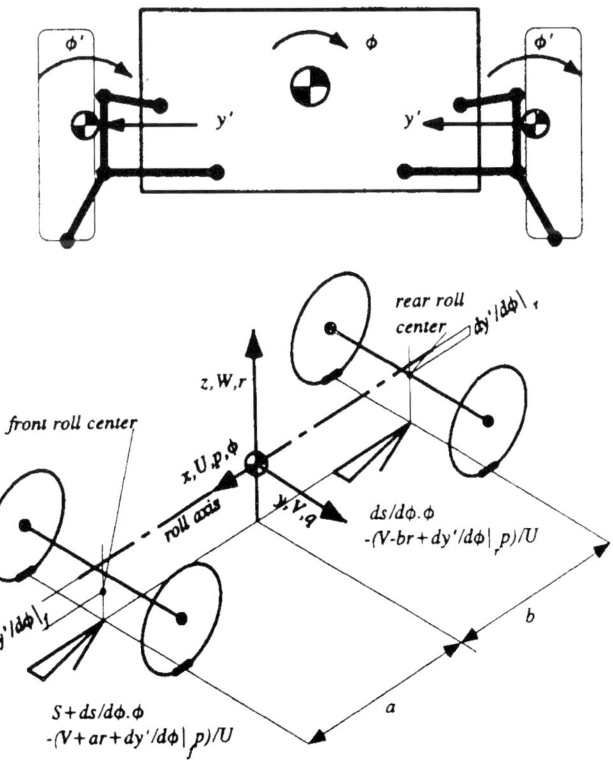

Fig 72: Roll effects

He uses the steady state roll angle of the suspension, derived from energy and work equations, and expressed in terms of the suspension derivatives, to examine roll attitudes and load transfer. The roll of the body causes potential energy to be stored in the springs and work is done by the tyres in developing lateral forces to keep the vehicle in a corner; and developing lateral motion as a result of the roll. As the vehicle moves on its curved path, the lateral inertial force acting at the centre of mass gives rise to an overturning moment, transferring normal load from inner to outer wheels. Because redistribution of normal forces causes the lateral forces generated by the tyres to change in a non-linear way, the balance of the vehicle may change, Ellis explains.

Roll-induced steering is possible at both front and rear axles, the relative roll stiffness of the suspensions affecting the proportion of the load transfer carried by each axle. Added to this are the effects of steering of the road wheels under roll-displacement, components of the lateral velocity at each contact patch due to tyre scrub also the camber of the wheels with roll displacement. Fig 72 (*top*) shows a 2-dimensional suspension for the case in which roll is the only motion and it is seen that the kinematic design of the suspension will constrain the motion of the road wheels in response to the roll of the vehicle. The kinetic energy of the system due to roll is

$$2T = Ip^2 + (m'V'^2 + I'p'^2)$$

but as $V' = dy'/d\phi.p$ and $p' = df/d\phi.p$, suspension derivatives referring to the centres of mass of the suspensions, then for the whole vehicle, Fig 72b, the effective inertia in roll is

$$I_x = I + \Sigma m'(dy'/d\phi)^2 + \Sigma I'(d\phi'/d\phi)^2$$

Current review of road-friendliness

Cebon joined with colleagues from Warwick and Nottingham Universities to give an up-to-date assessment of road-friendliness research[15] with particular reference to the common case of the articulated HGV. Recent thinking is that the dynamic component of tyre forces is the main factor in contributing to road damage. Not only is this significant compared with mean (static) tyre force but it is believed that different vehicles generate similar spatial patterns of dynamic loading. At a given speed, the tyre force time histories generated by a particular vehicle are repeated closely on successive runs over a given stretch of road, the phenomenon of spatial repeatability. This means that certain points along the road will be subjected to the peak dynamic tyre forces generated by many vehicles, thus significantly increasing the damage incurred at those points, the authors explain. Assessment procedures should therefore follow the criteria shown in the table of Fig 73.

As well as the dynamic loading effect on road damage discussed above, a road stress factor has also been used based on the so-called fourth power law:

$$E[P9t)]^4 = (1 + 6\underline{s}^2 + 3\underline{s}^4)P^4_{stat}$$

where $P(t)$ is tyre force; $E[\,]$ an expectation operator; P_{stat} the static tyre force; and $\underline{s}$ the co-efficient of variation of dynamic tyre force (the dlc). However, it must be noted that the stress factor does not account for the spatial distribution of tyre forces and is no more than a 'plausible rule-of-thumb'.

The need to have road damage criteria that relate to specific points along the road has involved summing the measured dynamic forces of each axle raised to a power n, applied to each location along a road, or sensor in a road measuring mat. This is known as a weighted aggregate force model, the nth power aggregate force A_k^n measured at location k being given by

$$\Sigma_{j=1}(P_{jk}^{\,n}) \qquad \textit{for } k = 1,2,3,\ldots, N_s$$

the power of n being chosen to represent the type of road damage being considered (4 for fatigue damage and 1 for rutting). The dlc and aggregate force methods have been compared in tests on a trunk road, Fig 74 showing the dlc averaged over all suspensions in each class of axle group for all trailer axle groups

'Road-friendliness' assessment tests should:

1. account for the effects of tyre and axle arrangements and static load sharing as well as dynamic loading performance;
2. use testing equipment which is not dependent on the arrangement of axles and tyres, and therefore does not bias vehicle design towards particular configurations;
3. be based on assessment criteria which reflect the potential of the vehicle to inflict fatigue and rutting damage to representative pavement structures;
4. be conducted for representative conditions of speed, road roughness and road construction;
5. account correctly for wheelbase filtering;
6. account for the effects of all axles of the vehicle on the pavement;
7. resolve/sample correctly both high- and low-frequency dynamic tyre forces;
8. utilize statistically significant samples of tyre force–time histories;
9. account for the effects of spatial repeatability of dynamic pavement loads;
10. account for the dynamic interaction between the suspensions of tractors and trailers.

Fig 73: Assessment procedure criteria

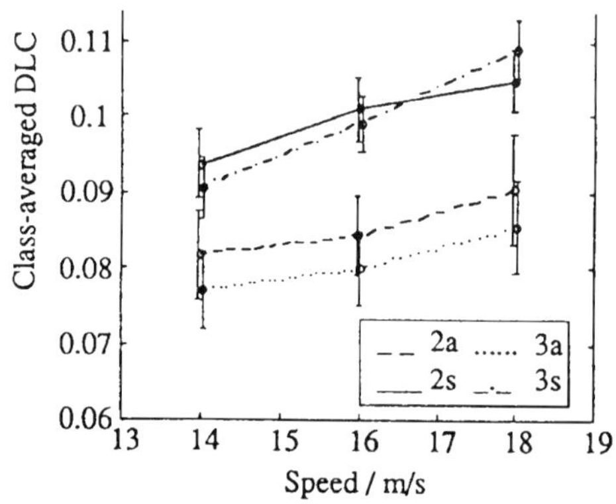

Fig 74: Dlc results (2a and 3a refer to 2/3-axle air suspension and 2a/3a to 2/3-axle steel suspension respectively)

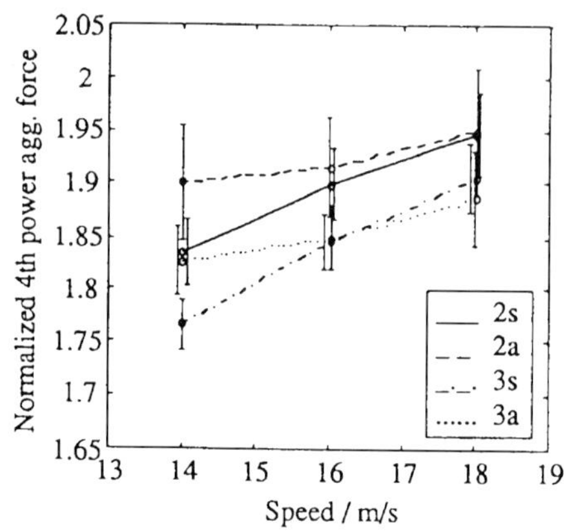

Fig 75: Fourth power results, error bars showing standard deviation

for which suspension type was identified. Fig 75 shows the fourth power results for the same trailer axle groups, normalized using the static weight of the axle group to enable different vehicles to be compared directly. The two criteria rank the suspension classes in different orders; this is due to spatial correlation between the tyre forces in an axle group which is not handled properly by the dlc.

Generally it has been observed that dynamic wheel loads increase with increasing road speed and road roughness; that walking beam and pivoted spring tandem bogie suspensions generate the highest loads, leaf spring suspensions generate less, and air and torsion bar suspensions generate the smallest levels of dynamic loads. An important feature of dynamic loading is that all suspensions and tyres on a vehicle are involved in the sprung mass modes of vibration and so effect of tractor-trailer interaction is important. A trailer performing well with one tractor may perform badly with another. A practical assessment test is therefore likely to require a standard tractor for testing with every trailer and a standard trailer for testing with all tractors.

Chapter 4: Electrical and electronic systems

The maturity of automotive electronics is seen in the desire to see the controller as seamlessly integrated with the vehicle system. This is evident in new developments in engine knock sensing, suspension, steering and drivetrain control. Vehicle system development is also being driven by telematic traffic management systems and the goal of vehicle headway control is not far away. Engineers from different disciplines are seeking knowledge of the digital systems used in electronic circuits and applying the knowledge to driver-vehicle man-machine control systems for creating information links with the surrounding road infrastructure and the driver becoming a predictable control link within refined electronic steering systems. The greater sophistication in hybrid, electric and IC engine, drive for vehicles is seen in the more accurate matching of performance to the actual drive cycles which such vehicles are likely to experience. The use of map-controlled drive management is emerging and some of the latest control techniques are seen in the first production hybrid, the Toyota Prius. Future possible hybrid development is considered by some to be with the use of supercapacitors to work in conjunction with energy storage systems based on batteries. Electronic control is even seen to be advancing into vehicle test engineering with the appearance of automated handling systems for track use.

Automotive electronics maturity

The seamless electro-mechanical vehicle is the current aim of automotive electronics engineers who seek to bring an end to inefficiencies caused by grafting on electric controls to systems originally designed for mechanical operation. This approach is seen here applied to engine, suspension, steering and drivetrain control. Meanwhile the principal growth area, telematics, is beginning to see firm proposals for OE navigation systems for less expensive vehicles in the adaptive cruise control and headway sensor described here alongside a proposal for a guidance system to allow vehicles to follow roadway white lines. Major proprietary system releases reveal the latest in engine management and rival systems for business driver communication with his/her office.

According to Chrysler Advanced Technologies specialist Dr Chris Bodoni-Bird[1], the emerging technology of mechatronics is set to bring about the seamless electro-mechanical vehicle which can learn from Nature in its development. The theme is reduction in mass by using seamless electro-mechanical integration by either exploiting intelligence and/or developing multi-functional componentry. This will also suit the increasing demand for new features within a fixed space envelope. Biometics is the study that sees how man-made systems can emulate Nature. In the key field of sensors and actuators perhaps man can learn from the sensing capabilities of certain bird species, for magnetism and light polarization, and electric eels and rattlesnakes for electricity and heat, he suggests. Exhaustive historical analysis of patents has shown that systems tend to evolve from mechanical through chemical and electro-mechanical to electro-magnetic (optical), and vehicle sensor development is said to reflect this. Because electronic systems also have the capability for integration, sensors can be shared between vehicle functions, as seen in Fig 1. Because, however, mechanical systems are often cheaper and difficult to replace, institutionally and technically, the compromise discipline of mechatronics is seen by the author as set for growth.

A way of providing an on-board control system without the addition of hardware is described by researchers at NGK Spark Plug[2] who show how ion formation around the spark-plug can be used as an engine control technique. While the ion density represents the condition status of combustion, analysis of the condition is made possible by measuring voltage/current between plug-gap and ignition due to the magnitude of ion formation. In so doing the need is removed for sensors of EGR or knock and a plug-gasket type pressure sensor.

While it is expensive to provide an ion detector for each cylinder, here the authors propose a technique in which the spark plug electrodes are used for detection. But to avoid damage being caused by high tension voltages a point on the ion response curve, Fig 2, must be used after high-tension discharge. However, some increase in circuit complexity is involved, Fig 3. To measure ion density a coil is used to generate several 100 V pulses consecutively with the spark ignitions, the electrical circuitry providing

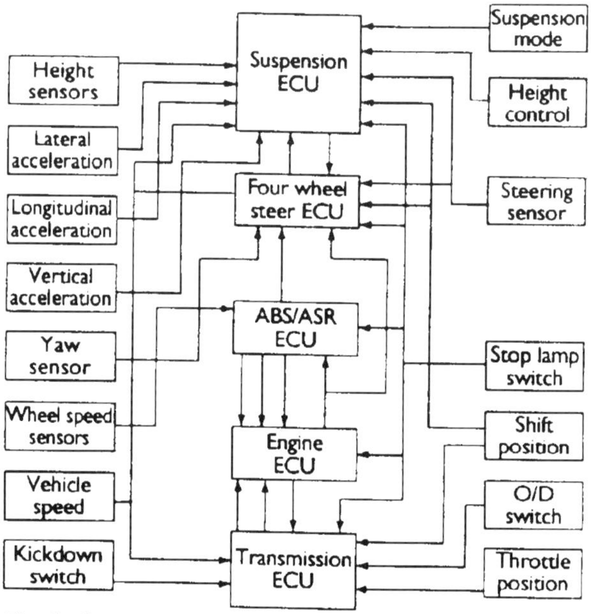

Fig 1: Systems integration effect on sensors

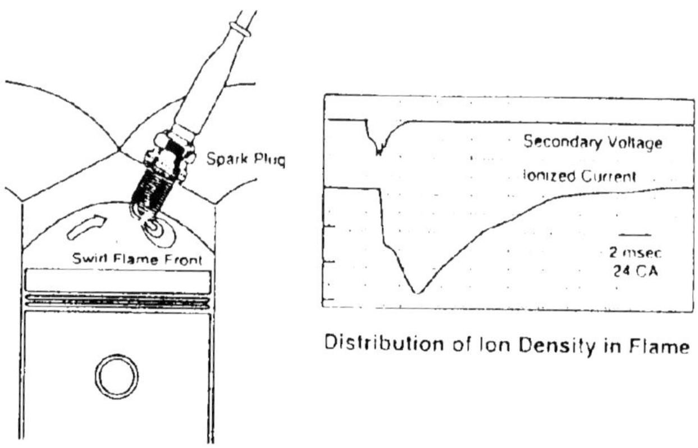

Fig 2: Ion current response

the means to measure voltage decay with time, which is proportional to ion density, Fig 4. The ECU analyses combustion by examining the pressure-wave pattern (magnitude and pulse-width) using ion-density and decay-period data. Input ignition signals, firing pulses, misfiring signals and the decay curve (and periods) are seen in the lower part of Fig 3. Amount of fuel injected is controlled so as to reduce the difference between fluctuation rate of the decay period and its standard fluctuation rate. Knock can also be detected because ignition and ion density waves are synchronized during detonation. But the decay period measurement technique alone cannot be used. The ion density voltage pulsation must be measured continuously. A capacitor is used to accumulate current generated by ion density voltage pulsation, and current flowing through an associated resistor in series with it, in order to sense the onset of knock.

Smart materials for suspension control

Conventional mechanical suspension systems are potentially convertible to active control with the use of electronically monitored smart materials, in the opinion of researchers[3] at TRW Inc. Effective control of the motions of sprung and unsprung masses of a spring-suspended vehicle can only be achieved by interconnecting hardware responding and adapting to changing disturbances. Optimum performance is obtained, it is argued, when system energy can be decreased and increased as necessary. Achieving variable spring rate with conventional active suspension systems requires high amounts of energy, with serious drain on the vehicle's capability.

By using smart materials, controllable by an external energy field, performance can be made to change in a predictable and repeatable manner. Among the smart solids, piezo-electrics exhibit a controllable energy transformation between electrical and

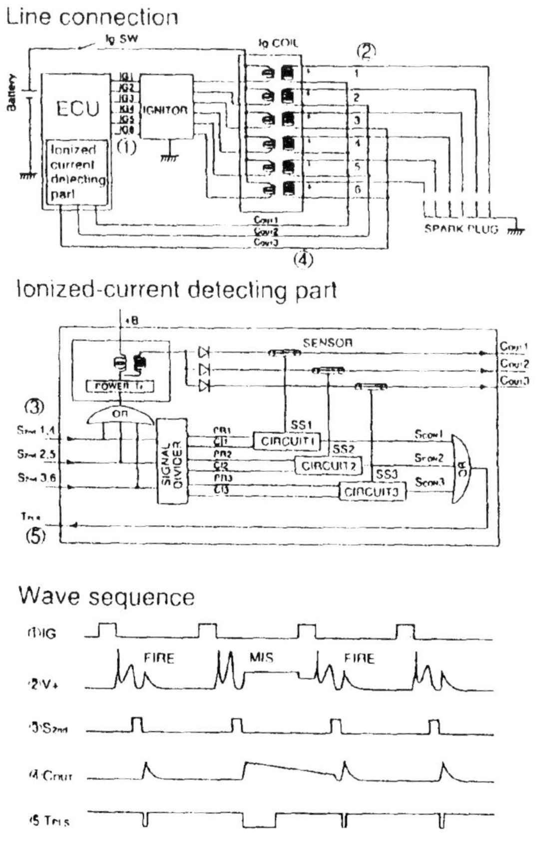

Fig 3: Secondary voltage ionised current method

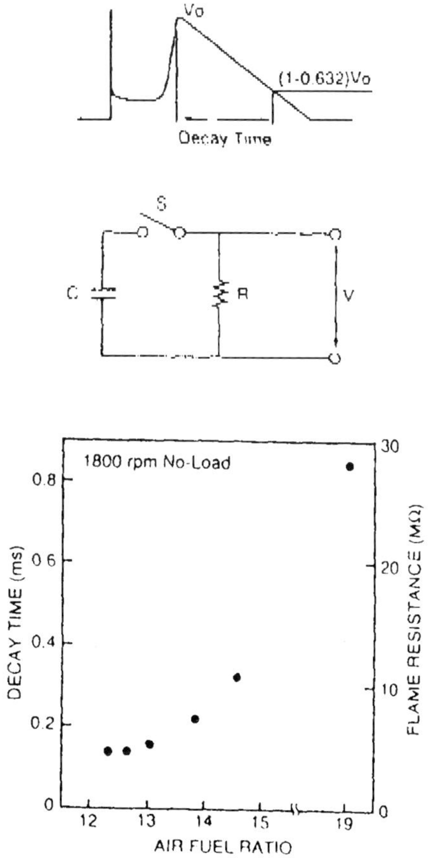

Fig 4: Determining flame resistance around plug gap

mechanical modes, hence their use in noise and vibration control, the authors explain. Smart liquids include those of Nature such as elastically deformable red blood cells suspended in plasma which may be reshaped with reduced diameters for easier flow through small blood vessels.

In suspension applications, rheological fluids are attractive and their behaviour is represented in a simple way by Fig 5: constituent particles which are randomly distributed in a zero-energy field, become aligned with the field under excitation. Considerable changes in shear force between the plates occurs during excitation by the field, hence viscosity is controllable. Both electro- and magneto-field excited types have been studied for many years.

So-called electro-rheological magnetic (ERM) fluids can develop very high levels, above 50 psi, of fluid shear stress while operating on vehicle battery voltage levels or moderate currents. This fluid also has the ability to set up shear forces independent of velocity, so no fluid flow is required. With such a fluid, a variable suspension damping characteristic is possible while reverting from current, space-wasting types to the more compact rotary configurations of earlier years. A typical parallel lower control arm suspension is shown in Fig 6. The electronic control allows an algorithm to process information for the control of the ERM fluid shear forces, hence the torque which the RACD can generate. Its cross-section in Fig 7 shows that energy is absorbed purely in shear mode between sprung and unsprung masses.

In controlling the fluid, the output signal from the ECU is transformed from an analogue voltage into a frequency, in the V/F converter, and used in a pulse-width-modulated power driver causing a current level to flow to a set of electro-magnets. Fig 8 shows a schematic of the ECU, the left side depicting

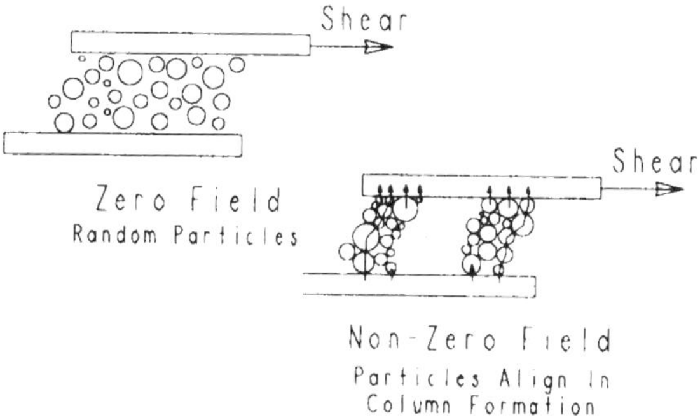

Fig 5: Particle orientation in rheological fluid

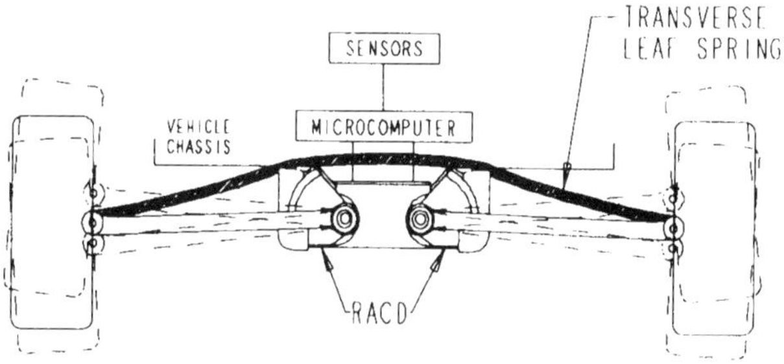

Fig 6: Parallel lower arm configuration

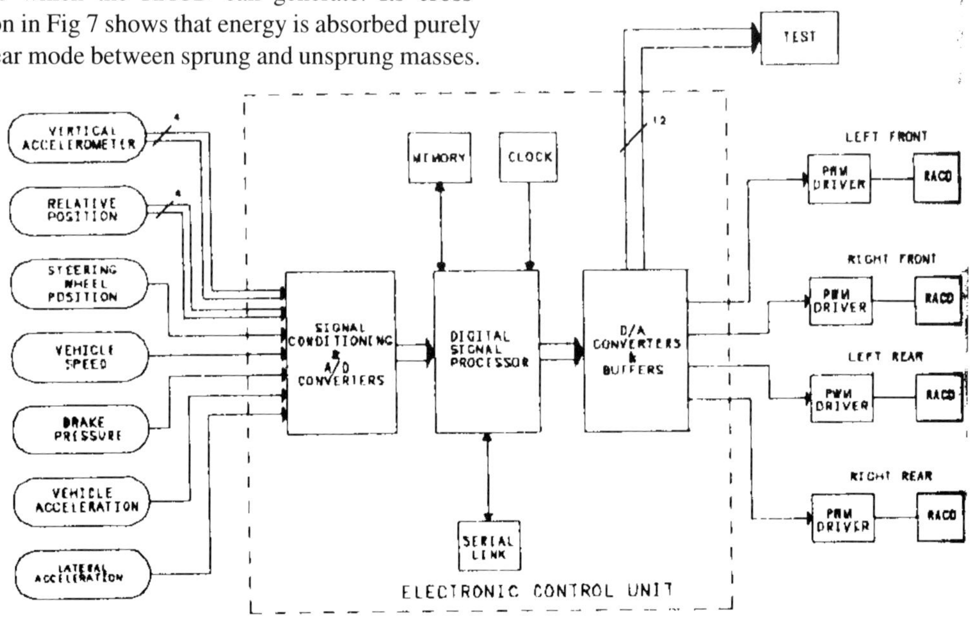

Fig 8: Electronic control unit

the inputs from the sensors which are converted for processing by the DSP and then reconverted to analogue for transformation to the appropriate pulse-width-modified signal. The ECU can tune the vehicle through software without any hardware change. This could involve mere days of work compared with the months currently involved in tuning the dampers for a new vehicle platform.

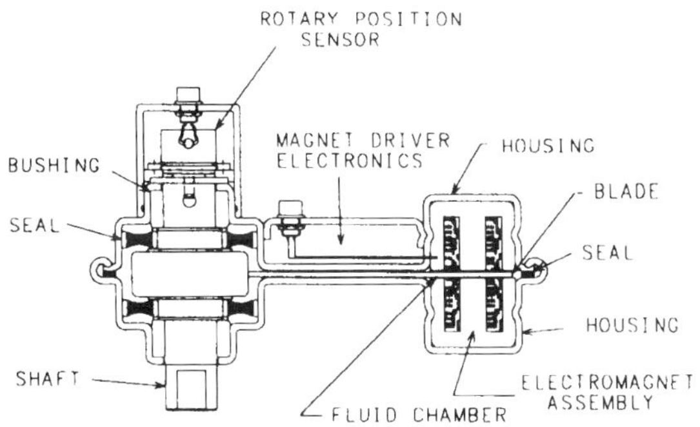

Fig 7: Cross section of RACD damper unit

The multi-degree of freedom matrix equation to control the dampers according to both vehicle and road dynamics is represented by the algorithm of Fig 9. For the future, more complex algorithms and integrated electronics are forecast by the authors. Microprocessor instructions could be executed at the rate of millions per second, transforms between time and frequency domains becoming possible and hence more complex control strategies permitted.

Electronic control of electric steering

Following Honda's successful pioneering of electric steering on the NSX car, company engineers[4] have reported a new-generation system for future application of intelligent steering, Fig 10. Its lighter and more compact construction allows its substitution for hydraulic PAS on a wide range of light vehicles.

The torque sensor, Fig 11, on the steering column relates steering input torque from the driver

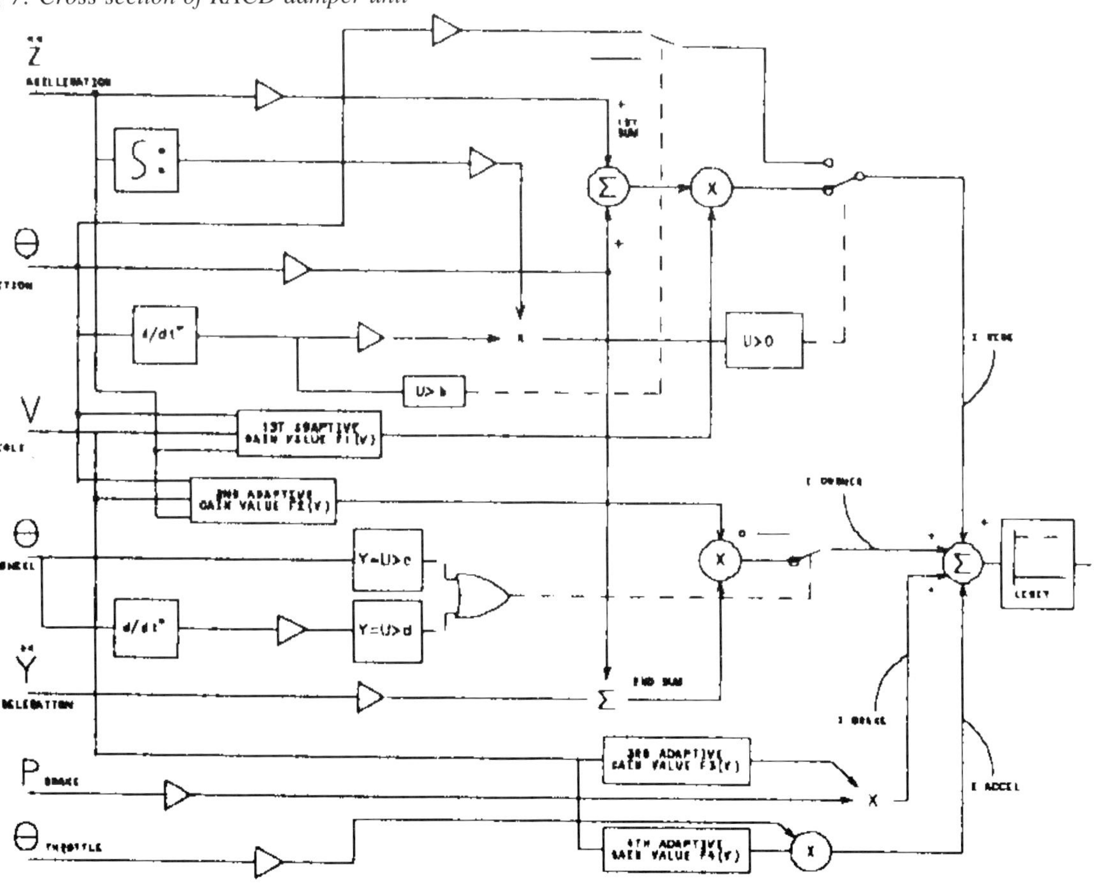

Fig 9: ERM/RACD control algorithm

with axial displacement and senses change of inductance in its field coil. The steering servo motor driving the rack has current feedback control and as well as the input from the column torque sensor also has a vehicle input related to instantaneous road speed; a PWM drive with power MOSFET is involved.

The seamless design of the torque sensor, integrated into input shaft, allows driver input and road kick-back to be evaluated. On the input shaft, steering wheel turning speed is derived from differential value of the torque as well as the direct value. On the output shaft, motor assist-torque data, and the reaction torque from the road, is obtained as a micro-vibration trace. Information on the inertia of the rotor is also obtained from the output shaft and the detrimental increment of inertia is cancelled out electrically to prevent its transmission to the steering wheel.

In the ECU for the new generation system, the microprocessor performance has been raised from 8 bit 12 MHz to 16 bit 16 Mhz. ASICs are now employed to customize the input and output circuits, with a resultant 35% reduction in part-count. Because the steering motor requires some 10 amperes for control, MOSFET switching and metal strip conductors are exploited with specially developed connectors. Motor and control-unit are now adjacent to one another so as to minimize cable losses.

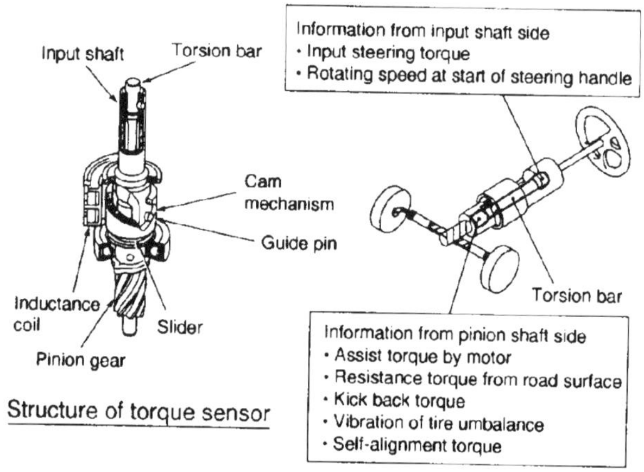

Fig 11: Torque sensor

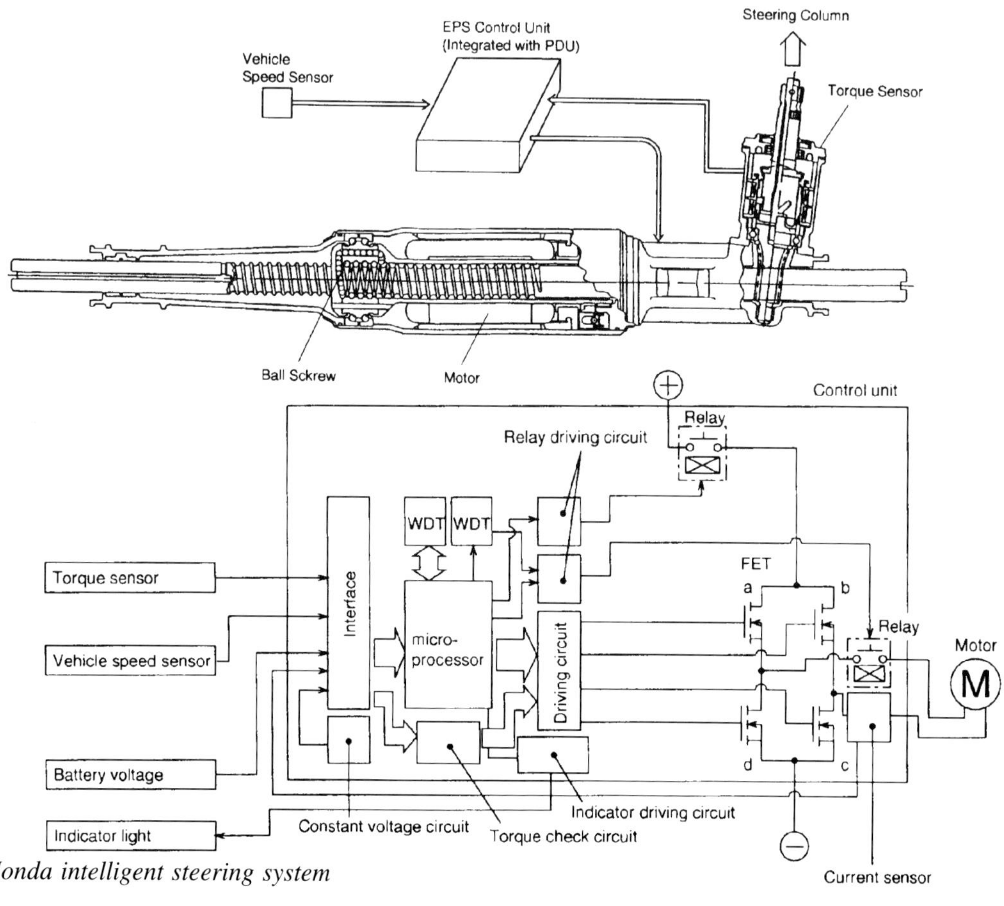

Fig 10: Honda intelligent steering system

Drivetrain control

The farther future in mechatronics is seen by researchers at Vairex Corporation[5] as the electric drivetrain. Full drive-by-wire will eliminate mechanical connections and replace them by electrically powered components and electronic controls, with the eventual reality of each wheel being capable of instantaneously, accurately and independently absorbing or developing torque between tyre and road with positive or negative value on any surface — with a dramatic improvement in traction, braking and handling performance. A particular advantage would be the introduction of electro-magnetic braking of a type that is more efficient than regenerative systems being pioneered on electric and hybrid propulsion vehicles. With regen braking, maximum braking

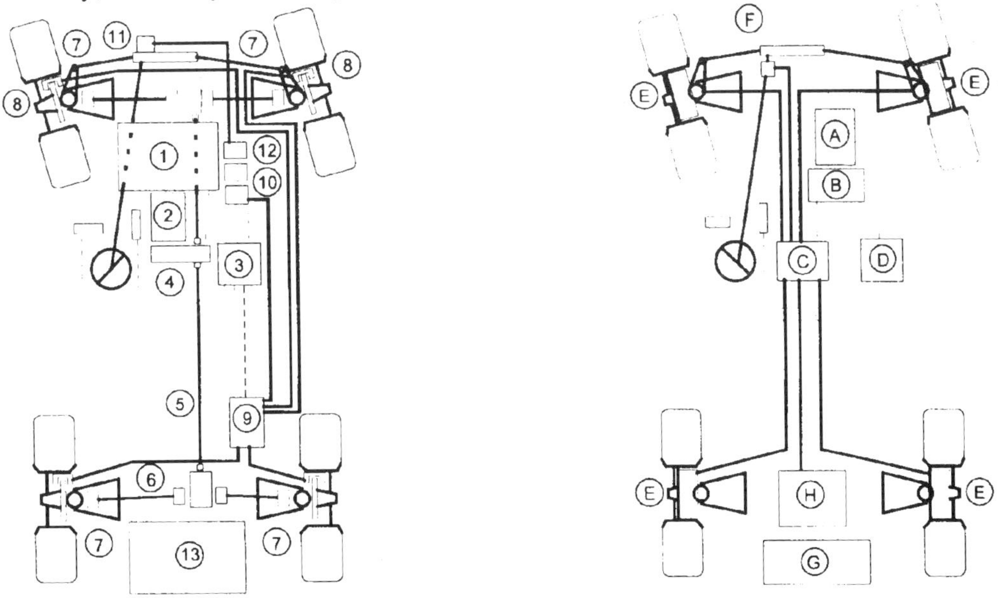

Conventional Drivetrain	Electric Drivetrain
1) 150 kW IC Engine	A) 50 to 75 kW IC Engine
2) Automatic Transmission	B) 75 kW Alternator
3) Logic Module	C) Power Module
	D) Logic Module
4) Transfer Case	
5) Driveshafts	E) 35 to 50 kW Wheel Motors
6) Differentials	
7) CV Joints	
8) Disc Brake Assemblies	
9) ABS/SC Hydraulic Module	
10) Brake Hydraulic Pump/Accumulator	
11) Power Steering Rack	F) Steering Rack
12) Power Steering Pump	
13) 40 L Fuel Tank	G) 20 L Fuel Tank
	H) Battery/Capacitor Module

Fig 12: Conventional (left) and electric (right) drivetrains compared

force is limited by the rate at which the energy storage system can absorb and dispense power; without the use of ultracapacitors, therefore, only supplementary braking is possible electro-magnetically. Also braking effectiveness is reduced to the degree that to the degree that zero braking force is reached slightly before the vehicle reaches zero speed. If reverse power rather than a load is applied to the generator, braking force down to zero speed can be generated and held there. The braking requirements of an average saloon car, say the authors, would require wheel motors, on each of the four wheels, with individual capacity of 35–50 kW. Fig 12 shows the author's electric drivetrain proposal alongside a conventional drivetrain. This would suit a 1750 kg vehicle, having all-wheel-drive, electronic torque split, traction control, automatic transmission, ABS braking and stability control. For safety, a conventional friction park-brake would probably be required.

Navigation system advances

The 1998 SAE Congress in Detroit was the occasion for the reporting of important telematics developments bringing closer the prospects of intelligent transportation systems on public highways.

Researchers from Hitachi[6] described an adaptive cruise control (ACC) system using a wheel torque management technique and involving a longitudinal control method. An electronically controlled throttle valve, electrically controlled brake and automatic transmission work co-operatively to provide smoother acceleration, better shift performance and better fuel economy in ACC mode (at a desired headway distance detected by a micro-wave radar sensor), while, in manual driving, wheel torque is proportional to throttle pedal displacement.

The management system seen in the lower part of Fig 13 selects signals from throttle-control, transmission and brake to send commands which depend upon the requested wheel torque. Conventional cruise control is obtained by adding a speed controller to the wheel torque manager, and on top of this, headway control computes a command speed and sends it to the speed controller which calculates the required wheel torque. Block diagrams for the speed and headway controllers are shown in Fig 14.

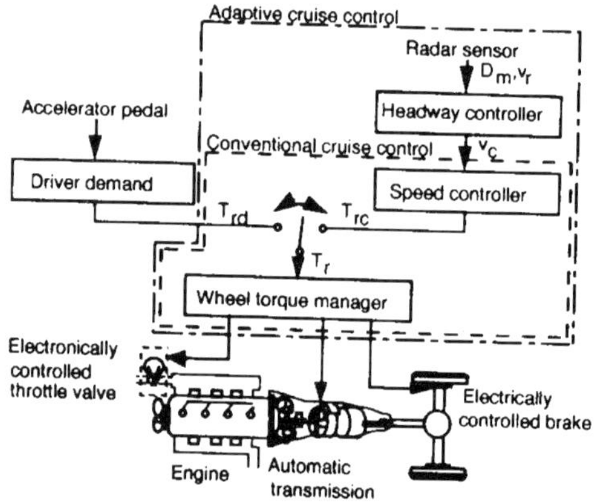

Fig 13: Constituents of ACC

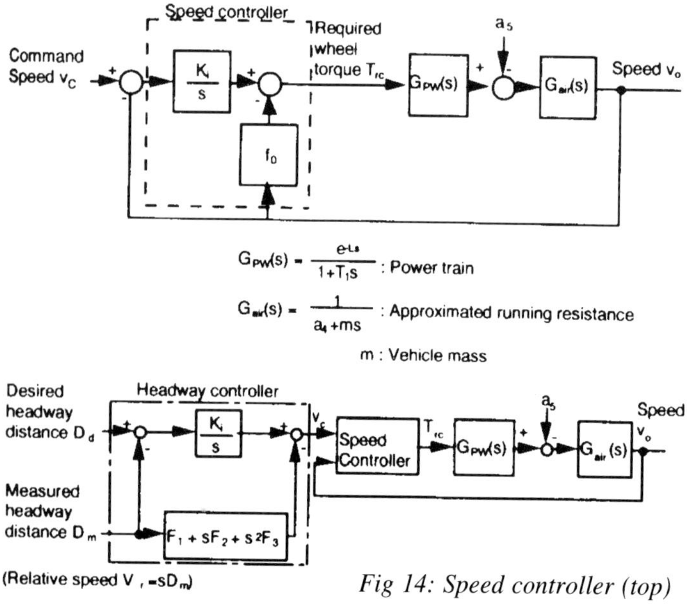

Fig 14: Speed controller (top) and headway controller

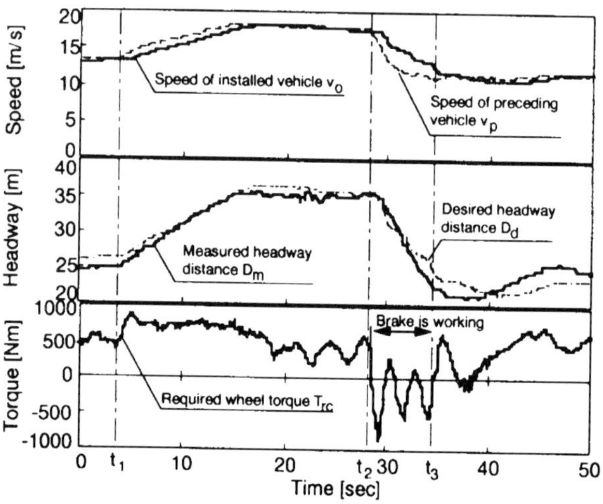

Fig 15: Experimental results of ACC

Tests carried out with the system installed in a vehicle showed the ACC performance as in Fig 15. From ACC start-up at $t = 0$, the preceding vehicle increases its speed v_p at time t_1 then decreases its speed at time t_3 by braking and at time t_4 its speed is kept constant. The wheel torque shows its peak at the beginning of acceleration and keeps it a constant as speed increases; in the deceleration phase, at about -2.5 m/s^2, the torque becomes negative under electrical braking.

Headway sensor design

The development of a suitable radar sensor for ACC has been described by engineers at Bosch[7] which is due for European introduction in the current model year in its initial form as a high-cost option on top-of-the-range vehicles. For more general application a less expensive version is under development, the first stage of which is an integrated sensor and control unit described in the paper. This would be incorporated in a typical ACC system depicted in Fig 16.

The main functional advantage of integration is improved communication between sensor and controller and that not only the prime target preceding vehicle is considered but up to 8 objects in the path of the vehicle (valuable when a vehicle ahead of the preceding vehicle brakes suddenly). Basic specification of the sensor is as follows:

Range	2-120m
Relative speed	-60m/s ... +60m/s
Angle	...+/-5°
Resolution cell	0.6m , 1.7m/s
Measurement rate	10 Hz

Freeways and inter-state highways with a minimum curve radius of about 500 m require a lateral detection angle as shown. A monostatic three beam frequency modulation continuous wave (FMCW)-type radar is employed, working at 76–77 GHz, all three beams sending and receiving simultaneously. Characteristic data for the radar are:

Frequency allocation	76 - 77.0 GHz
Total average output power	< 1mW
Bandwidth	220 MHz
Number of beams	3
Beam width (measured at the -6dB point of the radar-diagram)	4°
Overlap of two beams („Squint Angle")	2°
Total angular range	+/- 5°

A block diagram for the integrated system is seen in Fig 17.

Road guidance for drowsy drivers

To provide vehicles with a guidance means based on white road-line recognition, a system based on lane region extraction and line edge detection has been described by Matsushita engineers. The proposed method uses both low and high spatial frequency image data, respectively, to detect these two attributes of the white lines. The company has shown that detection from 5 to 40 metres in front of the vehicle is accurately performed at a detection rate of over 99.4%. As well as protecting drivers who become drowsy the system also is used to distinguish

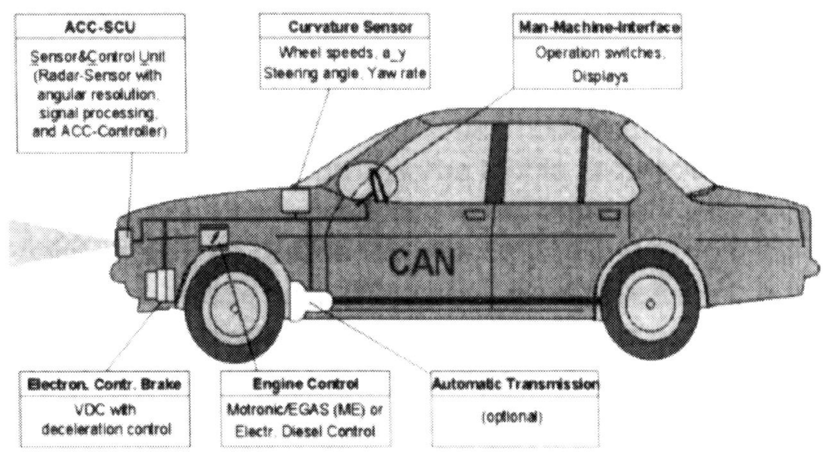

Fig 16: ACC system components

present lane from adjacent lanes in automatic cruise control. The two separate modes of image detection described above are intended to overcome the problems of intermittent white line fading and the problem of the detection camera vibrating in sympathy with the vehicle. Fig 18 distinguishes between the two modes. Lane region extraction uses a knowledge-based brightness rule to distinguish the inside lane area and/or a low spatial frequency image in order to smooth the brightness of the lane region. The system checks there is a smooth transition between the dark area in front of the vehicle and the bright area further away. Edge data are determined using each threshold in each scan direction between 50 lines and 200 lines, the following contour points being removed: those which are more than 20 dots away from the end-of-lane region (extracted from the low-frequency image); all non-continuous points and those which do not fit within the road width. Resultant white lines are calculated using a spline of 3 degrees, the detection process illustrated being implemented on a network of transputers. Camera-angle error compensation is based on distance between the lines being constant, contour points of lines in the image being transferred to points on the world co-ordinate mapped image seen in Fig 19.

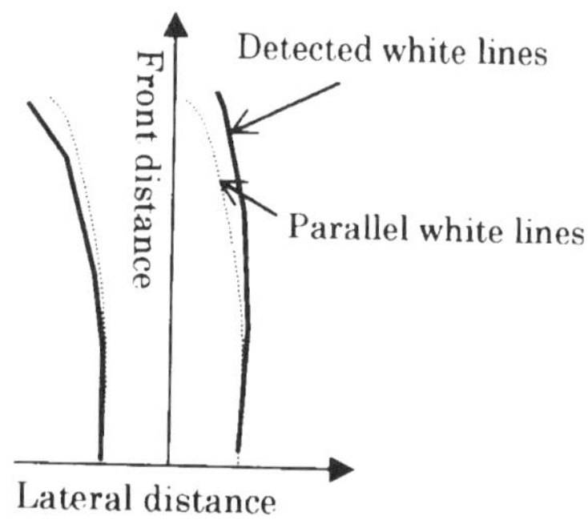

Fig 19: Mapped image

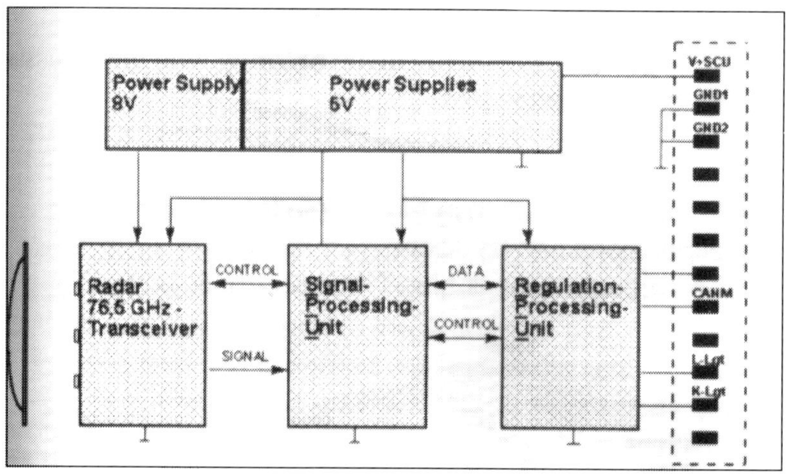

Fig 17: Block diagram of ACC-SCU

Fig 18: White lines detection process

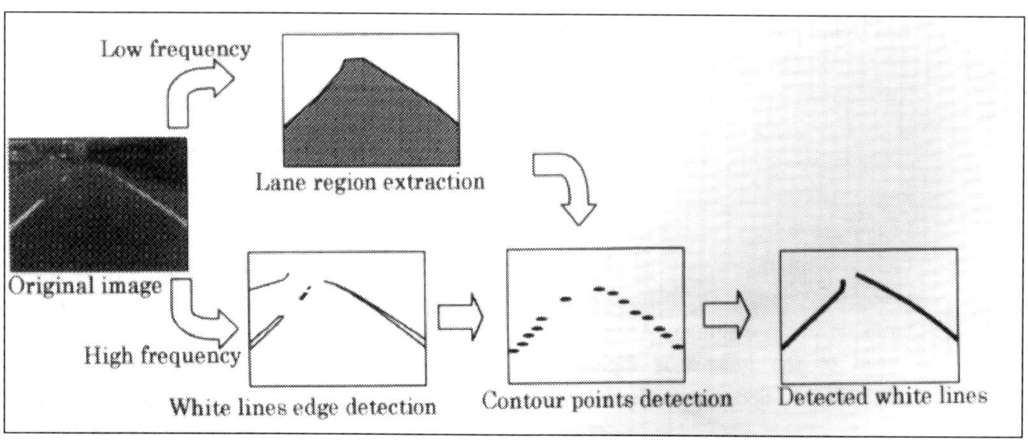

Digital circuits for computation

Logic gates (switching circuits) are known as building blocks for electronic circuits used in computer control. A recent work[8] on engineering mathematics provides an unusually accessible description of the building of such circuits, in the author's chapters on Boolean Algebra and Digital Systems.

The Boolean variable, he explains, involves the realization of the two logical states, 0 and 1, analogous to the open and closed positions of a mechanical switch but in digital electronics represented by low and high voltages in *'positive'* logic. The logic gates, Fig 20, provide Boolean functions analogous to multiplication, division, inversion and their different combinations.

Further combination of the basic gates may be used to form functional logic circuits such as that in Fig 21a which might be used for a vehicle warning buzzer so that it sounds when the key is in the ignition (A) and a door open (B) or when the headlamps are on (C) and a door open (A). Here there are two AND gates and an OR gate. The first two give the output A.B + C.A which is the input to the OR gate which has output A.(B + C).

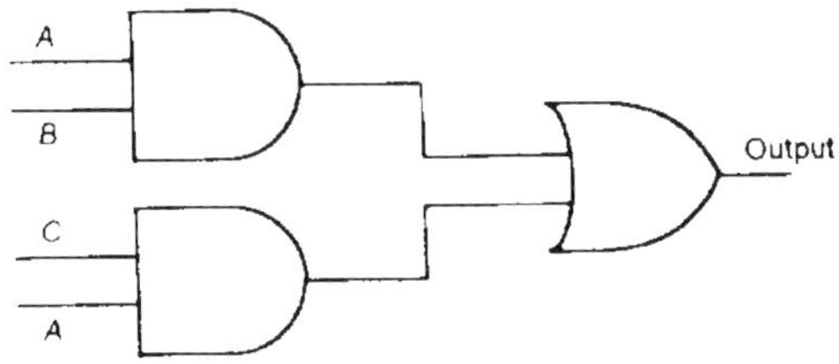

Fig 21a: Warning buzzer circuit

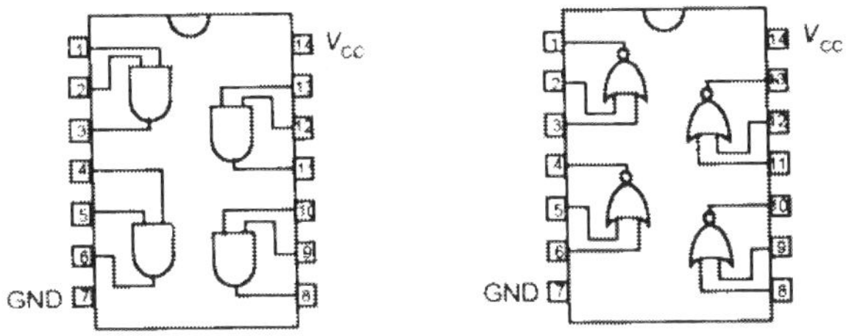

Fig 21b: 7408 and 7402 integrated circuits

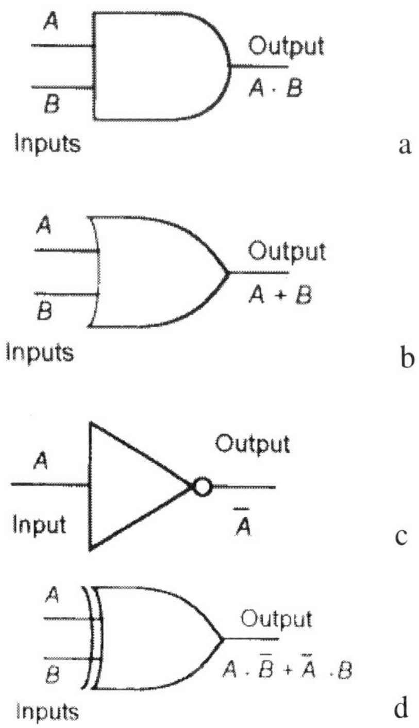

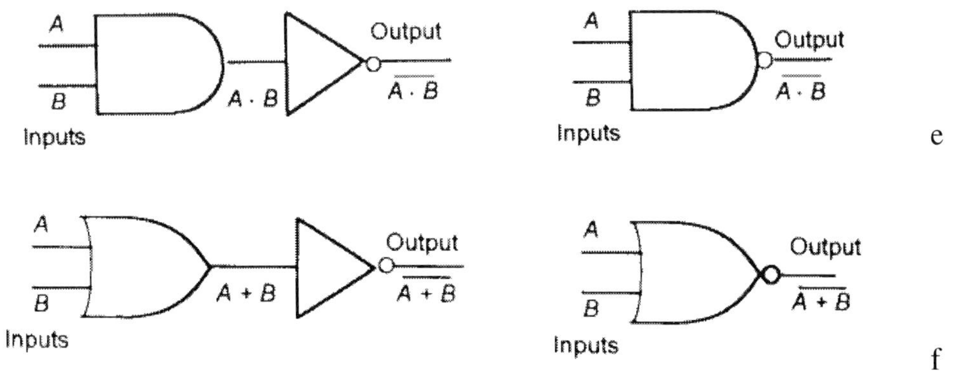

Fig 20: Logic circuits, a. AND; b. OR; c. NOT; d. XOR; e. NAND; f. NOR

In multiple combinations they come as relatively inexpensive integrated circuits, Fig 21b illustrating the 7408 and 7402 circuits. The first has four two-input AND gates supplied as a 14 pin package, with power supply connected to pins 7 and 14, supplying the operating voltage for all the four AND gates. A notch is cut between 1 and 14 to show at which end of the package pin 1 starts. In the 7402 package there are four two-input NOR gates. The use of 7408 as a clock enable circuit is shown in the view in Fig 21c, where signals are passed on to a receiving device only when a switch is set to an enable position. The clock emits a sequence of 0 and 1 signals (pulses), Fig 22. The AND gate will give an output when both input signals are 1.

Computer control usually involves implementation of arithmetic operations for which the binary numbering system is used having binary digits (bits) 0 and 1. Fig 23a shows a half-adder circuit giving the sum equal to 1 when either first or second input numbers, but not both numbers, is 1, obtained with an XOR gate. The carry-out bit is to be 1 when first and second numbers are both 1, obtained with an AND gate. A full adder circuit is seen in Fig 23b; the sum output arriving from the three inputs, first, second and carry-in number. This provides a 1 output when there is an odd number of 1 inputs, obtained by two XOR gates. The carry-out output is 1 when any two of the inputs are 1, obtained with three AND gates and an OR gate.

The half and full adder circuits can be combined to realize multi-bit adder circuits. Thus, for a 4-bit adder, Fig 24 is an appropriate circuit, in which the full and half adders are represented as blocks for simplicity. By connecting the carry-out from one 4-bit adder to the carry-in for another one, an 8-bit adder circuit can be obtained (used for adding positive numbers together), Bolton explains.

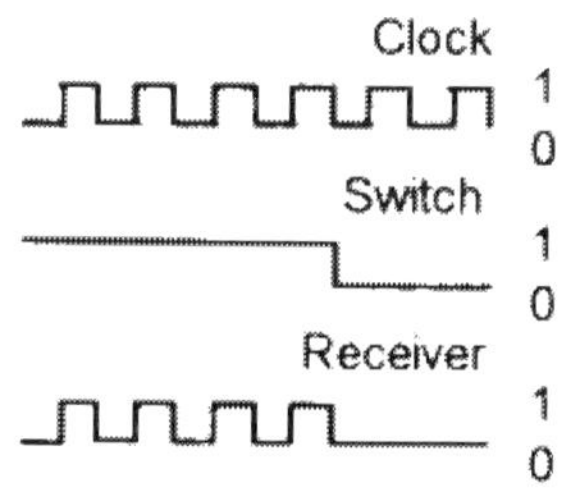

Fig 22: Clock enable signals

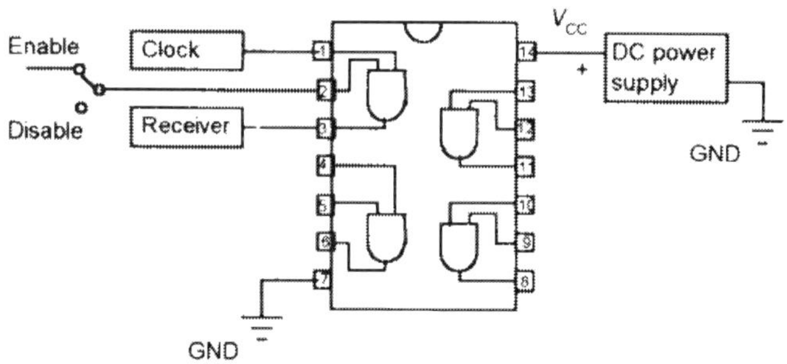

Fig 21c: Clock enable circuit using 7406 IC

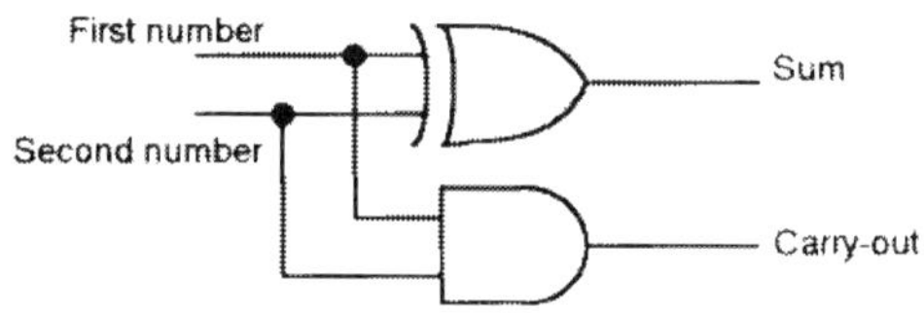

Fig 23a: Half-adder

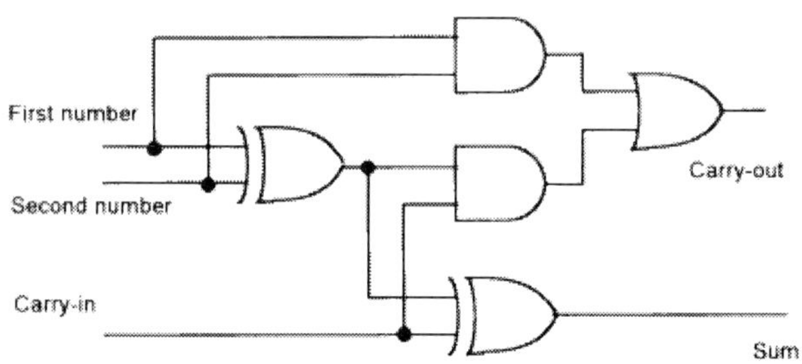

Fig 23b: Full adder

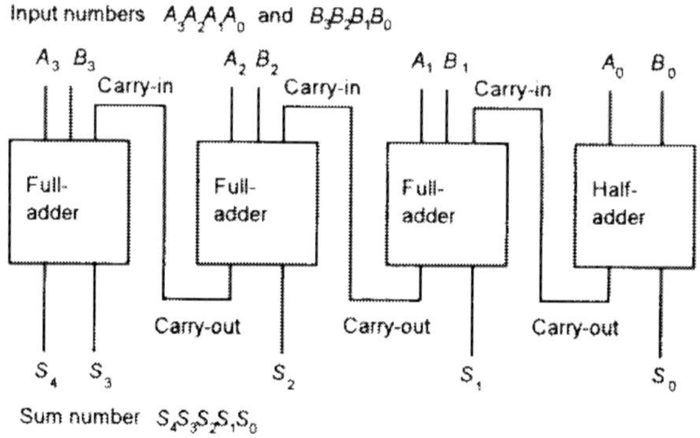

Fig 24: 4-bit adder

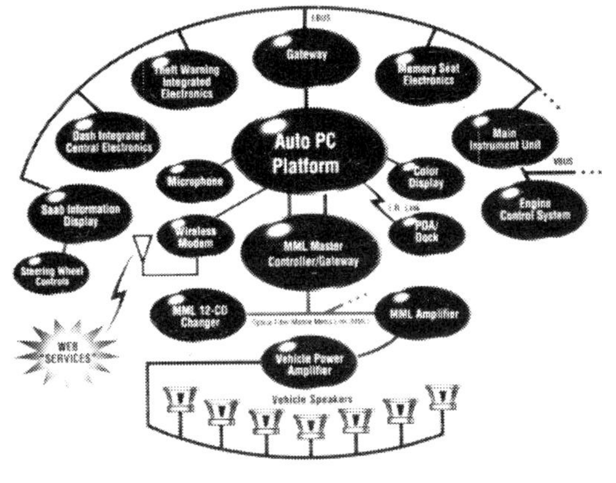

Fig 25: Personal Productivity info-technologies

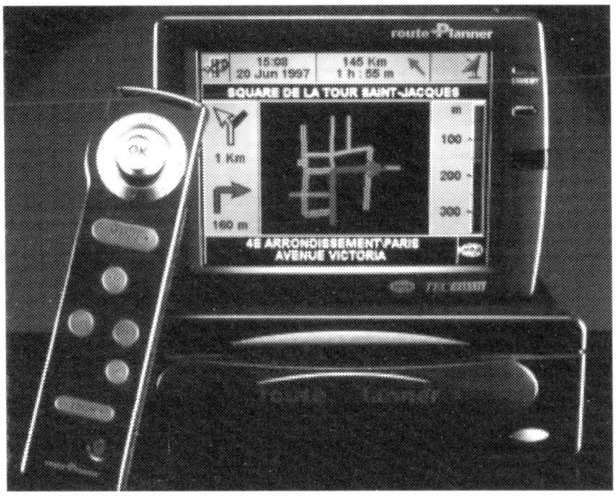

Fig 26: TEC mobility system

Proprietary control system advances

The Delphi Personal Productivity vehicle is a combination of information technologies, Fig 25, for permitting business drivers to remain in constant communication with their offices. First demonstrated in a Saab car it has involved co-operation between the suppliers and Microsoft/Mecel. By means of Windows CE, installed in an on-board computer, the system links with office-based Windows PCs. A speech interface called Auto PC allows hands-free communication for the driver with voice commands being used to send E-mail messages, for example. The facility also uses speech synthesis to convert incoming messages for the driver. There is, too, the option of using steering-wheel mounted controls instead of audio communication. The Delco Mobile Media Link employed is a high-speed fibre-optic serial data link capable of transmitting data at 98 MB/sec. It also has sufficient band width to allow reception of up to 50 channels of stereo audio and 15 channels of TV-quality compressed video. A cellular modem connection allows information like traffic updates to be input from an internet provider.

Following the introduction of the Magneti-Marelli Route Planner in-car navigation system, the next generation TEC mobility system under development, Fig 26, will allow the driver to access Pentium computer power through ConnectCar-PC via a voice

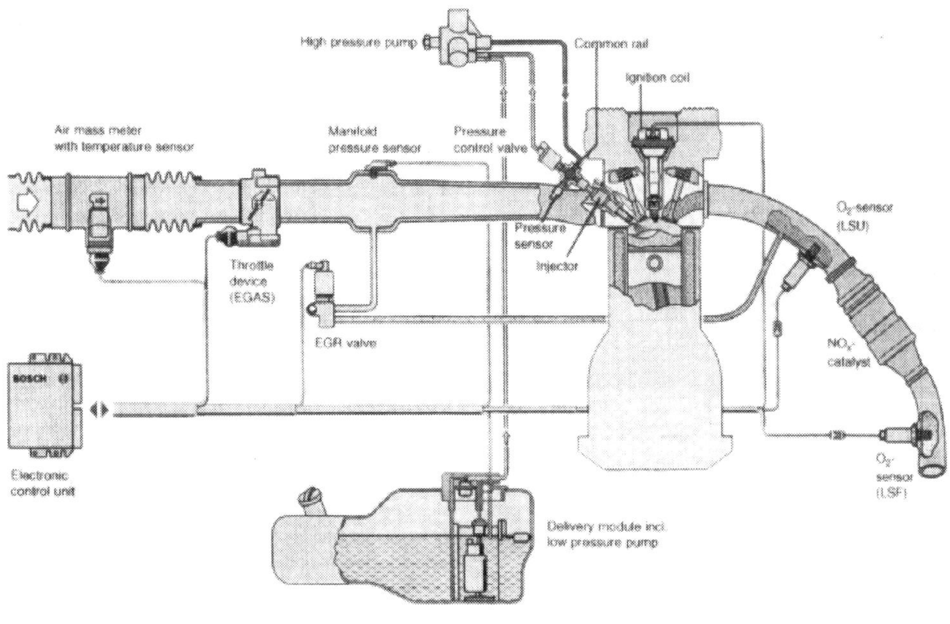

Fig 27:Motronic MED 7 engine management system

recognition interface. As well as helping the driver to navigate, he/she will be able to place and receive telephone calls and even browse the internet or send an E-mail.

The Bosch Motronic MED 7 engine management system, Fig 27, is designed for direct-injection gasoline engines and calls on a wide variety of sensor inputs before computing engine torque response to throttle-pedal displacement. Some 15–20% fuel savings are claimed for DI engines using the system compared with conventional state-of-the-art gasoline engines.

Delphi's E-STEER electronic steering incorporates a steering gear assist mechanism , Fig 28, and electronic controller to provide responsive steering assist, in several configurations, Fig 29. Sensors measure two primary inputs—vehicle speed and driver torque (or effort) on the steering shaft at the assist unit. These two primary inputs, along with other system variables and inputs, are continuously fed into an electronic control module, which analyses the data and determines the direction and the amount of steering assist required. The controller then generates a command to the variable-speed, 12-volt electric motor. Using intricate control algorithms, the controller's command varies the torque of the electric motor, which drives a gear mechanism to provide the required assist. A permanent magnet brushless motor is used, which allows the smallest package size, the smallest mass, and the lowest rotor inertia (for more responsiveness). The motor is also quieter, more reliable, and more powerful than many types of electric motors; it uses the vehicle's battery as its power source. This can be especially important to drivers who experience an engine stall while turning a sharp corner or rounding a steep mountain curve.

Traditional hydraulic systems can be sluggish until fluids warm up, especially in very cold climates. The Delphi system provides cold-weather rapid start. With its engine independence and fluid-free design, the system is less sensitive to cold temperatures even

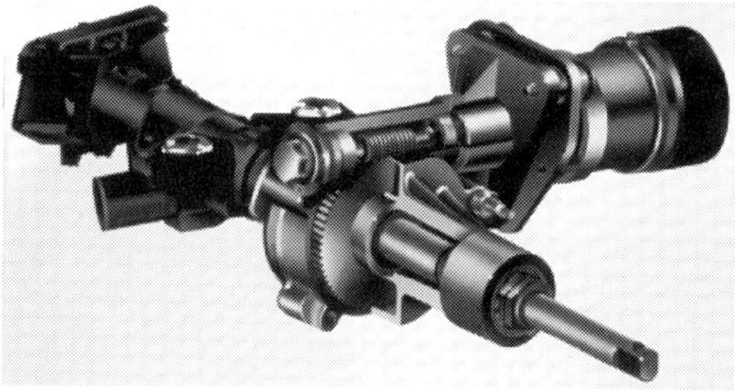

Fig 28: Assist mechanism

Fig 29: Different configurations

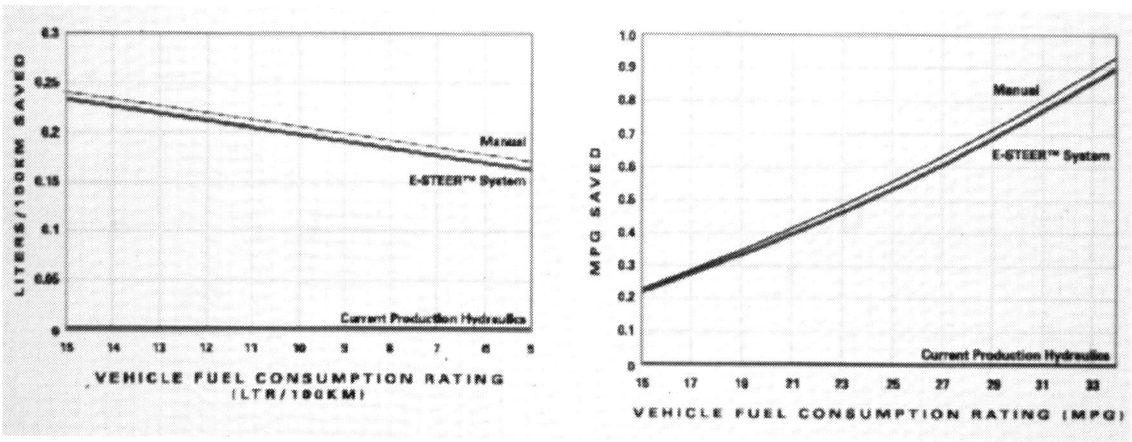

Fig 30a: Vehicle fuel consumption rating — comparison between manual steering and E-Steer;

at –40C The system also eliminates the large energy loss typically involved in those first few minutes of warm-up. When a major European manufacturer conducted a fuel economy test, it found virtually no difference in fuel usage between this EPS and manual steering, Fig 30a.

This EPS is an 'on-demand' system, running in a monitoring mode until assist is needed. Typically using less than 0.5 amp at idle and with the average usage between 1 to 2 amp, the system has extremely low parasitic energy losses for maximum efficiency. In addition to reducing engine drain, the system may also reduce overall mass. The system also provides improved stability for greater passenger comfort, control, and safety. There is smoother transition from one effort level to another—without compromising power steering assist between parking manoeuvres and highway driving .

System design with more options Speed-sensitive variable effort steering is an increasingly popular option, but the need for a controller makes it a costly addition to conventional hydraulic systems. Delphi's E-Steer can calculate the required assist level, based on vehicle speed and driver torque input, using its existing software and controller, Fig 30b. This makes speed-sensitive variable effort steering (and other options) available for very little added cost.

A further benefit of electric power steering is that assist can be provided in both directions, so it is no longer necessary to compromise suspension design in order to achieve the required returnability characteristics. The system's optional Assisted Return feature allows the returnability to be optimized independently, so that when the driver turns and then releases the wheel, E-Steer returns it safely to centre. The company has also developed a unique Steer Damping option that helps to provide a smoother feel and reduce the effects of issues such as free control. A conventional hydraulic steering system would require time-consuming and expensive mechanical modifications, but with E-Steer this can be achieved with software.

Fast system tuning The operating characteristics of any steering assist system must be carefully tuned to match the vehicle design objectives. With conventional hydraulic systems, tuning requires the removal of the steering gear and the manufacture and fitment of modified valves and torsion bars. This process limits testing to two or three configurations per day, with the whole process typically taking from one to six months. E-Steer allows very fast tuning without mechanical intervention, allowing the desired feel and performance to be achieved in a small fraction of the time previously required. All that is required is a laptop computer with Delphi's proprietary tuning package.

Reliability In addition to engine independence, there are specific reliability features built into the system. During development, company engineers set up testing situations to simulate possible faults in the system. With this knowledge, they built in several reliability features. The system continuously runs detailed self-checks and diagnostics, ensuring that all areas of the system are functioning properly. Redundancy (back-up or 'repeat' system elements) has been engineered into the system to provide reliability.

Packaging flexibility With no pumps, hoses, or belts on the engine, this simplified packaging provides maximum design and engineering flexibility. The system controller can be incorporated with the column, the rack, the pinion, or in a stand-alone

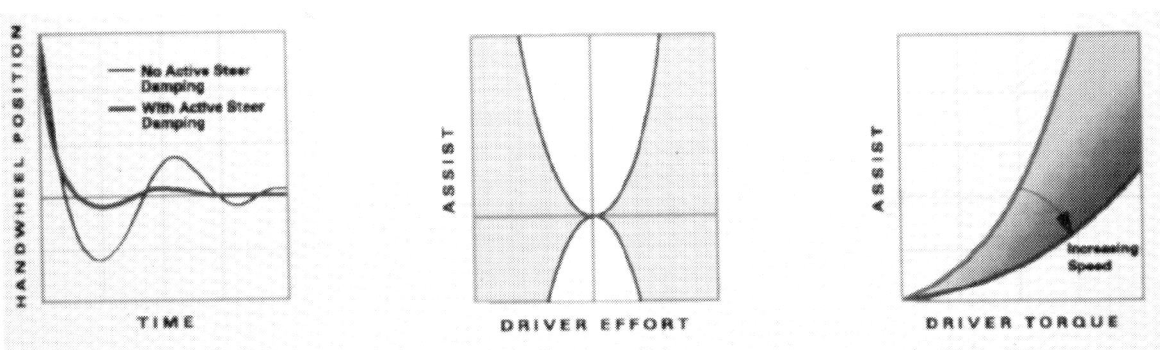

Fig 30b: Performance parameters: left: steer damping; centre: tuning capability; right: variable effort

location. The controller can be in the engine or the passenger compartment. The system frees up the part of the engine where a belt and pulley to drive the pump would otherwise be placed. This makes room for other features like pollution control or an air-conditioning compressor. And there are no more hoses snaking through the engine compartment. The system can help reduce the number of parts the car manufacturer has to track, order, and store. For example, two models in the same platform, a 4-door sedan and a 2-door coupe, can use the same physical steering system. With the advanced tuning technology of the system, one could custom tune a luxury feel for the sedan and a sporty feel for the coupe instantly.

For the future Delphi Steering anticipates the development of several options for future electric power steering systems. Potential combinations include selectable steering to go with a selectable chassis where the driver could push a button for a sportier feel, for example. Hand-held diagnostic tools could be used at the dealership, allowing consumers to request changes to the feel of their steering. As a cost-reduction strategy, the system could be very useful in advanced systems integration. With up-level integration, algorithms/software for other systems could be combined—like security, collision avoidance, steer by wire, or automated highway systems.

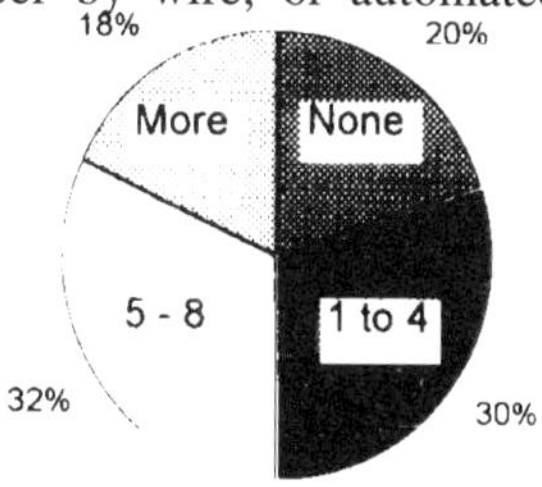

Fig 32: Use of vehicle per day

Hybrid drive prospects

A neat description of the problems of hybrid drive vehicles has come out of the results of the three-year HYZEM research programme undertaken by European manufacturers. According to Rover participants[9], controlled comparisons of different hybrid drive configurations, using verified simulation tools, are able to highlight the profitable fields of development needed to arrive at a fully competitive hybrid drive vehicle and demonstrate, in quantitative terms,

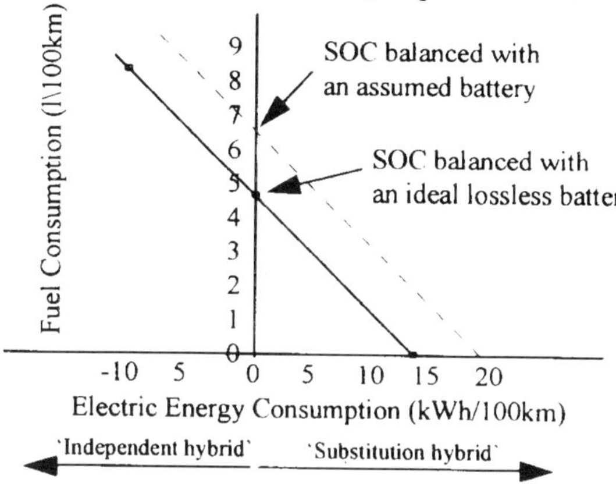

Fig 31: Characterizing a hybrid powertrain

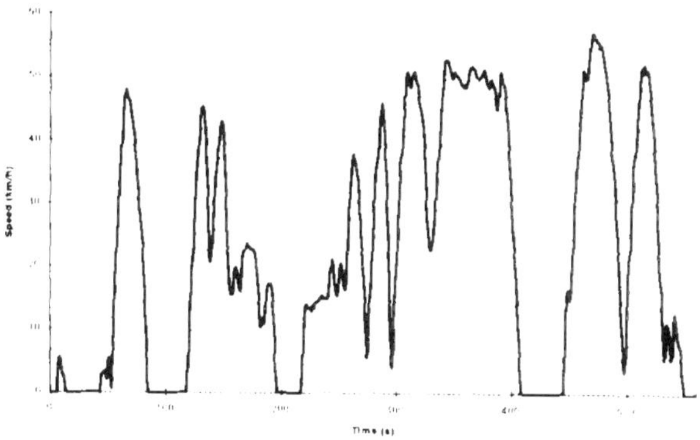

Fig 34: Synthetic urban drive cycle

Fig 33: Daily distances and trip lengths

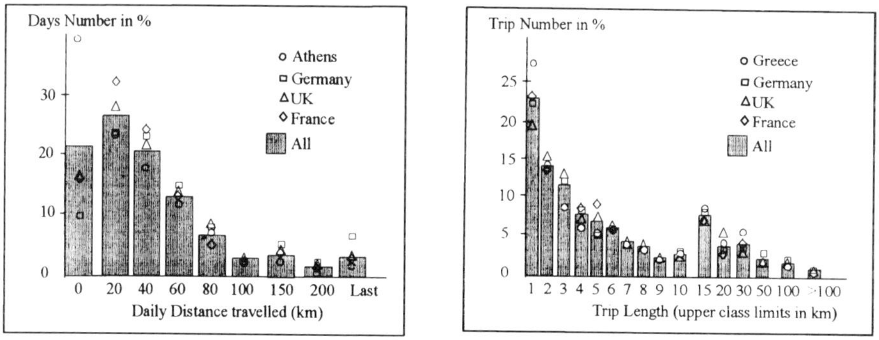

the trade-off between emissions, electrical energy and fuel consumption. Only two standard test points are required to describe the almost linear relationship: fuel consumption at point of no overall change in battery state-of-charge (SOC) and point of electrical consumption over the same cycle in pure electric mode. A linear characteristic representing an ideal lossless battery can also be added to the graph, to show the potential for battery development, Fig 31.

Confirmation was also given to such empirical assessments that parallel hybrids give particularly good fuel economy because of the inherent efficiency of transferring energy direct to the wheels as against the series hybrids' relatively inefficient energy conversion from mechanical to electrical drive. The need for a battery which can cope with much more frequent charge-discharge cycles than one for a pure electric-drive vehicle was also confirmed. Although electric energy capability requirement is less stringent, a need to reduce weight is paramount in overcoming the problem of the redundant drive in hybrid designs.

A useful analyis of over 10,000 car journeys throughout Europe was undertaken for a better understanding of 'mission profile' for the driving cycles involved. Cars were found to be used typically between 1 and 8 times per day, Fig 32, and total daily distances travelled were mostly less than 55 km. Some 13% of trips, Fig 3, were less than 500 metres, showing that we are in danger of becoming like Americans who drive even to visit their next door neighbours! Even more useful velocity and acceleration profiles were obtained, by data recoding at 1 Hz frequency, so that valuable synthetic drive cycles were obtained such as the urban driving one shown in Fig 34.

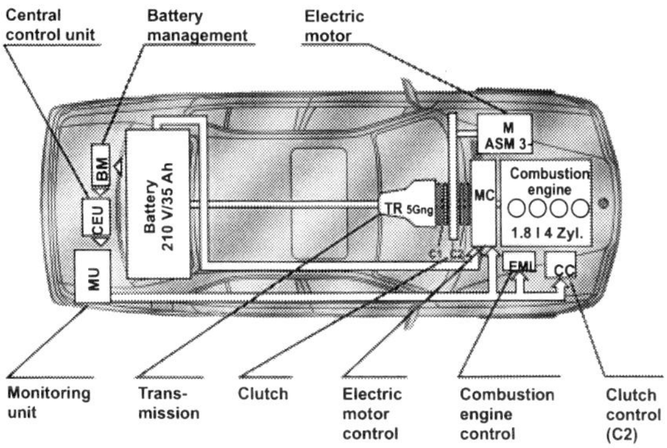

Fig 35: BMW parallel hybrid drive

Fig 36: Parallel hybrid drive mechanism

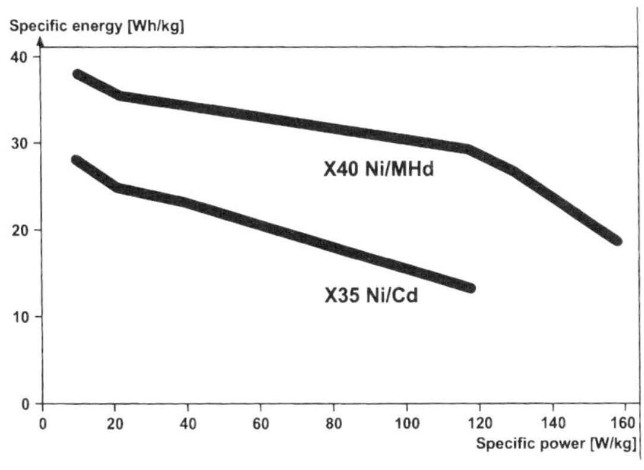

Fig 38: Ragone diagram for the two battery systems

	518i Hybrid	518i standard
Combustion engine 1.8-ltr. 4-cyl	162 Nm (119 lb-ft), 83 kW (113 bhp)	162 Nm (119 lb-ft) 83 kW (113 bhp)
Electric motor/ alternator	Siemens asynchronous 200V, max 280A, 95 Nm (70 lb-ft) (165 Nm/122 lb-ft max), 18 kW (26 kW max)	
Transmission ratios	5.09/2.8/1.76/1.24/1.0	5.1/2.77/1.72/ 1.22/1.0
Final drive	3.25	3.46
Ni/MH battery	DAUG-Hoppecke Cell type: X40, 180 1.2V cells/40Ah/1.06 kg, total energy content 9 kWh, maintenance-free Dimensions: 803/465/ 418 mm^3	
Ni/Cd battery	DAUG–Hoppecke, cell type: X35, 168 1.2V cells/35Ah/1.15 kg Energy content 7 kWh, maintenance-free Dimensions: 859/400/410 mm^3	
Weight, unladen kg (lb)	1940 (4278)	1418 (3127)
Performance weight kg (lb)	2139 (4716)	1618 (3568)

Fig 37: Vehicle specification

BMW researchers[10] have shown the possibility of challenging the fuel consumption levels of conventional cars with parallel hybrid levels, by using map-controlled drive management. The 2-shaft system used by the company, Fig 35, uses a rod-shaped asynchronous motor, by Siemens, fitted parallel to the crankshaft beneath the intake manifold of the 4-cylinder engine, driving the tooth-belt drive system as seen in Fig 36; overall specification compared with the 518i production car from which it is derived is shown in Fig 37. The vehicle still has top speed of 180 kph (100 kph in electric mode) and a range of 500 km; relative performance of the battery options is shown by Fig 38. Electric servo pumps for steering and braking systems are specified for the hybrid vehicle and a cooling system for the electric motor is incorporated. The motor is energized by the battery via a 13.8 V/ 50 A DC/DC converter. The key electronic control unit links with the main systems of the vehicle as seen in Fig 39.

To implement the driving modes of either hybrid, electric or IC engine the operating stategy is broken down into tasks processed parallel to one another by the CPU, to control and monitor engine, motor, battery and electric clutch. The mode task determines which traction condition is appropriate, balancing the inputs from the power sources; the performance/output task controls power flow within the total system; the battery task controls battery charging. According to accelerator/braking pedal

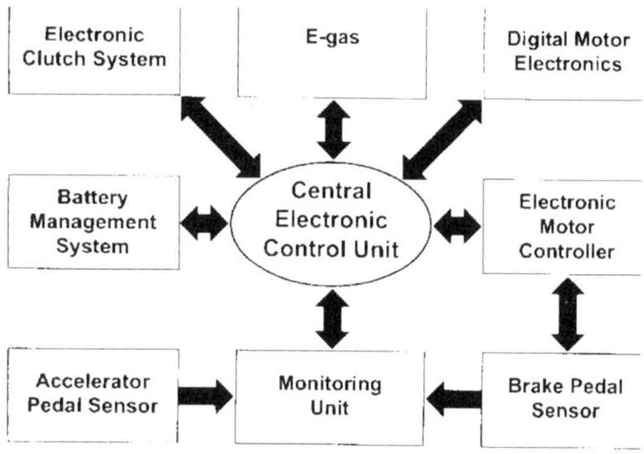

Fig 39:Vehicle management

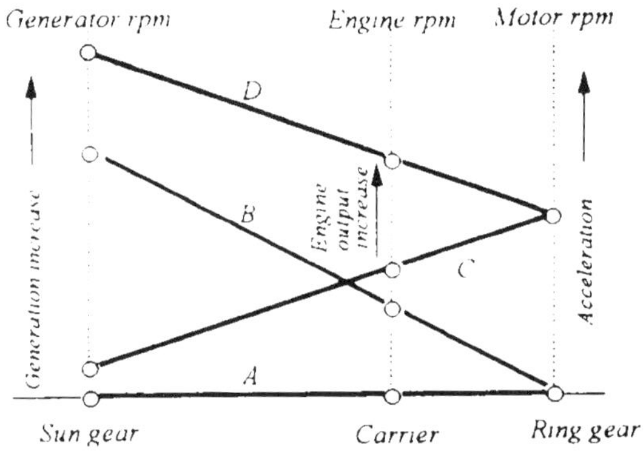

The three vertical lines in the diagram show the shafts in the planetary

Fig 41: Power interaction diagram

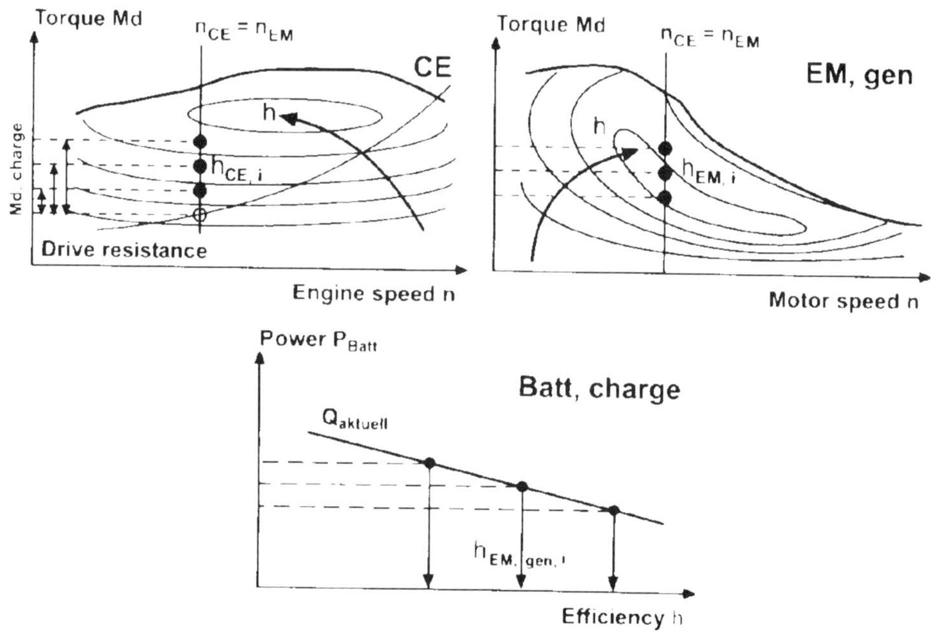

Fig 40: Optimized re-charge strategy

inputs, the monitoring unit transfers the power target required by the driver to the CPU where optimal operating point for both drive units is calculated in a continuous, iterative process. Fig 40 gives an example of three iterations for charge efficiency, also determined by the CPU, based on current charge level of the battery.

Production control-system

In the Toyota Prius production hybrid-drive car, described in a separate article in this issue, for the power-split device which is a key part of the system, company engineers[11] have provided the diagram of Fig 41 to show how the engine, generator and motor operate under different conditions. At A level with vehicle at rest, engine, generator and motor are also at rest; on engine start-up the generator produces electricity acting as a starter to start the engine as well as operating the motor causing the vehicle to move off as at B. For normal driving the engine supplies enough power and there is no need for generation of electricity, C. As the vehicle accelerates from the cruise condition, generator output increases and the motor sends extra power to the drive shaft for assisting acceleration, D. The system can change engine speed by controlling generator speed; some of the engine output goes to the motor via the generator as extra acceleration power and there is no need for a conventional transmission.

The control system schematic for the vehicle is in Fig 42, the THS (Toyota Hybrid System) calculates desired and existing operating conditions and controls the vehicle systems accordingly, in real time. The ECU keeps the engine operating in a predetermined high torque to maximize fuel economy. The corresponding schematic for the ECU is in Fig 43. It is made up of 5 separate ECUs for the major vehicle systems. The hybrid ECU controls overall drive force by calculating engine output, motor torque and generator drive torque, based on accelerator and shift position. Request values sent out are received by other ECUs; the motor one controls the generator inverters to output a 3-phase DC current for desired torque; the engine ECU controls the electronic throttle in accordance with requested output; the braking ECU co-ordinates braking effort of motor-regeneration and mechanical brakes; the battery ECU controls charge rate.

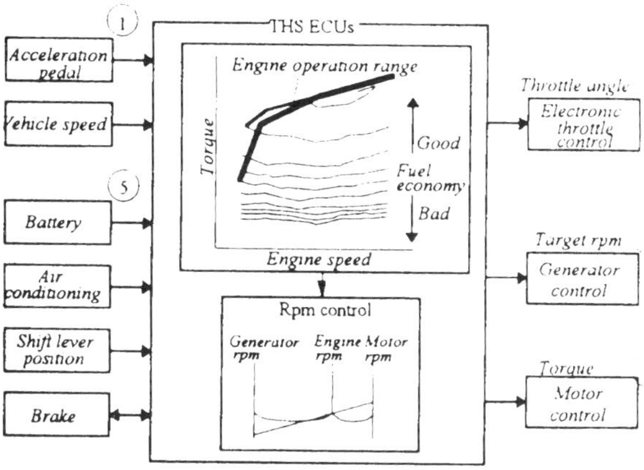

Fig 42: THS control system

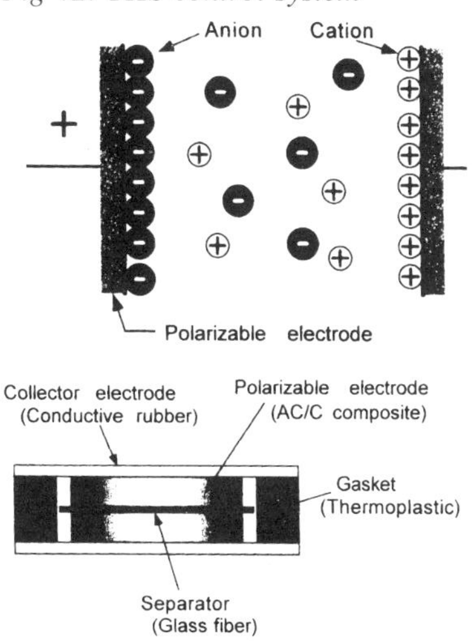

Fig 44: EDLC model (a), above, and cell (b), below

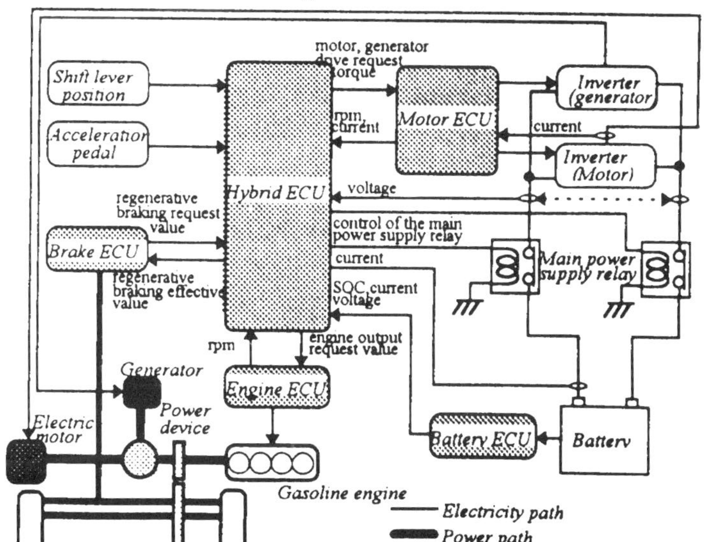

Fig 43: ECU schematic

Energy storage initiatives

According to researchers at NEC Corp[12], the supercapacitor will be an important contributor to the energy efficient hybrid vehicle, the absence of chemical reaction allowing a durable means of obtaining high energy charge/discharge cycles. Tests have shown for multi-stop vehicle operations a 25–30% fuel saving was obtained in a compact hybrid vehicle fitted with regenerative braking.

While energy density of existing, non-automotive, supercapacitors is only about10% of that of lead-acid batteries, the authors explain, it is still possible to compensate for some of the weak points of conventional batteries. For effective power assist in hybrids, supercapacitors need working voltage of over 100 V, alongside low equivalent series resistance and high energy density. The authors have produced 120V units operating at 24 kW fabricated from newly developed activated carbon/carbon composites. Electric double layer capacitors (EDLCs) depend on the layering between electrode surface and electrolyte, Fig 44(a) shows an EDLC model. Because energy is stored in physical adsorption/desorption of ions, without chemical reaction, good life is obtained. The active carbon electrodes usually have specific surface area over 1000 m^2/g and double-layer capacitance is some 20–30 $\mu F/cm^2$ (activated carbon has capacitance over 200–300 F/g). The EDLC has two double-layers in series, so it is possible to obtain 50–70F using a gram of activated carbon. Working voltage is about 1.2V and storable energy is thus 50 J/g or 14 Wh/kg.

Fig 44(b) shows a cell cross-section, the conductive rubber having 0.2 S/cm conductivity and thickness of 20 microns. The sulphuric acid electrolyte has conductivity of 0.7 S/cm. Fig 45(a) shows the

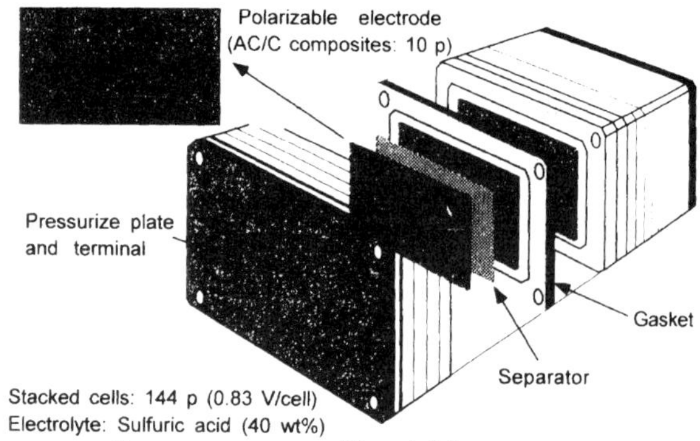

Items	Characteristics
Working voltage (V)	120
Capacitance (F)	20
ESR (mΩ)	78
Maximum current (A)	200
Weight (kg)	24
Volume (L)	17
Size (W x D x H mm)	390 x 270 x 160
Power density (kW/kg)	1.0
Power density (kW/L)	1.4
Energy density (Wh/kg)	1.7
Energy density (Wh/L)	2.4

Fig 45: High-power EDLC schematic (above)and specification (below)

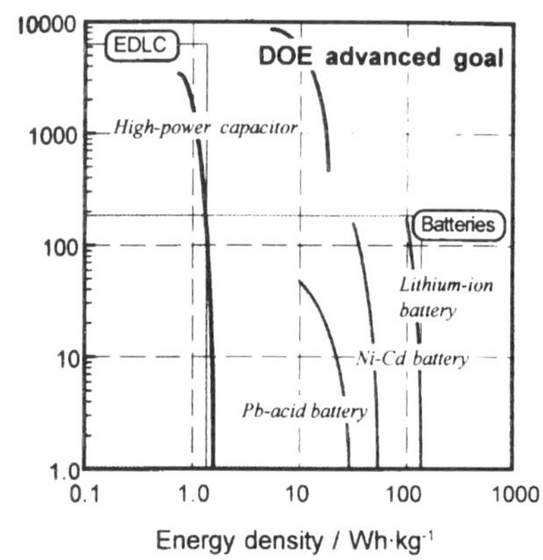

Fig 46b: Power density, y-axis in W/kg, vs energy density for high-power EDLC

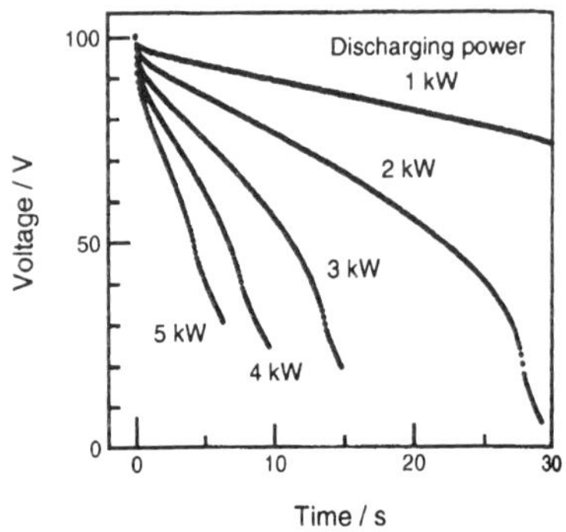

Fig 46a: Constant power discharge characteristics

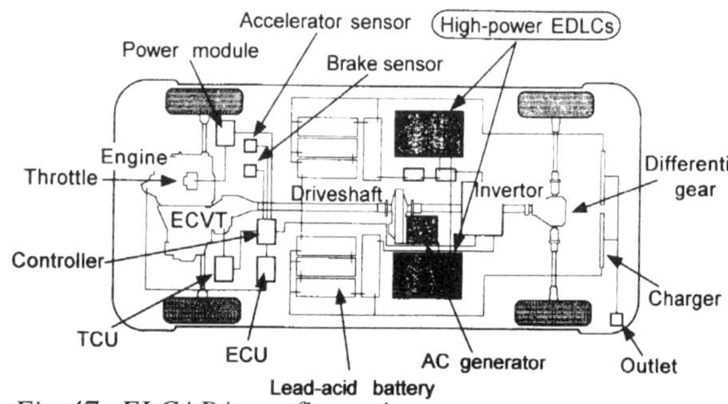

Fig 47: ELCAPA configuration

high-power EDLC suitable for a hybrid vehicle and the table in Fig 45(b), its specification. Plate size is 68 X 48X1 mm^3 and the weight 2.5 g, a pair having 300 F capacity. Fig 46(a) shows constant power discharge characteristics and Fig 46(b) compares the EDLC's energy density with that of other batteries.

Fuji Industries's ELCAPA hybrid vehicle, Fig 47, uses two EDLCs (of 40 F total capacity) in parallel with lead-acid batteries. The stored energy can accelerate the vehicle to 50 kph in a few seconds and energy is recharged during regenerative braking. When high energy batteries are used alongside the supercapacitors, the authors predict that full competitive road performance will be obtainable.

A high energy battery receiving considerable attention is the lithium ion cell unit, the development of which has been described by Nissan and Sony engineers[13] who point out that because of the high cell voltage, relatively few cells are required and better battery management is thus obtained. Accurate detection of battery state-of-charge is possible based on voltage measurement. In the battery system developed, Fig 48, cell controllers and a battery controller work together to calculate battery power, and remaining capacity, and convey the results to the vehicle control unit. Charging current bypass circuits are also controlled on a cell to cell basis. Maximizing lifetime performance of an EV battery is seen by the authors to be as important as energy density level. Each module of the battery system has a thermistor to detect temperature and signal the controllers to activate cooling fans as necessary.

Nissan are reported to be launching the Ultra EV in 1999 with lithium-ion batteries; the car is said to return a 120 mile range per charge. For the farther future lithium-polymer batteries are reported to be capable of giving 300 mile ranges.

There is also less strength in the arguments against EVs with news of cheaper solar cells. Rapid thermal processing (RTP) techniques are said to be halving the time normally taken to produce silicon solar cells, while retaining an 18% energy conversion efficiency from sunlight. Researchers at Georgia Institute of Technology have demonstrated RTP processing involving a 3 minute thermal diffusion, as against the current commercial process taking three hours. An EC study has also shown that mass-production of solar cells could bring substantial benefits and that a £350 million plant investment could produce enough panels to produce 500 MW annually and cut the generating cost from 64p/kWh to 13p.

A move towards fuel cells for electrical power generation is also gaining ground. Ballard Generation Systems has announced the start-up of a PEM fuel cell, natural gas fuelled, with 250 kW capacity and using Johnson Matthey electrodes and platinum catalysts. Compared with on-board fuel cells, the larger stationary generating ones can better exploit the high operating efficiency in relation to the relatively high capital cost. The US Dept of Energy argues that over one year 1MW of electricity generated by fuel cells will avoid 45 tonnes of sulphur dioxide and 19 tonnes of nitric oxides being emitted.

Ballard are also in joint agreement with Daimler-Benz on in-vehicle fuel cells and an A-class Mercedes-Benz, thus powered, is scheduled for around 2004. Mass production to bring down the cost of fuel cells is the objective.

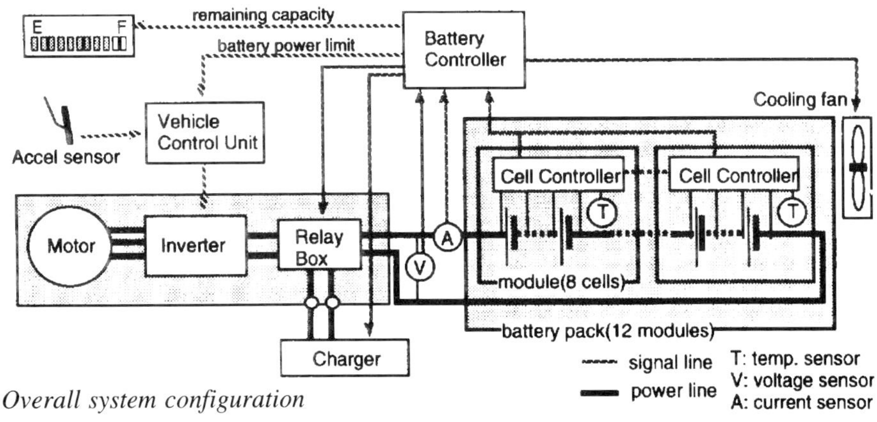

Fig 48: Overall system configuration

Automation of handling tests

Recent publicity surrounding deaths of road users due to accidents caused by OEM's road-test drivers, and the need to reproduce laboratory conditions of repeatability into track testing, will hasten the acceptance of the ABD steering robot, announced by Anthony Best Dynamics. Essentially the device comprises a motorized assembly (which replaces the steering wheel) whose stator is linked to a pair of load cells measuring torque reaction; these are, in turn, coupled to an adjustable arm clamped on to a pillar member between the vehicle floor and suction caps on the windscreen.

The computer-controlled steering robot provides steering inputs to vehicles during track testing. Historically, this type of testing has been carried out manually, but it has been difficult to ensure that accurate and repeatable inputs are generated. The system developed by ABD overcomes these problems, making it easy to obtain meaningful comparisons between different vehicles and vehicle configurations. The robot, whichcan be used in cars, vans and HGVs, is capable of performing the range of tests described in the international standard ISO7401 as well as many special tests, and has a specification born out of detailed discussions with a number of potential customers who all had considerable experience in this type of testing. The computer and control technology employed is a spin off from ABD's successful Suspension Parameter Measurement Machine (SPMM) installed in both the UK and the US. The SPMM is a cost-effective four wheel-station machine which can measure the suspension, kinematic and compliance characteristics of a wide range of vehicles. The new steering robot uses technology transferred from the SPMM to apply steering inputs to vehicles so that their transient response characteristics can be studied, providing accurate and repeatable inputs even in low frequency sinusoidal tests.

General objectives

At the heart of the robot is a direct drive DC torque motor with incremental encoder feedback. This provides the motive force in a small package which fits within the diameter of a standard car steering wheel. The steering robot therefore only requires a DC electrical supply, which is normally taken from the vehicle's own electrical system.

The steering robot gives very precise, smooth control of the steering without the need for gears or clutch, in a high performance package with minimum weight and inertia. The direct drive design also enables the driver to steer the vehicle easily when the unit is de-energized, as there is very little torque required to back-drive the motor. This enables the unit to be used to learn driver inputs, which can then be replayed by the motor, as well as providing a very simple and inherently safe construction.

The stationary side of the motor is connected to the vehicle body via a special universal torque reaction mechanism utilizing a built-in torque transducer, which is installed in the passenger side of the vehicle between the windscreen and the floor, Fig 49. The system is designed to be installed easily, to provide no access restrictions for the driver, to cause minimal loss of field of view and to be very rigid. The torque reaction system accommodates steering column run out without generating side forces or degrading the torque measurement. The motor assembly is normally fitted to the steering column in place of the vehicle's standard steering wheel, as this provides a very compact installation with minimum inertia. However, an adaptor plate can be supplied to enable attachment to the rim of the standard steering wheel for situations where it is difficult to reproduce the splines on the steering column or where only a small amount of installation time is available.

The steering robot is controlled by a portable industrial PC running a Windows NT based software package. The software enables a database of vehicles and tests to be built up with great flexibility in the test specifications. A range of sinusoidal tests are provided as standard; single period, continuous and swept sine, as well as step and pulse inputs. Alternatively, test profiles can be played out from points stored on a disk file which have been generated elsewhere (special tests or random inputs for example) or learnt from driver inputs. During operation, tests are selected and initiated using a touch screen, which also provides status information as the test progresses. Whilst the robot is operating the driver holds a joystick which gives him the ability to adjust the steering drift of the vehicle and to attenuate the playout amplitude. The robot is automatically de-energized if the joystick is released to enable the driver to regain manual control of the steering.

Fig 49 : Steering robot installed

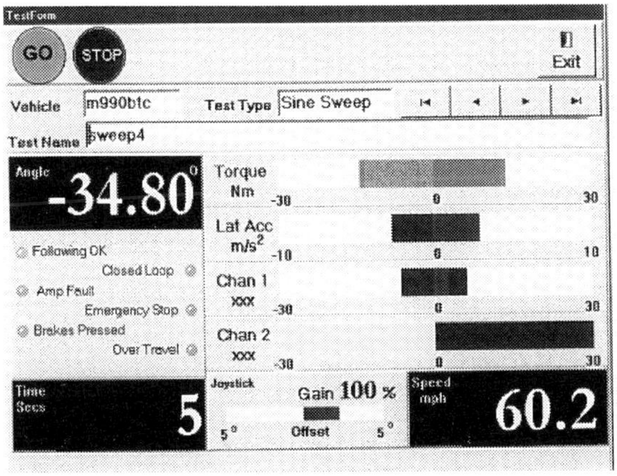

Fig 51: Computer test screen

The computer system also has the ability to capture and store run-time data from the robot's own transducers as well as up to twelve user-defined analogue inputs and three encoder inputs. The system also provides digital outputs to trigger external data capture, and analogue outputs of motor torque, angle and velocity.

The new test systems ABD has developed are a response to an increased thrust in the industry to reduce time from concept to production of a new vehicle, with increased use of computer analysis and prediction techniques. Initial design parameters may be based on experience but, increasingly, use is being made of measurements on previous models and on competitor 'target' vehicles. Steering and suspensions are refined with the aid of models for which measured data is essential; also managers need to ensure that development work is carried out only on vehicles that are known to be built to specification. All this can be achieved only by making the relevant measurements.

A particular specification

The following specification describes a robotic machine for applying low frequency inputs to a vehicle's steering system to enable repeatable measurements to be made whilst testing on a track. Fig 50 shows a schematic diagram of the robot. On the right of the figure is the servo-motor and angular position encoder which mounts directly to the vehicle's steering column. The torque reaction frame also functions as a stand for a flat, touch-screen, display (not shown). This is the user interface for the computer system, which is housed in a separate case located elsewhere in the vehicle (and not shown in the figure). A second, stacking case houses the motor amplifier and signal conditioning for the torque measuring system. Manual steering of the vehicle is still possible because the servo-motor is a direct-drive type and will back-drive easily when the power is removed. The motor has a steering wheel mounted around it to enable the vehicle to be driven normally. The motor is servo-controlled using position feedback from an encoder. This task is undertaken by a dedicated motion controller. The user interface, communication with the motion controller, and other non time-critical functions are handled by a Pentium-based industrial computer.

A telescopic pole between the vehicle floor and windscreen functions as the torque reaction frame for the motor. The two sections of the pole form a pneumatic actuator which is extended during fitting by using a low pressure air supply. Once firmly in position, a cross bar may be adjusted to allow the force links to be fitted to the drive unit at the correct angle. The two force links use charge devices to measure their axial loads. The steer torque is calculated from these loads by the computer. The torque reaction frame also acts as a mount for a removable control box which houses the colour, TFT, display with a touch screen. This provides the principal control interface during testing on the track. Another control box, mounted to the driver's window with suction pads, holds a two-axis joystick incorporating a button for activating closed-loop control of the motor. The spring centred left-right plane of the joystick adjusts the DC steering position, enabling the driver to cancel any drift of the vehicle during a test, while the forwards-backwards plane adjusts the attenuation of the test amplitude. The forwards-backwards plane has a built in friction brake to prevent accidental movement when altering the steering position. A third control box on a flying lead contains an emergency stop button, to cut all power to the motor, and a button which can be used to initiate tests (as an alternative to using the touch screen). This box can be placed in any convenient position within the driver's reach. The computer system, motor amplifier and signal conditioning hardware are housed in two separate rugged cases. These are designed to stack one on top of the other and have strong handles to enable them to be strapped securely within the vehicle. Interconnection cables between the boxes are arranged on one side of the boxes and are protected by hinged covers. The connections to the control devices, motor, transducers and power supply are all on the opposite side. The connectors are protected when not in use by removable lids. The control cases can be mounted at any convenient position, typically a front or rear seat. The main power supply is provided via a large DC to DC converter. This is typically located in the foot-well of the vehicle.

The steer angle, and torque, are measured and recorded by the steering robot and output on +10V analogue (DAC) channels for sampling by other equipment. In addition, velocity or acceleration can be output on a third DAC channel. A further 12 analogue input (ADC) channels and 3 encoder channels are available for monitoring any other signals during a test. The scale factors (EU/V) and offsets for all these channels are adjustable to maximize the dynamic range. The software features a database which records all the vehicles that have been tested, and catalogues the tests performed and the results. Fig 51 shows a typical screen for selecting the test vehicle. The results files contain the steer angle, velocity and acceleration, the torque values, and the spare analogue and encoder channels all sampled at 100 Hz. The files are available in ASCII format with a header describing the test and each channel of data. A facility to copy files to a network, or other drive, is included to facilitate transferring the results to other post-processing software. A number of standard tests,

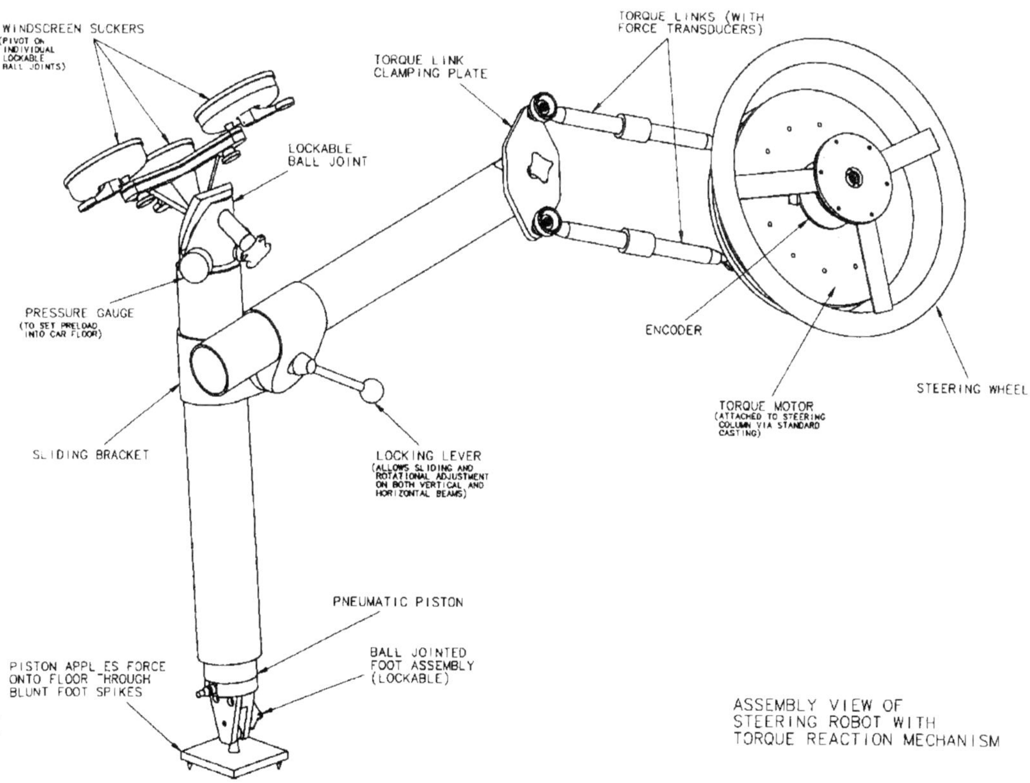

Fig 50: [illegible] reaction

or 'templates', are available to make setting up tests for new vehicles easier. It is possible to add extra templates if required.

During a test, the steer angle and test time remaining are displayed digitally on the touch screen in large text. The steer torque is also displayed but in the form of an analogue bar graph. In addition up to four of the spare channels can also be displayed, one digitally (primarily intended for a speed display if available), and three as bar graphs (for lateral acceleration). A target limit can also be set for each spare channel. Reaching the target level during the test causes the colour of the display to change to give a clear visual indication to the driver. The bar graphs also have a peak hold facility to show the maximum levels achieved in each direction during the test.

Performance modes

The software performs the following types of steering test: in the continuous sine test, for a fixed frequency sine output, the user can select the frequency, amplitude and test time. The playout is automatically ramped at start and finish to help maintain a straight vehicle heading. In the sine sweep test, the user can select the test time, start and finish frequencies and a start and finish amplitude. A facility to generate a sweep with constant peak velocity is also available. In this mode the software calculates the instantaneous amplitude necessary to maintain the peak velocity constant throughout the test. The sweep is logarithmic in frequency. There is also a single period sine test for lane change testing. The amplitude, frequency and direction can be specified. Pulse tests, in addition, allow step or ramped inputs to be applied as a half cycle (move to an angle, dwell, move back to zero). The overall time, step angle and ramp rate can be specified. In User data test mode the software outputs data contained in a file. The software will check the file to make sure that the robot is capable of generating the speed required. The data will have previously been saved to the file by using the 'learn mode', or may be generated externally and imported from an ASCII file. The software also has a learn mode to record steering movements input by a driver and save the results in a file. The file can then be played out as a test

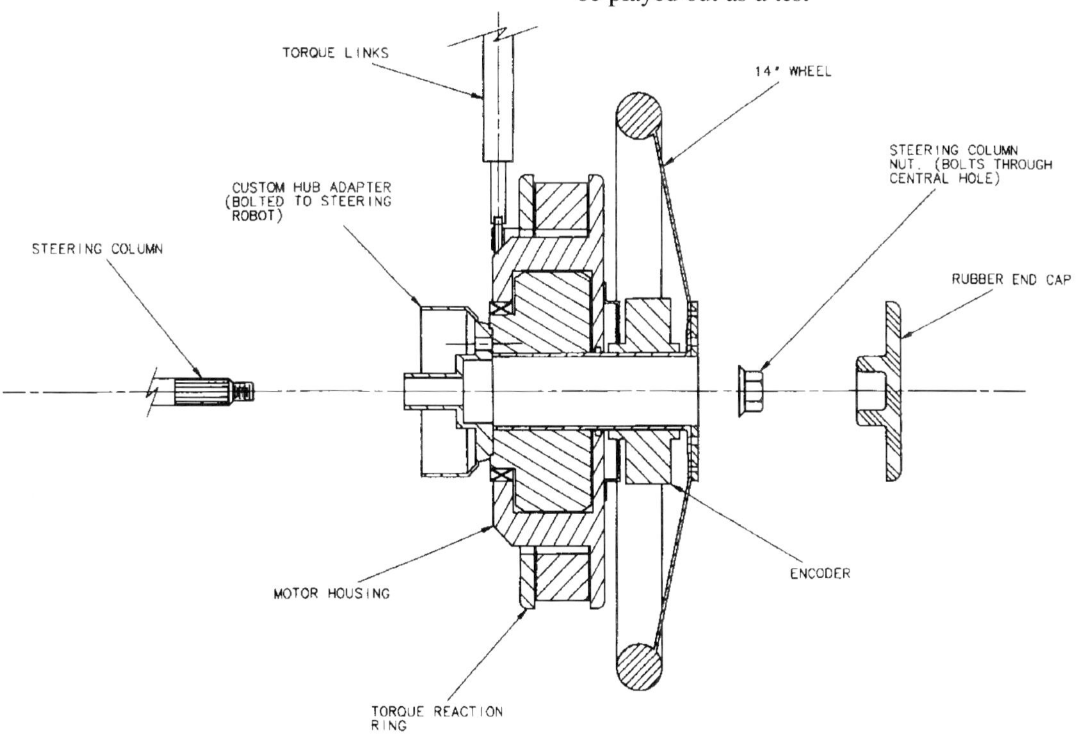

OUTLINE VIEW OF STEERING ROBOT WITH 14" WHEEL

Installation and operation

A hole through the motor assembly makes attachment to the steering column easy, Fig 52. Installation of the torque reaction system is very simple and requires no tools other than a low pressure air supply. The computer system can be installed at any suitable place in the vehicle within 2 metres of the driver, and the display and joystick are easily attached.

The database would normally be set-up for a new vehicle and any tests defined whilst the vehicle is in the running-shop. A normal PC keyboard (and mouse) would be used at this stage. Simple modifications can be made to the test setups later while on the road, by using the touch screen. Once at the test site, the driver would set the correct direction and velocity, and would press and hold the joystick button to engage the motor. The robot will then be in control of the steering. Any drift of the vehicle may be cancelled by the driver by moving the joystick in the left-right plane. Pressing the touch screen button (or remote button) will start the first test in the sequence, and cause a digital output to change state to signal to external data capture systems. The playout will begin after a preset delay (defined in the test setup). A second digital output will signal the start of the playout. During the test, the joystick up-down plane controls attenuation of the test amplitude. Releasing the joystick button at any time will disengage the motor and allow manual steering. Pressing the touch screen during a test will stop the test and reset the steering to straight ahead. Tests may be modified and repeated via the touch screen or by using the keyboard.

During the test the display will show:

1) The steering wheel angle (large text)
2) Direction of first turn in the test
3) Warning that the robot is not following the test profile to a user-set tolerance
4) Percentage test amplitude (100% clearly indicated)
5) Test time remaining
6) Test and robot status
7) Bar graph displays of ADC and encoder channels

The data from a series of tests would be saved to disk for post processing later, but results can be viewed graphically on the touch screen immediately a test is complete.

Chapter 5: Vehicle development

Among the recent vehicle introductions, four stand out as marking a major innovative input to vehicle development. The Mercedes A-class is a significant advance in small car packaging as well as breaking new ground in secondary safety. Ford's Focus brings the company's 21st century hard-edge styling to its most mainstream model and also marks a completed market repositioning of a class of vehicle starting as the Mark 1 Escort to a vehicle refined enough to compete in the neo-luxury sector. The Land Rover Freelander brings the price of comfortable 4-wheel drive vehicles down to realistic levels and is a breakthrough in body option versatility. Project Thrust SSC presents the ultimate in competition vehicles and gives a taste of the technologies required at the limits of vehicle-to-ground adhesion.

Mercedes A-class

The world's pioneering motor manufacturer, and possibly still the most prestigious, had a momentous year in 1998, and seemed to emerge as the senior partner in Daimler-Chrysler. Its earlier attempts to break into compact car making have also now borne fruit in the re-launched moose-proof A-class, Fig 1, which has become possibly the class-leader in city cars. The revised suspension system is said to make the A-class the safest car in its class. The modified rear suspension, Fig 2, now has the special tie rod, shown, incorporated and all models benefit from ESP, the company's Electronic StabilityProgramme. Chassis was also slightly lowered, anti-roll torque increased, rear track extended slightly, dampers/ tyres replaced, suspension stiffened and greater understeer built into the handling dynamics.

Fig 1: Mercedes A-class

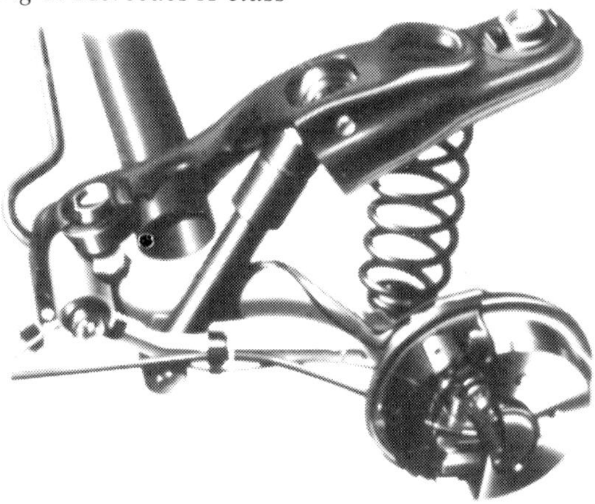

Fig 2: Revised rear suspension

Secondary safety systems

The sandwich floor concept allows the engine and drive assembly of the A-class to be shunted back underneath the body in the event of a frontal impact rather than penetrating the passenger compartment. This ingenious layout, Fig 3, means that the new Mercedes model virtually achieves the high safety standard of the E-class. Even in the event of a side impact the 'sandwich' concept of the double layer floor, Fig 4, provides clear advantages, because the occupants sit about 20 centimetres higher than in other cars, with the result that the impact occurs below the occupant cell. The A-class complies not only with the future EU frontal impact crash test standard but also with strict safety regulations on side impacts in the USA and the European Union. In terms of safety a car measuring only 3.57 metres long with a kerb weight of only about 1000 kg has a number of basic disadvantages to overcome — especially if it is built on conventional lines. The low vehicle mass, short front-end structure and limited space available for crumple zones are fundamental handicaps that until now have left little room in this vehicle class for further progress in vehicle safety. This situation has been revolutionized with the A-class.

The main aim in the safety development of the car was to make maximum use of the short front crumple zone in a head-on collision and to clear all components that obstructed the crumpling process out of the way. Because of the rigid structure of engine and gearbox, they play practically no part in

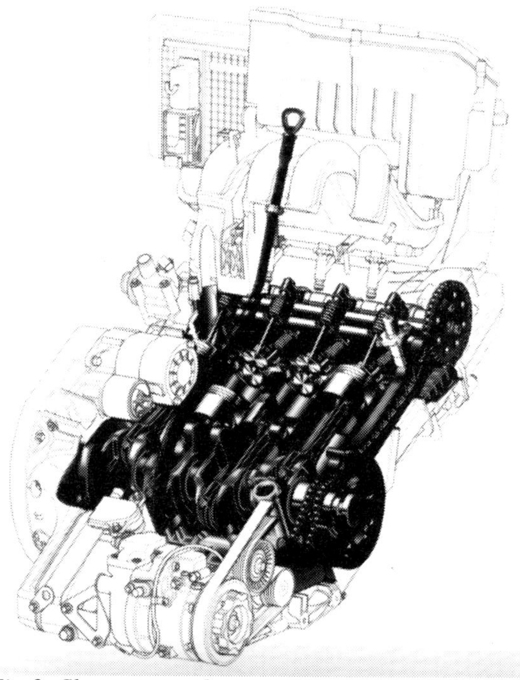

Fig 3: Slant-mounted power unit

the deformation, and in conventional compact cars they can be pushed back in a single block that can intrude into the interior and injure the occupants.

Here the innovative arrangement of engine and gearbox reduces the risk of their forming a single block and therefore makes an essential contribution to occupant safety. In the A-class, the engine and gearbox are positioned at an angle partly in front of the occupant cell and partly underneath it, so that in the event of frontal impact they can slide downwards rather than straight back. This is possible due largely to three unusual features: the unique inclined construction of the engine and gearbox, the shape and installation of which are specifically designed to accord with the crash principle of the A-class; the development of a frame-type integral support to accommodate the engine and gearbox (Fig 5), which in a crash can disengage at two of its eight fixing points thereby allowing the engine and gearbox to slide downwards; thirdly the design of the stable front floor panel (pedal floor) as a surface along which the engine and gearbox can slide in the event of a crash.

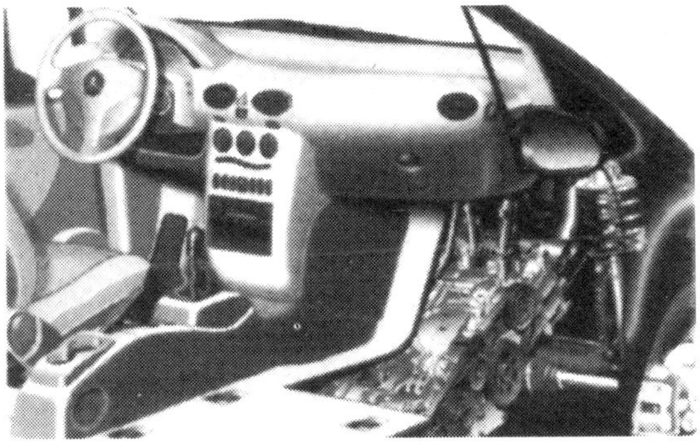

Fig 4: Double-level floor

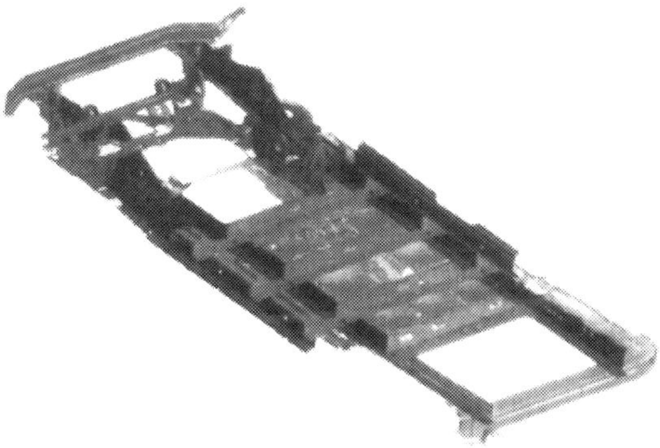

Fig 5: Vehicle chassis-frame construction

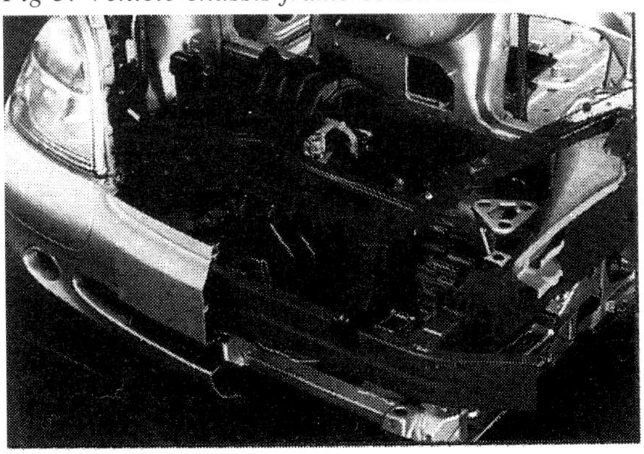

Fig 6: Deformation zones

This maximum possible deformation length available at the front end of the Mercedes A-class is due solely to the fact that the rigid engine and gearbox are displaced under the passenger cell in a serious head-on collision. Without this capacity for evasion and the ability of the engine and gearbox to 'duck' out of the deformation area, the front end of the A-class would have to be about 25 cm longer in order to give the occupants the same degree of protection. The fact that the short crumple zone also does not subject the occupants to severe stresses as a result of the higher deceleration values is due to the retaining systems specially matched to the front end structure.

Other advantages of the new body concept are the straight side members. They absorb high deformation forces right from the start of body deformation and thereby permit much more favourable deceleration characteristics than in cars of conventional design. As a result, the occupants share in the deceleration of the vehicle earlier and over a longer period, which reduces the stresses to which passengers are subjected. The triggering threshold of seat belt tensioners and airbags on the driver and front passenger side can also be more precisely defined thanks to the longitudinal rigidity of the body structure. The front end structure has deformation zones, Fig 6, on three levels. The impact energy in the front area of the A-class is widely distributed over three planes: the frame-type integral support, which serves to accommodate the front axle, engine and gearbox; the straight side members leading to the aluminium and steel front module; the second upper side member plane on a level with the headlamps.

Since most frontal impacts are offset crashes with unilateral stressing of the front end structure, there are strong transverse connections between the impact zones designed to ensure that the deformation areas on the opposite side, remote from the impact, are also involved. This produces an homogeneous front end structure which will prove effective in all types of frontal collision. The front wheels also have a role to play in energy absorption. They form an additional load path and in an impact are braced at the stable nodal points of A-pillar and sill. This effect is particularly important in the frequent front end crashes with overlap on one side — the so-called offset collisions.

M-B engineers have matched the strength of the occupant cell to the high front end deformation force. The raised flat floor, which is welded from below to a grid-like support structure, gives the occupant cell a high degree of stability so that it is essentially not deformed either in a frontal impact or in an offset crash with 40 or 50 percent overlap. Airbags and belt tensioners and a polystyrene insert in the footwell, which form part of the standard equipment on all Mercedes cars, complete the list of measures to protect the occupants.

From the outset the 'Big versus Small' aspect was a prime consideration in the safety development of the A-class. In a head-on collision involving two cars of differing mass a suitable distribution of the deformation load between the two vehicles must be achieved so that there is no danger particular to the occupants of the smaller car. For some time now Mercedes-Benz has been developing its cars according to this principle and claims to be the first to test cars through crash tests against a deformable barrier. This test method, which will not be specified for all new cars in the European Union until October 1998, permits an especially realistic simulation of typical accidents involving oncoming traffic, and specifies a head-on collision at 56 km/h and 40 percent overlap against a deformable barrier. The A-class passes this test even at an impact speed of 65 km/h, corresponding to a significantly high stress loading.

Compatibility means that in a collision involving two vehicles they should activate one another's crumple zones and be capable of transmitting the deformation load uniformly to both bodies. For this to happen it is necessary for the front end deformation resistance of large and small cars to approximate to one another. As an example, in the E-class the front end structure of this Mercedes model is constructed so that in a crash the impact energy for the other vehicle is reduced — intelligent safety engineering means mutual protection. In a smaller car, however, the design principle is just the opposite. Because of the lower mass and the shorter front end the Mercedes engineers designed the A-class with an especially rigid front end so that in an accident with another, larger oncoming vehicle it can activate the other vehicle's deformation zones. In addition to matching the levels of force, the uniform distribution of impact forces over the entire height and width of the front end plays a particularly important role. On the A-class this purpose is served by the second impact plane with its stable transverse connections to the lower support structure. Even in a frontal collision with a larger car, the occupants of the smaller one are not at a major disadvantage in this case.

The strengths of the sandwich concept of the

Fig 7: Body-in-white

car are not confined to front-end crashes. They also apply to side and rear impact collisions: in the side-on crash the other car hits the A-class at the precise level where the body of the A-class is at its strongest, at the level of the load-bearing structure for the floor. The occupants are seated above this impact level and are therefore optimally protected. The load bearing structure of straight cross members, Fig 7, and side members is better able to contain and reduce the impact energy than in a car of conventional body design in which the impact mostly occurs above the floor structure.

The short crumple zones at the front of the A-class make it necessary to activate the belt tensioner immediately after the crash. As a result, the occupants not only take part earlier and longer in the deceleration of the vehicle, but they also have a longer distance for their crash-induced forward movement. The earlier the safety belt builds up its restraining force, the longer the distance available to the belt and belt tensioner in which to arrest the forward movement of the occupants. In this way, the forces unleashed on the occupants are low, despite the massive deceleration of the occupant cell. This also serves to compensate for the design disadvantages of short deformation zones in small and compact cars. The innovative restraint system in the A-class includes full-size airbags for driver and front passenger. The airbags fitted as standard have a volume of 64 and 130 litres. They are operated by a new propellant technology which offers certain advantages when recycling: the gas generator for the driver's airbag is filled with an acid-free solid propellant, and the passenger airbag contains an environmentally safe liquid gas mixture of helium and argon.

The car has inertia-reel seat belts with belt tensioners on the front seats: newly developed compact belt tensioners compensate for seat belt slack and ensure that in a crash the occupants are even better tied to the passenger cell by snugly fitting belts. The belt strap is tensioned by small steel balls in the seat belt retractors which, when the belt tensioner is activated, are set in motion and rotate counter to the belt strap shaft. In this way they first stop the belt strap and then coil it back up at high speed. This drive system has two outstanding advantages: high tensioning capacity and compact construction of the seat belt retractor.

The inertial reel seat belts for the front seat occupants are height-adjustable in three stages. The seat belt retractors of the front seat belts incorporate torsion bars which in a crash are slowly deformed from a certain force upwards, thereby reducing the locking action of the inertial reel seat belts. This reduces the risk of injury to the occupants who moreover, thanks to the force limiter, can sink very gently into the airbags. Belt force limiters and airbags are thus designed to function as part of a precisely co-ordinated system. The side airbags in the front doors form part of the standard specifications of the A-class. A newly developed gas generator with liquid gas filling is also used for these. M-B also fits belt tensioners as standard on the outer seats of the rear seat unit. These ensure that in a crash rear seat passengers also benefit in good time from the deceleration of the body structure. The height adjustment of the inertial reel seat belts functions automatically on the rear seats. As on the C- and E-class cars, child seats which fold out of the rear seats at the touch of a button are available as an option.

Primary safety system

For the first-launched vehicle, the front axle has a modified McPherson suspension system with coil springs, twin-tube shock absorbers and torsion bar stabiliszer, Fig 8. The suspension components are mounted together with the rack-and-pinion steering gear, engine and gearbox on a frame-type integral support that is bolted to the body at eight points. Unlike the original McPherson suspension, the anti-roll bar of the A-class does not take part in wheel location but is linked to the suspension strut via plastic suspension: the results include better elasto-kinematics and increased rolling comfort.

Ride comfort is also enhanced by the low-friction suspension struts and damper guide units, the Teflon coating of the front bearing of the torsion bar and the various noise-damping measures. All wheel location, suspension and damping components have been weight-optimized, such that engineers have achieved an exceptionally good balance between load-bearing ability and intrinsic weight: the effects of this balance only become evident in the sum of the car's many detailed solutions.

Another conceptual advantage that is felt in the handling of the car is the position of the steering gear in front of the centre of the wheel: this layout is only possible because of the sandwich double-layer floor concept, with the engine and gearbox located partly underneath the floor panel. Steering response is improved by this geometry.

A compact servo pump for the steering of the A-class, for the first time in a Mercedes is not mechanically driven but is powered by an electric motor with electronic control. Pump capacity can be adjusted precisely to suit demand and the instantaneous traffic situation. For example, when driving on the straight, the electric motor runs at reduced speed and consumes less energy. The three-spoke steering wheel of the car is a new development and is designed in particular to satisfy the needs of lightweight construction. The 370 millimetre diameter steering wheel is an aluminium pressure-casting, with a rim of polyurethane foam coated in plastic or leather, depending on the equipment level and customer option. The steering wheel material crumples in a controlled way in the event of high-force collision impact.

The company opted for trailing arm suspension with coil springs, torsion bar stabilizer and single-tube gas-pressure shock absorbers. The main advantage is that trailing arm suspension can be designed with all its components underneath the load floor, so that it does not encroach on the interior space. The shock absorbers and springs are located in an otherwise unused space in front of the wheel centre. With this geometry, the suspension components support the straight side members in a crash, thereby also improving occupant safety.

The dynamic qualities of the rear suspension are particularly due to special tie-bolts that control the elastic deformation of the side members and combine with them to form a rectangular linkage structure, Fig 9. Changes in the angle of the side

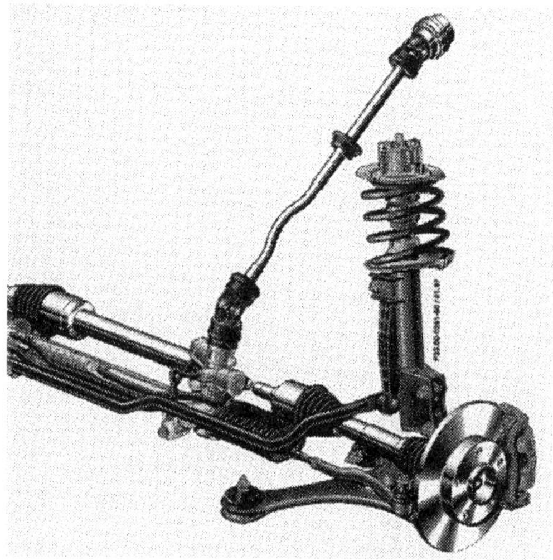

Fig 8: Front suspension

Fig 9: Suspension supports

member and wheel about the vertical axis are compensated with up to 75 percent efficiency. The suspension components, side members, coil springs and shock absorbers are mounted on a suspension subframe in a single compact block that is fixed to the bodywork on four sound-proofed rubber bearings, partly with hydraulic damping. The rear suspension unit is mounted with the aid of four vertical assembly bolts. None of the suspension components projects upwards above the level of the floor.

The technology and layout of the brakes, Fig 10, correspond to the particular features of the A-class drive system and vehicle concept. Engineers opted for floating-calliper disc brakes at the front and drum brakes at the rear, because this combination provides optimal braking safety, stability and endurance for a front-wheel drive car, say M-B. A brake booster and the anti-lock braking system provide added safety. The drum brakes on the rear wheels are characterized by a lightweight construction: the supporting body for the wheel brake cylinder and brake shoes, the anchor plate and brake cylinder are made of aluminium. In order to satisfy the high safety requirements, the hubs are made of steel and the brake cylinders of grey cast iron. An automatic adjusting system compensates for brake pad and drum wear. The parking brake in the A-class is operated manually, using a lever between the front seats. The control cables are maintenance-free, since they have automatic length adjustment.

ESP keeps the A-class directionally stable and reduces the risk of skidding when cornering by selectively applying the front and rear brakes. While driving, the ESP computer compares the actual behaviour of the vehicle with the calculated set points. If the car is deviating from the safe 'ideal line', the system responds at high speed with a specially developed logic, bringing the car back to the right track in two ways, firstly by the precisely dosed braking of one or more wheels and/or by controlling engine torque. By these methods, ESP corrects both driving errors and skidding movements caused by slippery or wet surfaces, loose chippings or other difficult road conditions that normally give the driver very little chance to keep the car in line by steering and braking action. The A-class is the first and only car in the subcompact and compact car class to have such an innovative driving safety system. The Electronic Stability Programme also includes the compnay's electronic Brake Assist System.

Fig 10: Braking system

Ford Focus

Ford's Focus, Fig 11, brings the company's 21st century hard-edge styling to its most mainstream model and also marks a completed market repositioning of a class of vehicle starting as the Mark 1 Escort to a vehicle refined enough to compete in the neo-luxury sector. The refinement of individual systems to achieve the transformation is described below.

Friction within the front suspension struts is reduced by off-set coil springs that eliminate bending forces acting upon the strut, Fig 12. Zero off-set geometry — via new A-arms supported by stiff, front A-arm bushes — provides precise lateral wheel control for enhanced precision and on-centre steering definition. Large, rear A-arm bushes permit longitudinal compliance for reduced impact harshness and improved ride performance. Negative scrub radius (–12 mm) enhances braking stability, particularly on surfaces in which available grip varies between the left- and right-hand tyres. Low steering friction levels minimize resistance and stickiness for improved feel, response and driver confidence. . Body-roll is controlled by anti-roll bars, which are 20 mm diameter front and rear. Large, pre-loaded bushes provide immediate linear control. Steering input induces immediate directional change (turn-in) with very little body roll. Ultra-stiff cross member, attached directly to floorpan, increases the precision of geometric wheel control and further stiffens the body. Friction within the steering system has been reduced 20 per cent and damping friction reduced, Fig 14.

Adoption of a fully independent, Control Blade multi-link rear suspension system, though similar to the system used on the Ford Mondeo estate, is redesigned for Focus. Here the Control Blade makes extensive use of stampings for key components, which reduces cost, unsprung weight — by 3.5 kg per wheel — and assembly time, while offering even greater geometric precision, Fig 13. It has ride comfort, steering precision, handling, braking and stability advantages, as well as NVH and package gains over the common twist-beam axle. There is greater longitudinal compliance without compromising lateral stiffness, which boosts steering precision, feel, handling and driver confidence. Intrinsic stability is provided by passive rear-wheel steer. The wider distribution and de-coupling of five separate load paths, compared with two for the twist beam design, helps to reduce transmission of noise, vibration and harshness levels to the passenger compartment. The control blade, spring link and toe link are all pressed-steel components. The camber link is the only casting. Compliant bush provides recession, which allows rear wheels to move both upwards and rearwards to absorb road bumps.

Fig 11: Ford Focus

An all new, highly rigid yet lightweight platform architecture (826 kNm/rad; 1070 kg for the 3-door — stiffest and lightest in class) provides best possible base for enhanced ride, handling, steering and active/passive safety. The combination is achieved through laser-welding, optimized panel gauges, extensive swaging, local reinforcement of critical points and optimized spot-weld distribution. Cross-member and powertrain pick-up points are as much as 150% stiffer than surrounding areas. The lightest model weighs just 1070 kg; both lighter and a full 15 per cent stiffer than all recent class entrants.

The rear side rails, for example, are traditionally made from four separate parts. For Focus, these parts have been replaced by a single laser-welded blank, in which the gauge reduces progressively in three stages from front to rear. The result is a structure that is stiffer, with more rigid mounting points for the suspension, has improved crash performance and saves more than 1 kg per rail in weight. Similarly, the gauge of steel in the B-pillar is 2.25 mm thick at the top but tapers to just 1.1 mm at the base, providing a stiff upper section for a controlled and superior performance in the event of a side-impact crash.

Zero offset geometry gives total separation of load paths, with lateral forces being fed through the stiff forward A-arm bush (1) while longitudinal forces are controlled by the more compliant rear A-arm bush (2) shown in Fig 13. Large-diameter coil springs are offset, to feed suspension loads directly to them, eliminating bending forces on the struts for reduced friction and superior responses, Fig 14.

Fig 12: Large diameter coil springs are offset so that suspension loads are fed to them without bending forces applied to the struts

Fig 13: Zero-offset geometry allows lateral forces to be fed through bush (1) while longitudinal ones are controlled by bush (2)

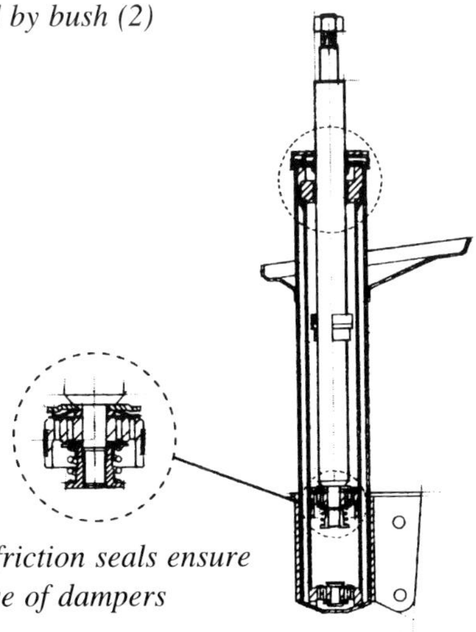

Fig 14: Low friction seals ensure quick response of dampers

Stability is enhanced by maintaining optimum braking performance to the rear wheels in all conditions, whether the vehicle is heavily laden or carrying only the driver. On non-ABS-equipped vehicles, this is provided by a pressure cut-off proportioning valve. For the 1.8- and 2.0-litre petrol and diesel estate, a load-sensitive pressure limiting valve is used.

When ABS is specified, brake proportioning is achieved using the electronic brake-force distribution (EBD) system. It uses the ABS sensors to monitor the speed of front and rear wheels and adjusts the pressure distribution to help the rear wheels retain traction. EBD is self-compensating for all load conditions and imperceptible to most drivers. It is designed to reduce stopping distances and improves vehicle stability in conditions that fall short of triggering the ABS. ABS also forms the basis of the car's Traction Control System (TCS) and Electronic Stability Programme (ESP). TCS is a development of Ford's dual-mode system with low-speed brake intervention coupled to a cut in engine power, above 31 mph, delivered via the EEC-V engine management module.

ESP acts whether or not the driver operates the brakes. In tests, it maintains vehicle stability at much higher speeds than would be possible without it — thereby placing less demand upon driver skill and reducing stress in extreme conditions. ESP uses software to control engine power and brake individual wheels to restore stability when the driver has exceeded or is beginning to exceed the car's dynamic capabilities. Driver's intentions — direction of travel, measured via a steering angle sensor — are compared against vehicle behaviour. Vehicle behaviour is monitored constantly by three additional sensors for yaw rate (rotation), lateral acceleration (centrifugal force) and wheel speed, with the latter sensors being integral to the ABS. Yaw control is achieved via the vehicle's ABS, with braking being directed to individual wheels through the ESP modulator. ESP uses information from three sensors and the four ABS wheel-speed sensors to determine the cornering behaviour. It interactively compares the data in a dynamic handling map stored in a special on-board computer. Additional ESP sensors monitor steering angle, lateral acceleration and yaw moment, which denotes the vehicle's tendency to turn about its vertical axis. As soon as any tendency for the vehicle to move off the chosen line is detected, engine power is reduced, and individual brakes are applied momentarily to bring the car back on course.

Powertrain With the Zetec SE low-friction engine, Ford used high-compression ratios, knock sensing and low idle speeds to deliver a 25 per cent real-world fuel economy improvement over past engines. The existing 1.4-litre Zetec SE unit, now fully re-calibrated for Focus to serve as a fuel-efficient 75 PS power unit, is supplemented by an all-new all-alloy 100 PS 1.6-litre Zetec SE engine. All engines require servicing only every 10,000 miles, with spark plug renewal every 40,000 miles and valve clearance attention only every 100,000 miles on petrol versions. All also feature a new torque roll axis (TRA) mounting system, which helps reduce powertrain shake on rough roads, results in a smoother idle quality and greatly reduces the degree of noise and vibration transmitted from the drivetrain into the bodyshell, Fig 18. Internally, the Zetec engines feature an alloy ladder frame between the base of the block and the crankcase, Fig 19,which boosts rigidity of the powertrain assembly by 30 per cent. The alloy block Zetec SE has a similar stiffening member added as a bearing beam within the crankcase skirt. These devices — together with new lightweight pistons and con rods — reduce second-order shaking forces by 20 per cent. Newly designed cam covers and low-noise ancillary drive systems generate significant improvements in running refinement and reduce perceived engine noise by 50 per cent.

Cables transmit fore/aft and left/right gear lever movements to the selector mechanism, eliminating a potential path for NVH to the interior, Fig 20. Cable shift also is smoother than a rod-operated system. Hydraulic clutch requires lighter pedal effort with less travel for a smoother take up. An updated version of the Ford B5 five-speed manual transmission has external ribs along the transmission and clutch housing to increase drivetrain stiffness for reduced noise and vibration. Internally, first and second gears are equipped with double-cone synchronizers for smoother downshifts. Selection of reverse is also slicker. Durability is enhanced by the incorporation of strengthened final drive bearings and a new synthetic oil lubricant, which also eases gear selection when cold. New sealed-for-life bearings and fluid seals protect from dirt particle ingress for extended transmission life.

Now available on Focus is a four-speed auto-

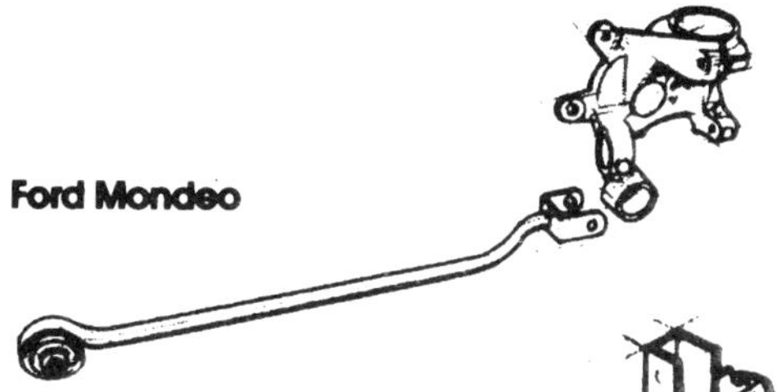

Fig 15: Earlier design

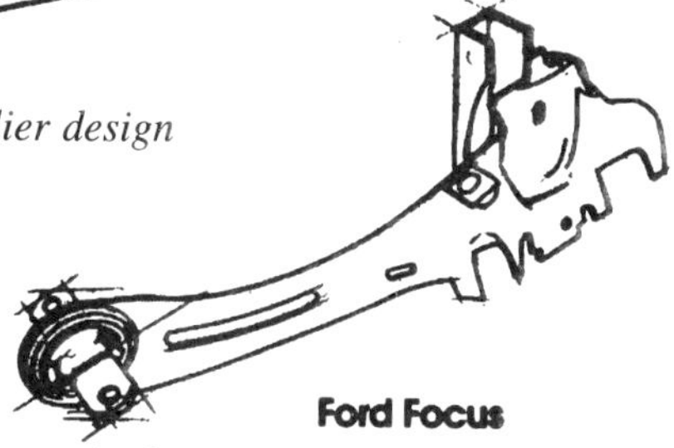

Fig 16: Current single-piece Control Blade

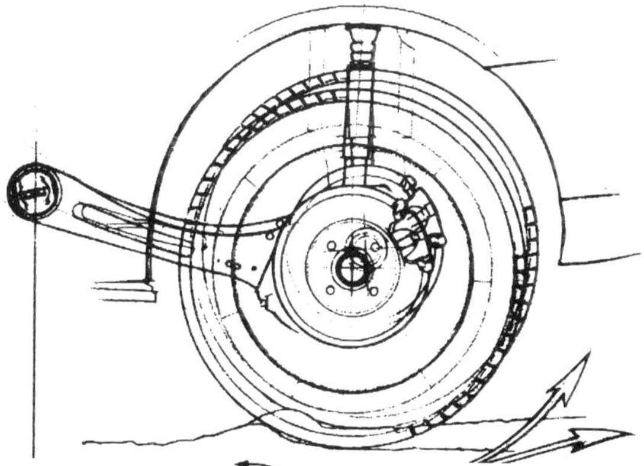

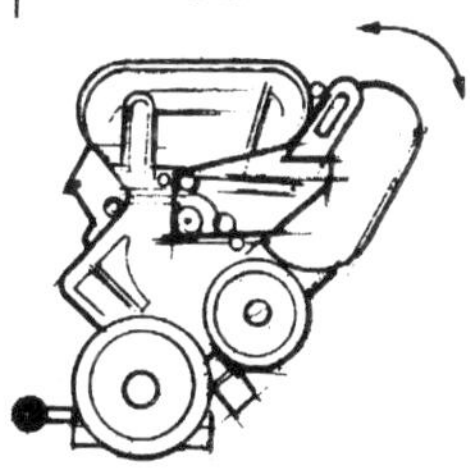

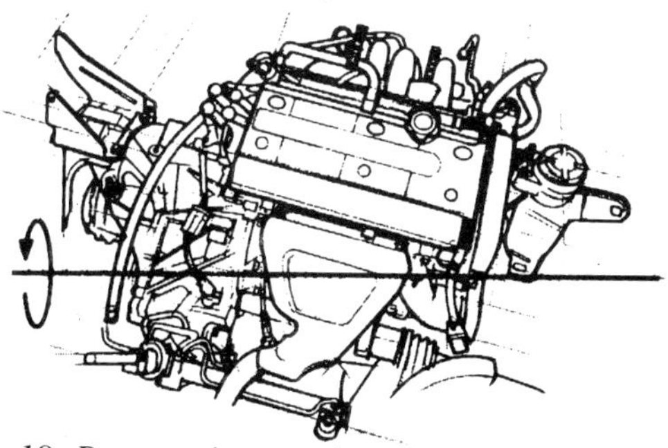

Fig 18: Powertrain mass is supported by two mounts in line with the virtual axis of rotation, a separate arm resisting torque

matic transmission which features an overdrive-top and lock-up torque converter. It has been designed specifically for use in front-wheel drive applications and is unusually light and compact. It is controlled by an electronic synchronous shift control (ESSC) module linked to the EEC-V engine management system. A centrally mounted quadrant selection lever has six positions (P,R,N,D,2,1), while overdrive can be switched by a separate, thumb-operated push-button at the side of the lever. The ESSC works in conjunction with the EEC-V using information read from 18 different engine and transmission sources to calculate the best possible shifting strategy for the driving and operating conditions. Fuel economy and performance are close to those of the equivalent manual transmission models, while the system's shift quality and speed of response set new standards for this class of vehicle.

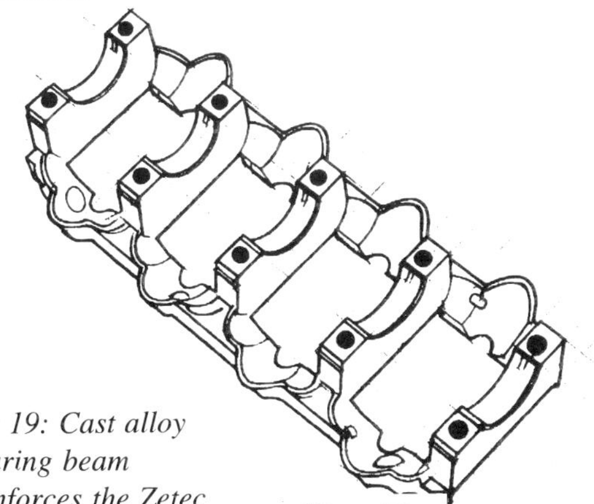

Fig 19: Cast alloy bearing beam reinforces the Zetec SE bottom end

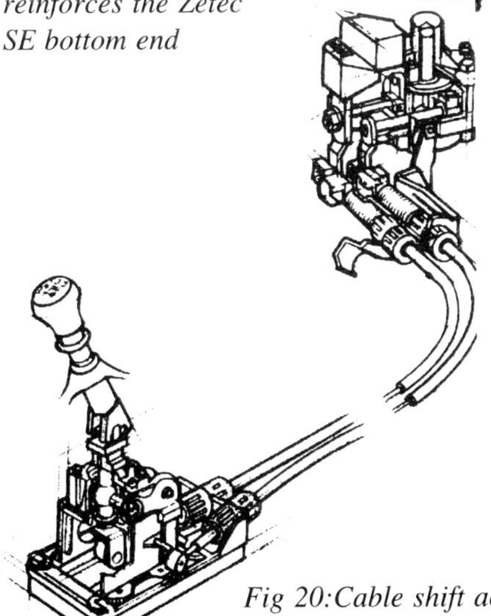

Fig 20:Cable shift actuation is chosen to reduce noise paths

Body structure and systems Focus meets more than 100 different real-world crash tests, modes and events, well beyond existing or currently proposed legislation, including all mandatory European and US destructive tests. At the front, beyond the bumper support beam, the car's cross member is made from ultra-high-strength steel, so it is robust and stable and will not tear upon impact, Fig 21. Resistance to tearing helps spread crash energy into the car's lower side rails, which is particularly effective in offset impacts. Additional to the progressive collapse of the lower side rails, which absorbs most of the energy ahead of the passenger safety cell, a secondary load path channels the remaining energy over the roof and around the doors to ensure safe evacuation of passengers.

At the rear, the objective is to minimize intrusion of the passenger compartment in a rear impact, while also ensuring that the doors remain operable and the integrity of the fuel system is maintained. This is achieved via the progressive collapse of the rear side rails, supported by the wheel housings. Stamped and laser-welded, the side rails are made up of three different metal gauges — changing in thickness from 1.6 mm to 2.6 mm (in the area of the fuel tank) and 2.4 mm — to provide progressive controlled collapse. The compound structure avoids over-gauging near the rear bumper and eliminates the need to add reinforcement further back. Vehicle weight is thus reduced, by 1 kg per side, and the number of parts and welding operations are reduced, improving assembly tolerances and providing robust control of the collapse performance. For side-impact protection, the entire side structure, which includes door beams and anti-burst door latches, is designed to absorb energy and reduce intrusion, especially above the door bars. To achieve this, the B-pillar is made from a laser-welded, high-strength steel blank, with a gauge thickness that reduces from 2.25 mm at the top to 1.1 mm at its base. Additional inner reinforcement, designed to increase rigidity, makes it highly resistant to 'kinking', so it can more effectively channel crash energy into both the rocker and roof rail, and support the side airbag during deployment.

All elements of the Focus restraint system work in harmony so that the whole structure is integrated and fine-tuned to provide maximum protection. This starts with the locking of the inertia seat belt reels which occurs approximately 6 milliseconds after impact. (All timings depend upon the nature of the impact, whether it be straight, angled, offset, against a solid object or another vehicle). Approximately 4 milliseconds later, before the head and chest begin to move forward, the airbag sensors trigger the front airbags and the small pyrotechnic charges in the pre-tensioners, which eliminate any slack in the seat belt and negate the effects of bulky or layered clothing. The front airbags take approximately 35 milliseconds to fully inflate, which is completed just before the occupant meets the bag, approximately 50 milliseconds after impact. To reduce chest loading and minimize injury, Focus is equipped with load-limiting devices within the inertia reel assemblies. These pay-out belt webbing after a predetermined load is exceeded. This process is progressive and occurs between 45 and 70 milliseconds after impact, ensuring that the cross belt pressure against the occupant's chest remains fairly constant.

The load-limiting device consists of a torsion bar within the traditional inertia reel, and the typical pay-out is 150 mm. The system is calibrated to help ensure that front seat occupants do not meet the airbag until the optimum point, which is just as it is beginning to deflate. The progressive pay-out also ensures that the upper torso is less likely to twist, an action that can compromise the efficiency of the front airbag. Seatbelt pre-tensioners are designed to pull the buckles down and take up any slack, optimizing their position on the torso and maximizing the restraint provided by the belt webbing.

Focus' new head-and-chest combination side airbags provide greatly enhanced head and chest protection in side impacts. In addition to acting as a barrier between the occupant and the side of the vehicle, the tall, 15.5-litre side airbag — stored in the seat back's outer side bolster — is designed to cushion the head, minimiszing lateral head injuries.

The car is equipped with the latest version of Ford's Safeguard Passive Anti-Theft System (PATS). Safeguard is a Thatcham-approved high-security engine immobiliszer operated via a miniature transponder within the key head. It transmits a unique coded signal, with several trillion possible codes, to a transceiver situated around the steering lock. Once the key is identified as correct, the engine is mobilized. Without it, the engine will not start and is effectively dead. PATS offers several advantages over competitive systems. The system arms itself,

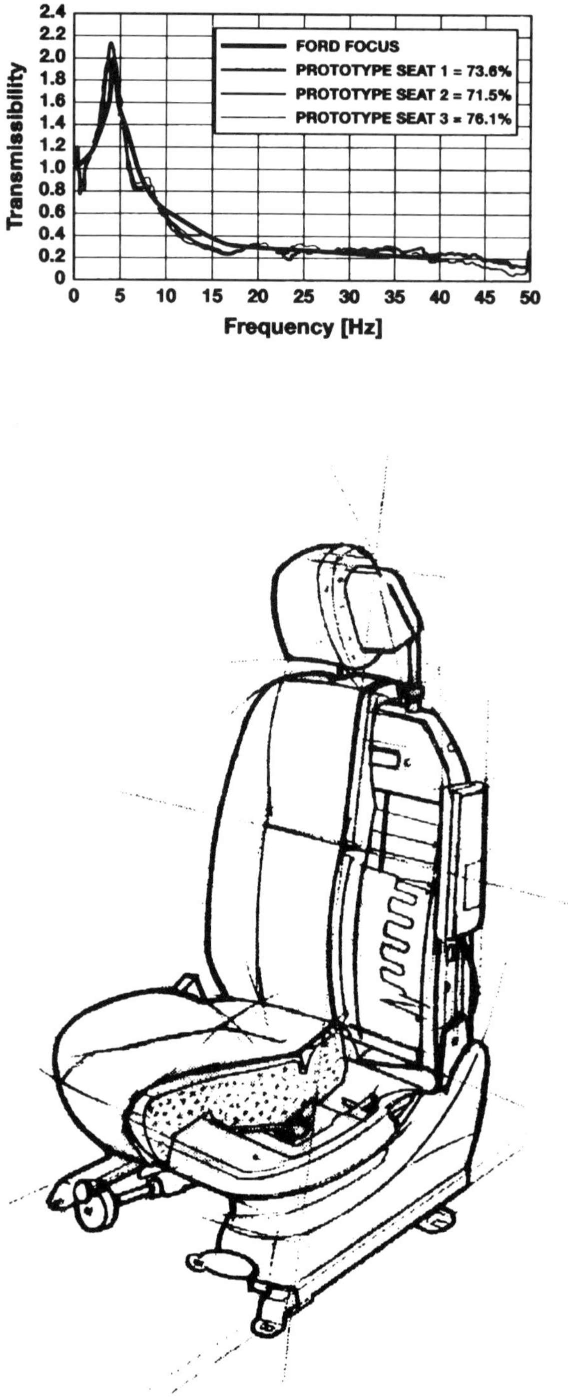

Fig 22: Seats are designed to attenuate vibration in the 10–50 Hz frequency range

eliminating human error through forgetfulness. The interception of the key code by a thief is virtually impossible because the transponder only has the power to pass the code by radio to the transceiver when it is within the electrical field of the transceiver. PATS also incorporates a special erasing procedure and smart 'learning' mode to maintain security after resale and eliminate the requirement for a master key. In addition, a new, key-operated hood-release mechanism makes a potential theft even more difficult. By eliminating access to the engine compartment via the traditional cabin lever and cable, to ensuree that a would-be thief could gain access to the engine or alarm siren only with the car's ignition key.

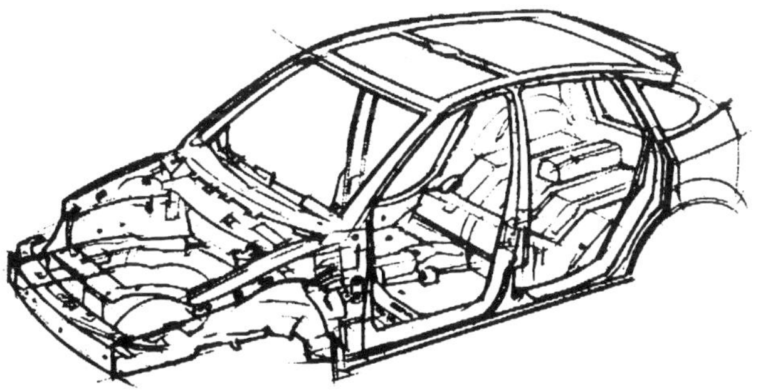

Fig 21: High specific-stiffness body shell

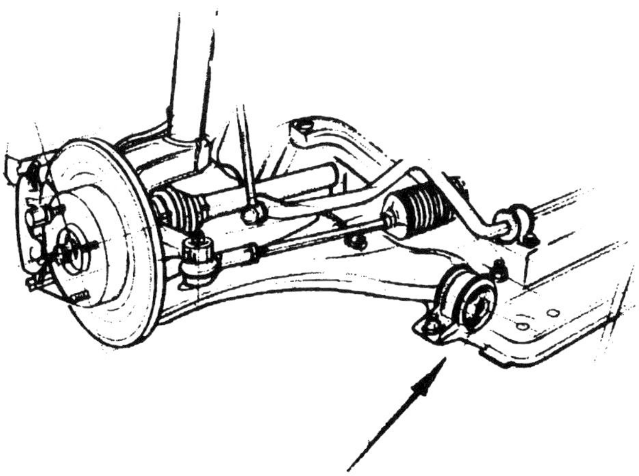

Fig 23: Large diameter rear A-arm bushes give longitudinal compliance without compromising lateral stiffness

The car's shrouded lock mechanisms and wiring present the first obstacle to forced entry, rendering break-in many times more difficult. A further obstacle is posed by a double-locking door system, which is linked to the car's central locking and optional remote-control facility. Once the double locks have been set, only the key or remote button can release the locks. A further deterrent is presented by a tamper-proof perimeter and interior scanning alarm. Once armed, it detects the opening of any doors, tailgate or hood, as well as any movement within the passenger compartment. The remote-control locking system has a 10-metre range and uses a miniature transmitter working on a rolling frequency system to eliminate possible interception by an electronic scanner.

The body structure and seating were tuned to reduce resonant peaks and move natural vibration frequencies outside the operating range, Fig 22. Deep swaging of body panels, especially the floorpan, help to maintain stiffness while permitting down-gauging for weight reduction. Where this was not possible, around the bulkhead for example, dual-function deadening pads were developed, with a viscous layer to dampen sound and a mass layer to absorb it. As well as tuned suspension mounts, Fig 23, torque roll axis engine mounts also are used to reduce powertrain noise, improve idle quality and reduce powertrain shake on rough roads. Such a mounting system separates the powertrain mass from its torque reactions by using two rubber-to-metal bonded mounts located at either end of the assembly, in line with the axis of rotational inertia, plus a dedicated torque reaction link. The front mount incorporates a hydraulic chamber so the modulus of the rubber can be optimized to reduce NVH transmission at high engine revs, without reducing idling smoothness.

Two direct acoustic paths from the drivetrain to the interior have been eliminated with the adoption of a hydraulic clutch mechanism and cable-operated gearshift. A unique flexible exhaust joint, situated behind the engine manifold, limits the transmission of drivetrain vibrations along the length of the exhaust system. It also has specially tuned exhaust hangers positioned at the natural vibrational nodes to provide further isolation. 'Helmholtz' resonators are used to filter or select individual frequencies so that an appropriately purposeful, yet subdued sound is obtained.

Land Rover Freelander

With prices pitched at under £16 000 the Freelander, Fig 24, has finally brought Land Rover within grasp of the profitable small sports-utility-vehicle market and first results are showing the vehicle to be performing well financially. Trade sources have projected higher residual values than the Honda CRV and Toyota RAV4. It is the first of the company's vehicles with a unitary construction body shell and transverse power unit and a well-planned engine compartment gives good service access, Fig 25.

Structure

The integrated underframe comprises high-strength steel box-section longitudinals, with eight crossmembers, all welded through to the floor pan. Box-sectioned outriggers from the crossmembers connect the side sills through to the centre tunnel member. Front and rear suspension sub-frames are bolted rigidly to the shell to enhance its torsional rigidity. The upper shell involves monoside pressings; there is a deep rear-heelboard serving as a further crossmember; there is a strong integral fascia panel, too; all these combining to enhance the stiffness of the superstructure. For the three-door variant the torsional figure is 13,500 Nm per degree and rises to 17,500 on the Station Wagon.

Within each lower A-post there are five diaphragm pressings to stabilize the box-beam and provide additional crush resistance under impact. Front-ends of the underframe longitudinals crush progressively on impact, after initial deflection of bumper armature and crush cans, a tapered box-section in the top of each engine bay side-panel also

Fig 24: Land Rover Freelander

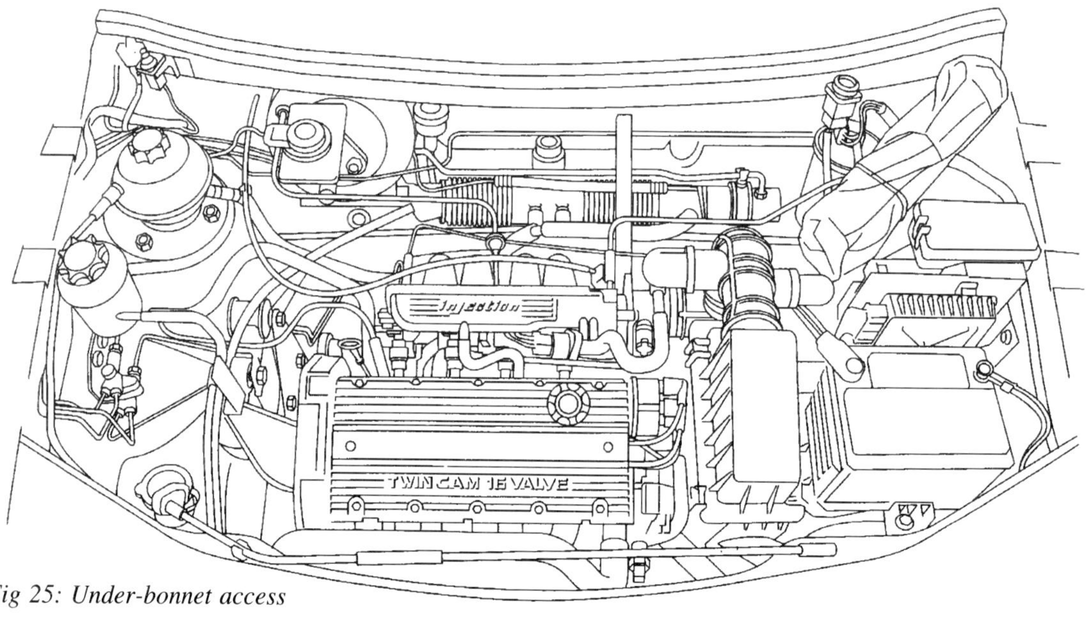

Fig 25: Under-bonnet access

absorbing impact. In more severe crashes, bending of the curved portions of the main longitudinals takes place. The suspension towers are braced back to the strong A-pillar area via box-sectioned members either side of the bulkhead. The rear bush mountings for the lower front suspension arms are designed to break free of the subframe, under severe impact, to minimize intrusions into the front footwells. From the bulkhead backwards the floor structure has a 'branching' layout to help spread impact forces as widely as possible, with a curved 'horn' box-member sweeping back from the longitudinal to the central tunnel area, the box-sectioned side sills also being brought into play.

Body systems

A spoiler-type electrically operated glass sun-roof has been engineered to tilt open at the rear and slide backwards above the roof panel, to maximize rear headroom, two-stage slides permitting 70% clear aperture opening against the usual 50%. At the front of the vehicle's roof there is a twin-panel targa structure, with detachable T-bar. Panels can be individually tilted open at the trailing edge, or taken out — at which point wind deflectors spring into position. Detached panels stow in a bag on the back of the rear seat squab. The complete targa unit is constructed as a cassette, having a one-piece SMC moulding which is bonded to the bodyshell with a polyurethane seal. Together with a number of the interior mouldings this is made by a gas injection moulding technique to increase specific strength and stiffness.

Roof modules for the 3-door vehicle come in either softback or hardback forms, both of which involve the same tail-door electrically operated drop-glass as the station wagon, Fig 26. Roofs are thus interchangeable for those customers wishing to invest in the alternative. On the softback variant, Fig 27, the vinyl roof folds upwards and furls within an integral tonneau at roof level so rear vision is unimpaired. PVC sidescreens are zipped in place and can be individually removed to leave the rear roof as a

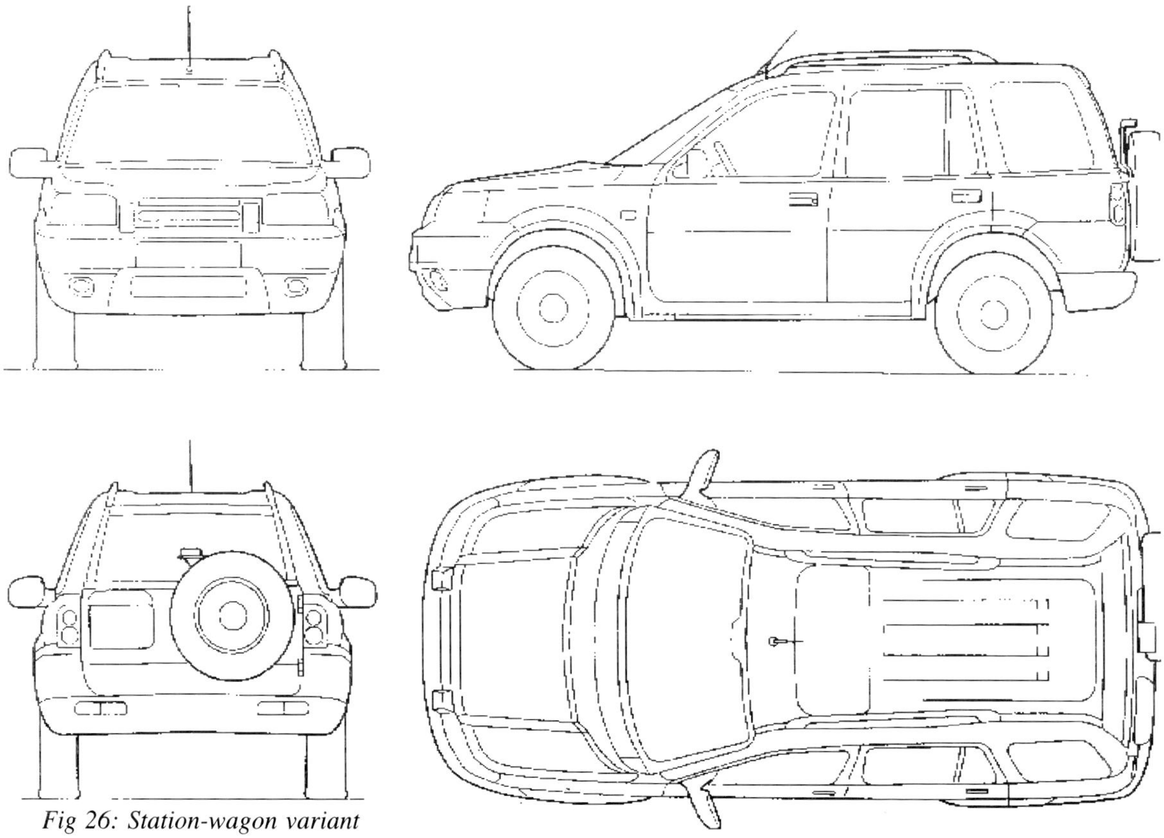

Fig 26: Station-wagon variant

canopy with strong air through-flow. Key body dimensions are tabulated in Fig 27.

A green-tinted laminated windscreen is bonded in place and all other window glass is Pilkington Optikool solar-control glazing, cutting solar transmission by 50%. Front wing panels are in an engineering polymer chosen for its damage resistance and 1.5 kg per wing weight saving. They are jointly painted with the main shell and have to withstand 180 C stoving temperature. Headlamp lens too are polymer-mouldings, a silicon-coated polycarbonate being used to allow a complex optical pattern which maximizes beam control. The front bumper and grille surround is a large polypropylene moulding mounted on an aluminium armature while that at the rear is mounted on a compression moulded glass-mat reinforced polypropylene member. Also in polypropylene are wheelarch mouldings, wheelarch liners, lower sill flange finishers, tail-door lower edge finishers and mudflaps.

Interesting interior detail is the facia-mounted rocker switch construction, Fig 29. The switch contact metal is integrated during the injection moulding and a common base moulding and contact plate are used. Switch warning lights are LEDs rather than bulbs. Another detail is in the driver seating, which has an adjustable lumbar support that does not stiffen the squab springing when adjusted outwards.

The side-hinged tail-door carrying the spare wheel also involves a specialized construction technique, for mounting the wheel without bridging inner and outer skins over the space required for the drop-glass. A large die-cast aluminium drum-moulding carries the wheel and spreads the mounting bolt loads over a very wide area. A neat support arm for the high-level stop-lamp is integrated into the drum. A gas-spring strut beneath the door has 'keeps' at half and fully open positions and facilitates operation by users of all sizes. The drop-glass has a grooved seal at its top edge and electrical interlocks are provided alongside a short-drop/lift control system to release the glass from the groove when the door is opened and closed.

Vehicle mechanical systems

A key quality of the vehicle is the provision of good off-road performance while maintaining on-road handling comparable with the best of conventional car practice. Rover Cars experience was drawn upon for the independent suspension system involving McPherson struts on all four wheel stations which provide vertical travels of 7 in front and 8 in rear. There are high levels of fore-aft compliance, around twice the normal 5 mm. Steer fight is avoided by carefully designed suspension bush configuration, aided by a steering rack with centre take-off, giving very long track rods. An anti-kickback valve in the power steering hydraulics eases the shocks of cross-country driving, Fig 30. While disc brakes are used at the front, the rear wheels rely on drum brakes to provide park-brake performance appropriate to the extremes of cross-country operation. The ABS has a special control system which builds in electronic traction control (ETC) and hill-descent control (HDC).

DIMENSIONS	
Overall length	4382mm (including 16" spare wheel)
Overall width	2074mm (including mirrors)
Overall width	1805mm (excluding mirrors)
Overall height	1757mm (including roof rails)
Wheelbase	2555mm
Track front	1534mm
Track rear	1545mm
Minimum ground clearance	193mm (to front 'axle')
Approach/break/depart angle	EC 30°/24°/34°
Front overhang	836mm
Rear overhang	991mm (including 16" spare wheel)
Maximum front leg room	1062mm
Minimum rear leg room	934mm
Front head room	998mm
Rear head room	1019mm
Front shoulder room	1420mm
Rear shoulder room	1350mm
Loadspace height	922mm
Loadspace length seats up	44[illegible]mm
Loadspace length seats folded	1059mm
Loadspace width	1400mm
Loadspace width between arches	978mm
Loadspace sill height	638mm
Loadspace to roof with seats up	563 litres
Loadspace to roof - seats folded	1313 litres
Loadspace to glass - seats up	371 litres
Loadspace to glass - seats folded	722 litres
Loadspace door aperture width min	924mm
Loadspace door aperture width max	1086mm
Underfloor locker	17 litres
Fuel tank capacity	13 gallons / 59 litres

Fig 27: Key dimensions

Fig 28: Softback variant

Fig 29: Interior features

ETC adds a further degree of limited slip control, to that provided between front and rear axles by the viscous coupling, but across each axle. HDC prevents tobogganing on sharp downhill brake applications by a hands/feet-off approach in which engine overrun at idle is detected and brakes gently applied to maintain maximum descent speed of 4.4 mph.

The powertrain involves either the K- or L-series engines, the K being specially tuned to provide power peak 250 rpm further down the speed range. The engine is also the first to use a returnless petrol injection supply system, with more sophisticated electronic control over the in-tank fuel pump. The L-series diesel has water-cooled EGR to reduce NOx emissions.

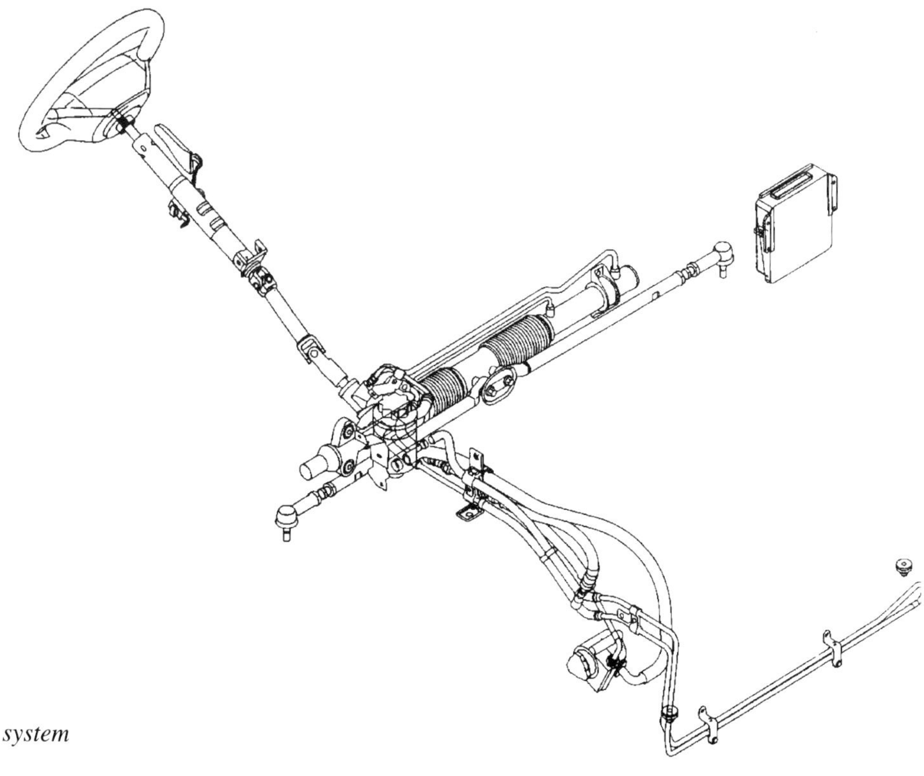

Fig 30: Steering system

Project Thrust SSC

G. T. Bowsher, Chief Designer (Mechanical) Thrust SSC Project, was responsible for spaceframe/structure/wheels/brakes/suspension/steering systems/engine mounts/ throttle systems. Seen in Fig 31 with the Dunlop-made wheel and brake, he described the engineering to a contemporary specialist journal[1].

He explained that Thrust SSC, although entirely new in concept, with twin jet engines and rear wheel steering , incorporates technology developed in Thrust 2, which regained the world land speed record in 1973. A major feature of the rolling gear is the adoption of solid aluminium alloy wheels and a power hydraulic braking system which uses two disc brake callipers acting on each specifically designed high speed carbon disc. The Thrust 2 experience gave valuable information on the safe 'footprint' size for transferring vehicle weight to a desert surface. It also made it possible to predict how wheel brakes might effectively be used to stop the vehicle. This is because the same rolling interface applies (aluminium alloy wheels and the low adhesion, friable surface of Black Rock desert where the very high speed runs were carried out). The shape and ground clearance of the new car dictated a wheel diameter of 34 inches (0.86 m). At the anticipated peak design speed of 8400 rpm this means that the radial acceleration at the wheel periphery is 34,000 g — effectively making 1 gm of material weigh 34 kg.

Dunlop undertook the manufacture of all wheels for the jet car, both the high speed solid alloy sets (with spares) and the low speed wheels which utilized the main undercarriage tyres of the Lightning

Fig 31: Chief Designer G.T. Bowsher with Dunlop wheel and brake

fighter. This proved a major contribution at an early stage, and at a time when credibility for the project was of great importance. To produce the high speed wheels solid L.77 forged aluminium blanks supplied by HDA Forging were machined to provide a rolling tread width of 10 inches (0.25 metre) for the front wheels and 6 inches (0.152 metre) for the rear, steering, wheels. This is in line with the weight distribution of the vehicle's 10 ton static weight.

A straight web design for both front and rear wheels became essential in order to minimize stress, but for the front wheels, with the wider tread profile and larger peripheral mass, it also became necessary to omit the traditional central hole and have a solid centred wheel; only then could the high speed stresses be kept within the limitations of the L.77 material. In terms of the wheel installation this did not matter, as the decision, already taken, to steer Thrust SSC by the

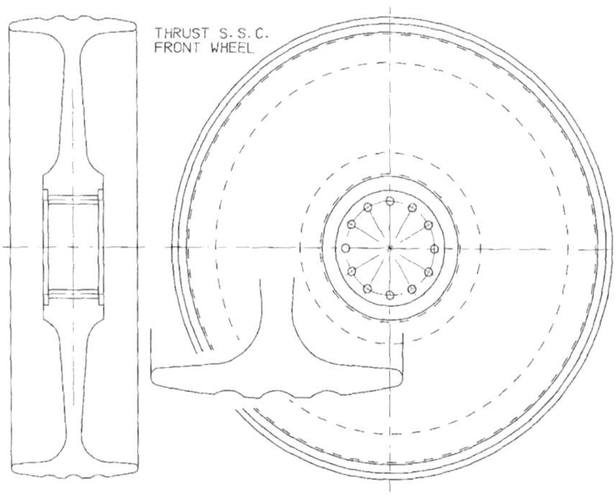

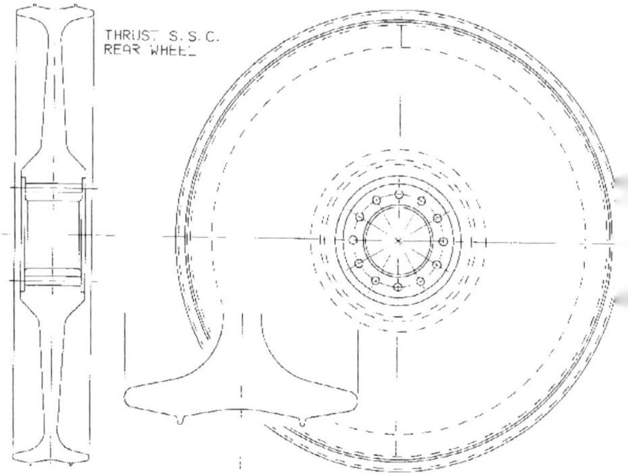

Figs 31/32: Front and rear wheels after final machining

rear wheels and not the front meant that the central hole was not essential: its principal use would have been as a location for machining. As the rear wheels were to be steered a central bearing/stub axle hole was a necessity, but this could still be accommodated within the material's stress limits because of the narrower tread width and lower peripheral mass. Three dimensional computational finite element stress analysis techniques were used in the design of the wheels, minimizing errors before a commitment to costly machining. Of interest, the rear wheels are slightly higher stressed than the fronts at any given speed, indicating the effect of the central hole.

The finalized wheel shapes are given in Fig 31 for the front wheels and Fig 32 for the rear wheels. Fig 33 shows the three dimensional stress contours of a front wheel at 8000 rpm., the darkest shade indicating the highest stresses (note those in the bolt holes) and the lightest the lowest stresses, with the other shades giving intermediate values. All wheels should offer minimal resistance to rolling and yet have a high lateral resistance in order to provide good vehicle control. Three circumferential grooves on the radiused profile of each front wheel tread provided for this, allowing a progressive take up of wheel load and for desert surface deformation to give the required lateral stability. Twin keels were a feature of each rear wheel, with the tread shape again profiled to give a progressive take up of load, these were intended to penetrate the desert surface and allow the generation of lateral steering forces, when steering took place, in addition to providing good lateral stability to the rear of the jet car. The wide spacing of the keels improved the damping capacity at the desert surface interface and helped prevent wheel 'shimmy', a feature of the first set of rear wheels, which exhibited only a single central keel of low damping capacity.

Although a parachute was the primary means of stopping the jet car from high speed, Fig 34,wheel brakes were essential to cover the lower speed range (200 mph to rest), when the parachute loses much of its effectiveness. In an emergency (if the parachute failed), the wheel brakes would need to stop the vehicle from over 300 mph.

In most respects, the wheel brakes for Thrust SSC were laid down as a design fourteen years before following the conclusion of the Thrust 2 project: the best features of that design were retained whilst other features were design-developed as a natural follow on and improvement. It was simply a question of waiting for a new vehicle. Each wheel brake comprises a wheel driven disc/rotor and a pair (one leading and one trailing) of brake callipers which hydraulically clamp the disc between suitable friction stators (brake pads). The brake callipers react the friction drag loads developed between the brake pads

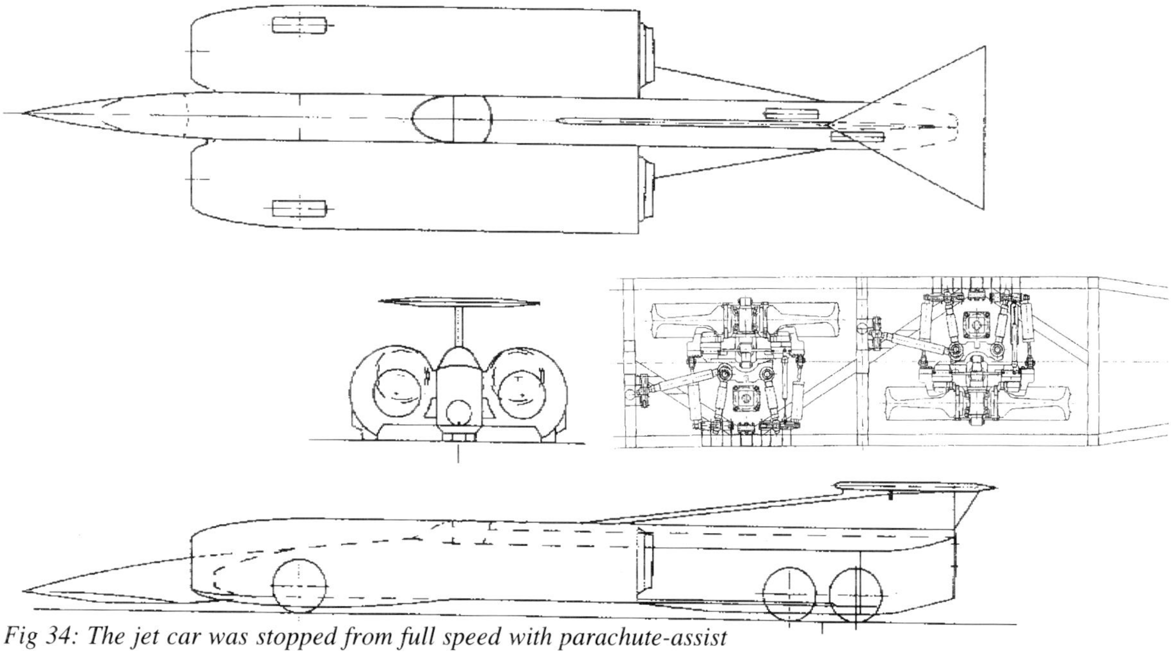

Fig 34: The jet car was stopped from full speed with parachute-assist

and disc and pass them into the vehicle's structure, whilst the brake disc absorbs the vehicles kinetic energy in the form of heat. The twin callipers equalize the brake loads in the system, particularly those reacting through the wheel bearings, and much improve the heat distribution in the brake disc.

A critical element in the brake design, the author explains, is the drive junction between the wheel and brake disc. That of Thrust SSC took the form of its predecessor's brake in having drive slots of a particular shape machined into the inner periphery of the disc, and engaging these with drive lugs machined into the wheel carriers. The slots, of almost 'keyhole' shape, allowed the disc to expand due either to high temperature or high rotational speed by sliding radially on the drive lugs whilst retaining its concentricity with the wheel, the precision of the slot/lug fit ensuring this. The shape of the 'keyhole' drive slot minimizes disc stresses, either thermal or rotational, whilst allowing a failure mode which retained the integrity of the disc as a major safety feature.

A degree of lateral compliance between the hydraulically actuated brake pads and disc is essential, partly to provide for evenness in the lateral force clamp, but also to allow the brake pads to move clear of the disc when not in use. In the Thrust SSC's predecessor's brake this compliance came by laterally springing the brake disc into a fixed position, and applying a pair of twin opposed piston callipers: the opposed pistons would move the brake pads into and out of contact with the brake disc, whilst the latter could still expand and contract radially as required. The webs of the Thrust 2 wheels had been cranked around the brake callipers, an arrangement completely unacceptable for the straight webbed wheels of Thrust SSC. For the new vehicles brake callipers, the body is fixed, and the disc allowed to move axially on the slot/lugs in order to provide the necessary lateral compliance in connection with single sided hydraulic pistons. Now, the hydraulic pistons would move the outboard brake pad into contact with the disc, then move the disc laterally into contact with the fixed inboard pad in order to provide the brake clamp. This arrangement minimized the space required for the fixed arm/fist element of the calliper, and fitted in well with the straight webs of the new wheels. In a laterally moving disc it is essential that all lateral forces are equal and equally disposed about the brake disc's axial centreline, this ensures that the disc does not jam in the drive lugs when actuated. The

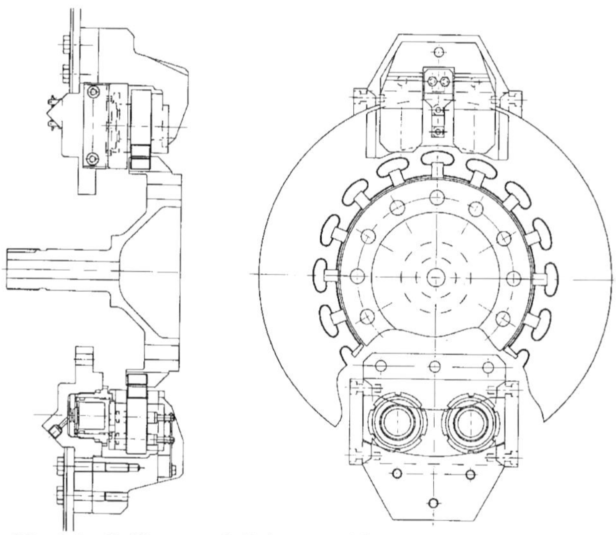

Fig 35: Calliper and disk assembly

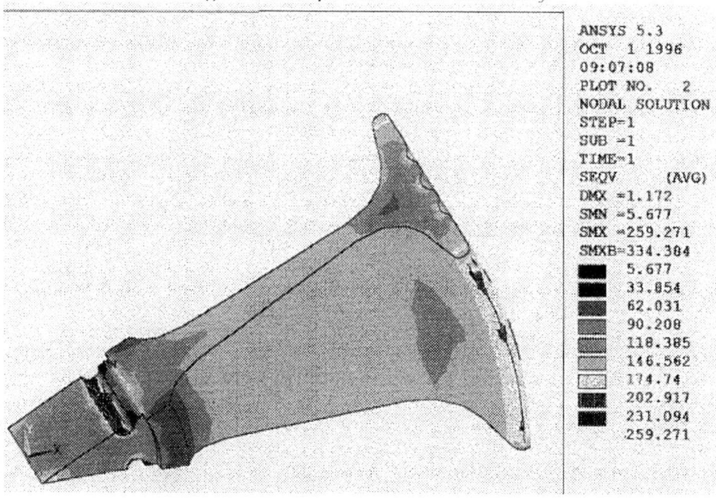

Fig 33: Results of FE stress analysis on front wheel at 8000 rpm

two callipers (one leading and one trailing), of the SSC brake arrangement provide for this, but each calliper has two pistons the pistons: are therefore diagonally connected on separate hydraulic lines, so that should an hydraulic line failure occur, the brake disc will still be evenly loaded in both the axial and rotational planes. The arrangement of the SSC brake is illustrated in Figs 35 and 36. Brake design was common throughout the jet car, with two brakes applied to each front wheel, and one for each rear wheel: brake effort therefore matched the dynamic weight distribution of the vehicle when decelerating on a low adhesion desert.

As both the wheels and brake discs were intended to operate at much higher speeds than in the former jet car, the compacted graphite iron of the original discs could not be used again. Several alternatives were considered, one being to minimize high speed stresses by reducing the mass density of the disc material. Of three possibilities in this area, one was the use of carbon/carbon fibre, used so effectively in aviation and race car brakes. It was with this in mind that Dunlop were approached in order to assess both the suitability of carbon as a disc material,

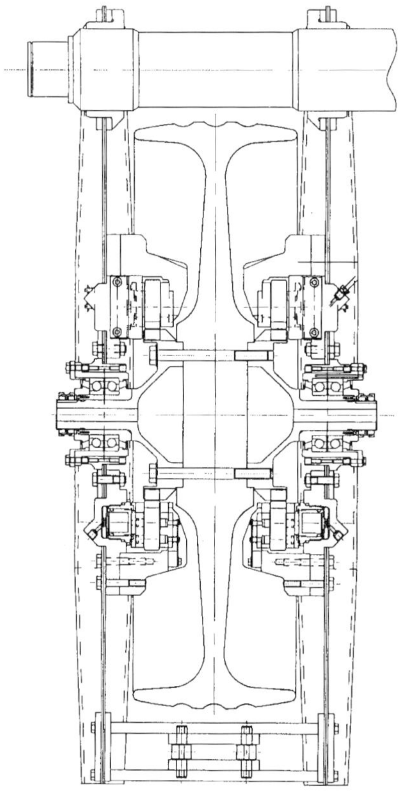

Fig 36a: Front brake

and their interest in supplying such discs.

In order to use this material, the brake pads would need to be from the same material, even though its thermal properties would ensure very high temperatures throughout the pad depth when compared with conventional materials. In view of the anticipated high brake pad temperatures two changes were made to the brake calliper design: the fixed calliper bridge/fist, originally specified in light alloy, became re-specified in steel, and the automotive style hydraulic pistons were changed to the Dunlop Aviation standard screw-in units, which were already fitted with a high temperature heat insulator. In addition, the brake discs were slightly reduced in outside diameter in order to use an available carbon disc blank, and the quantity of disc drive slots increased to sixteen.

All rotary components from the wheel/brake assembly needed proof testing before being passed for use on the jet car. Quite apart from any theoretical analysis of wheel/disc shapes, spin testing to at least 9000 rpm would ensure the use of the equipment at up to the 8400 rpm design speed. Spin testing took place at Pystock/Farnborough, under the good offices of the DRA, following balancing by Schenck, and all components passed without problem, particular emphasis on the post run inspections being directed to the relatively highly stressed wheel bolt holes. Because of the relatively large inertia of the wheels compared with the available motor power of the spinning machine, a spin test from rest to 9000 rpm and back to rest took 35/45 minutes, with a full 5 minute stress soak in the critical region of 8500–9000 rpm. An additional test, for one brake disc only, was to spin test to destruction: this occurred at 10,500 r.p.m.. For physical reasons, it was not possible to spin test a wheel to destruction.

In operation the brakes, because of large interface friction variation with generated temperature, required a learning curve in order to get the best out of them. This done, the brakes have been trouble free for the whole of the project, and pad/disc wear has been small enough for only one complete set of brake parts to be used for all 66 runs completed. Although small, a disc/pad damage differential became progressively noticeable between the front and rear brakes as the run numbers increased, with the rears taking the greater punishment, the likely cause was the heavily dust laden air, disturbed from the desert dust by the front wheels, passing with a scrubbing action between the rear pads and discs. Generally, brake temperatures peaked in the region of 450/500 C, but in a parachute failed situation temperatures of 800 C were reached, these temperatures were easily coped with. The high speed wheels were never operated to their true operational limits, and therefore never achieved the maximum stresses anticipated from both rotational speed and wheel loading. Tread damage from stone impacts during the second Jordanian test series became a potential problem as speed rose to the mid-500's: these damaged areaswould be stress raisers on an increasingly stressed surface at even higher speeds.

Wheel interaction with the desert surface became a dominant feature of the Black Rock operation as speeds rose above 600 mph, and good vehicle control became even more critical as the variations in desert hardness and surface texture over a 13 mile run length played their hand, particularly as an intermediate rainy period had changed the desert characteristics. In general, speeds up to 500 mph allowed the desert to imprint the shapes of the wheel treads, but through 600+ mph the desert beneath the wheels became very broken in their passing, possibly due to the formation of 'impact' shock waves. In the region of 650–750 mph the desert both beneath and either side of the car broke up considerably, probably due to the formation of large vehicle generated shock waves, those which were not airborne being dissipated by the desert itself. Beneath the wheels, to a depth which varied from 0.75 inch to 4 inches, the desert became pulverized into a fine dust: it was almost as if at high speed the jet car ran on a 'fluidized bed' rather than a solid desert, though sensors still indicated the dynamic variations in wheel load. In terms of desert firmness, miles 7–13 were best whilst miles 0–6 were softer with the measured mile being a combination of both: it was quite possible to feel the difference in desert drag over this length when driving a conventional vehicle.

Intermediate rain before the final supersonic runs changed these characteristics, with reducing hardness, and with a particular detrimental effect on miles 4–6, where the surface became particularly friable. A total of 66 runs completed the programme for the Thrust SSC jet car, from the first tentative 40 mph on the Farnborough runway to the final pair which created the world's first Supersonic Land Speed Record at 763.035 mph, (Mach 1.0175), and taking an intermediate LSR of 717.144 mph on the way. All runs were officially timed by the United States Auto Club under FIA rules. In the configuration used, with the Spey 202 engines the jet car would go no faster, as it could not penetrate the rapidly increasing supersonic drag. An engine change to the more powerful Spey 205 engines might have helped an increase in speed to 800 mph, but the true supersonic objective had been reached, and it was decided to leave that mark to those who follow.

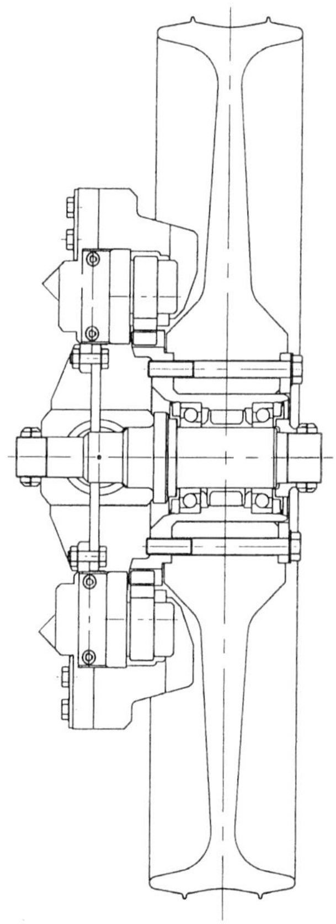

Fig 36b: Rear brake

Chapter 6:
Systems development: powertrain/chassis

A decade of patents specifications and company product releases provides an insight into automotive product development up to the period of new designs covered in the first five chapters. In the area of powertrain and chassis systems the trends to reducing both power losses and environmental pollution are clear in the examples which follow.

Engine developments

Performance 'on spec'

Cosworth Engineering's MBA high performance engine, first built as a technology demonstrator, is a high power-density unit which develops near maximum torque over a very wide speed range and incorporates many race-learnt techniques.

Having a dry weight of only 120 kg (overall dimensions 536 mm high X 20 wide X 490 long), the peak power of the MBA at 162 kW at 7000 rpm is commendable and the maximum torque of 256 Nm is developed at 4500; good in-gear acceleration can be obtained from 4000 rpm, in fact. Being produced independently of any client programme, considerable detail is now available on its design and construction. The aluminium alloy unit is configured as a 2.5 litre, 24 valve, 90 degree V-6 and is notable for such innovations as inlet port throttling and an unusual head and block structure. The block itself weighs only 26 kg with its 87 mm bore and 70 mm stroke layout. This 1.24 B/S ratio allows the fitment of large diameter valves for high specific performance but a lower ratio would be used for more emission-conscious variants. The block has a contra-rotating balance-shaft built into it which, running at engine speed, balances the primary couple inherent with this cylinder configuration.

This shaft also has a centrifugal oil-separator incorporated with it, Fig 1, as part of the advance crankshaft ventilation system. Oil is jetted to the underside of the pistons for cooling at high engine speeds. The head has pent-roof combustion chambers, giving 11:1 compression ratio, but the head-to-block joint is unique and involves angled head bolts.

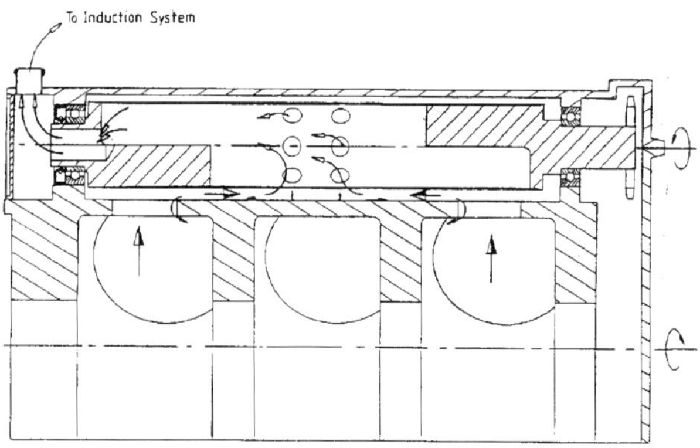

Fig 1: Vertical section through block showing balance shaft axis above that of crankshaft

Camshafts are pre-assembled into separate housings which simplify the main head casting. Valves are 35 mm diameter for the inlet and 30 for the exhaust and have an included valve angle of 40 degrees. The barrel apertures which form the port-throttles are also incorporated into the head.

Cylinder block side walls are held in tension by the angled head bolts while the cylinder wall is in compression by reaction to the bolt load. Solid fire rings seal each bore while O-rings seal the water and oil galleries. Cross-bolting of the main-bearing caps, Fig 2, helps to contain costs by eliminating the need for tight-tolerance machining. Mating horizontal surfaces on the block and iron caps are first machined flat then the bolt holes are drilled and tapped. After bolting down the caps, dowel holes are drilled in the two vertical clearance slots between each cap and the block. Cross-drilled dowel pins, and then bolts, are then fitted prior to line boring the assembly. A two-

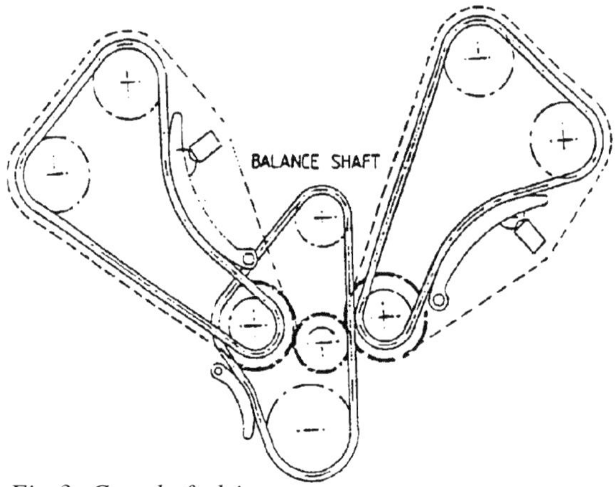

Fig 3: Camshaft drive

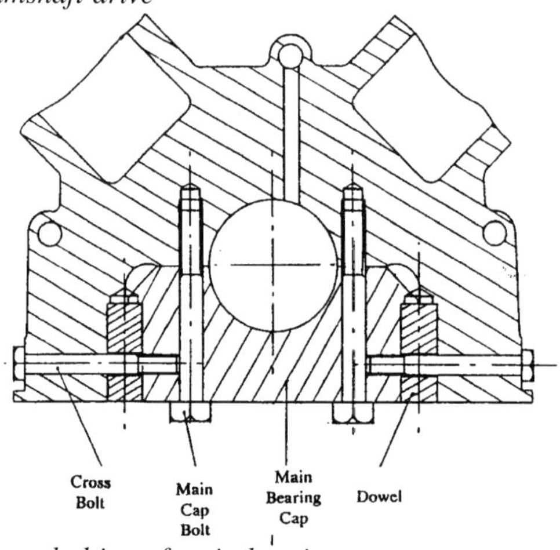

Fig 2: Cross-bolting of main-bearing caps

stage camshaft drive is featured with helical gear on the cam nose driving two primary gears which independently chain-drive one pair of camshafts. The two-stage reduction thus allows smaller camshaft sprockets than usual, which reduces the overall size of the engine. Chain life is also extended with this slower moving layout, made quiet by individual cases moulded in special grade of nylon, Fig 3.

The crankshaft has offset pins to provide even firing, main and big ends being 50 and 42 mm respectively. Connecting rods weigh only 0.49 kg and are 128 mm between centres; piston weight, too, is kept down to 0.4 kg. The induction system comprises a variable-geometry plenum which helps to spread WOT torque over a wide speed band. Two chambers into the plenum individually feed one bank of cylinders; they are linked together by two pipes with switchable valves controlled by the engine management system.

In a conventional plenum-throttled engine there is pressure differential at part throttle between inlet and exhaust ports. Conventionally there is simultaneous opening of inlet and exhaust valves towards the end of the exhaust stroke which allows residual exhaust gas to flow back into the inlet port, the resulting charge reduction lowering NOx emission. An external EGR system would accentuate this process with external pipes. The MBA has a patented internal EGR system which circulates exhaust gases rich in HC; this claims to reduce both NOx and HC emissions. A high performance engine usually requires long cam durations and a large valve overlap period and the resulting exhaust dilution is greater than on a conventional unit. On the MBA the resulting rough running that this would cause is obviated by port-throttling which limits the reverse flows. The depression created in the inlet port, during induction, recovers after the inlet valve closes, according to speed and load conditions. As the inlet reopens there is little pressure differential between inlet and exhaust ports. The independent throttle mechanism, Fig 4, on the MBA involves a port-throttle upstream of the inlet valves, the tracts being fed into a common plenum. Thus, at a given speed and load, degree of exhaust gas dilution is a function of inlet port depression. Internal EGR is therefore controlled by the relationship between the two throttles, Fig 5.

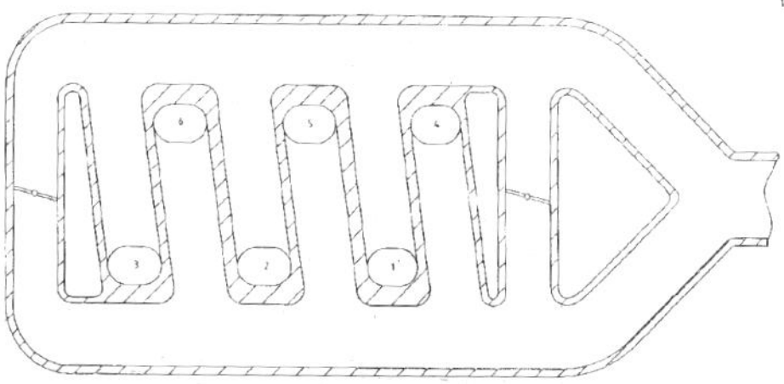

Plenum throttle

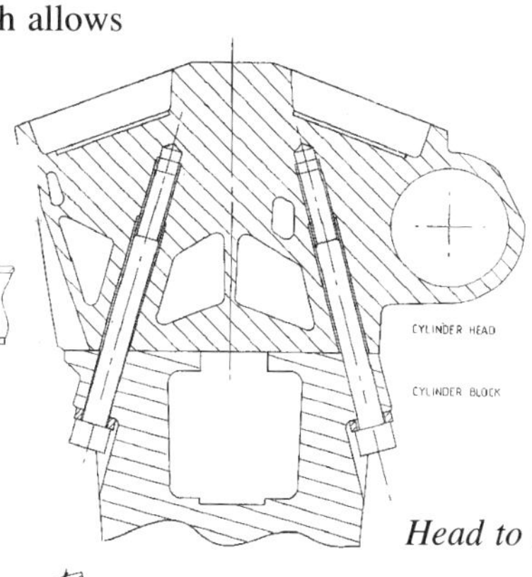

Head to block joint

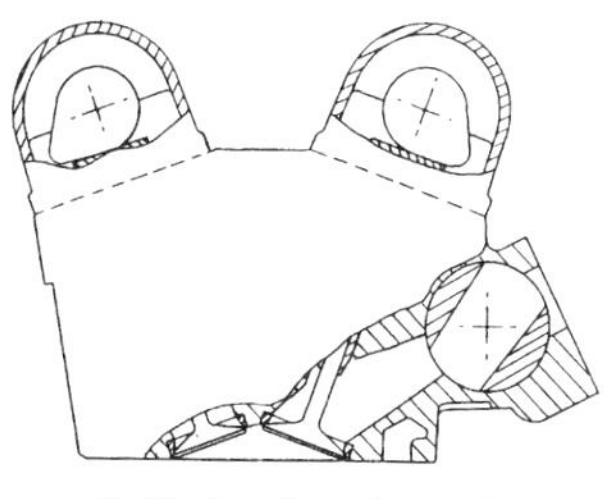

Cylinder-head section

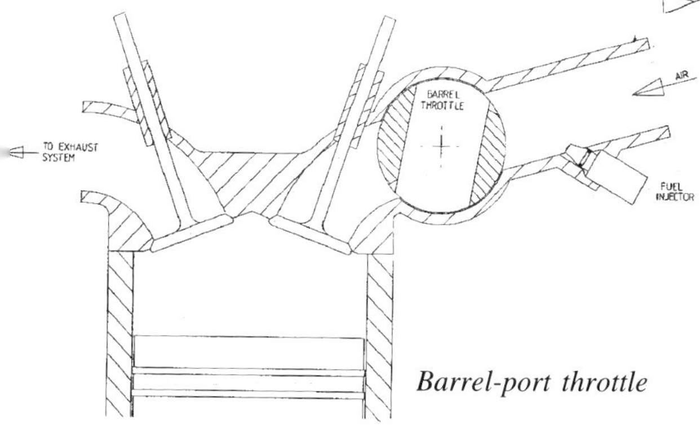

Barrel-port throttle

Fig 5: Cylinder head details (above left and right)

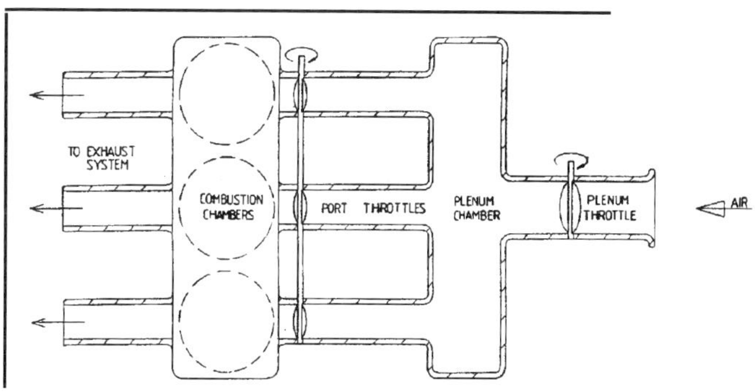

Fig 4: Plenum and port throttles

Advanced car and truck engines

Lotus Engineering's Jewel Project, Fig 6, seeks to meet some of the challenges of future engine requirements. It involves a monobloc design to improve heat flow from the head to the main water jacket, the cylinder head having hardened surfaces interfacing with the aluminium alloy monobloc. Resulting reduced temperatures for pistons and liners permit a safe raising of the compression ratio for greater engine efficiency.

A hollow crankshaft saves weight and lowers inertial forces — also helped by the use of metal matrix composite conn-rods. Corrosion of such parts as water pump and thermostat housings is avoided by the use of GRP composites; fuel rails, too, are injection-moulded for both lighter weight and better thermal properties for resisting fuel vaporization. Cam covers and outer walls of the coolant jacket are also of composite construction. A rubber joint isolates the monobloc casting from manifold and sump to reduce noise transmission.

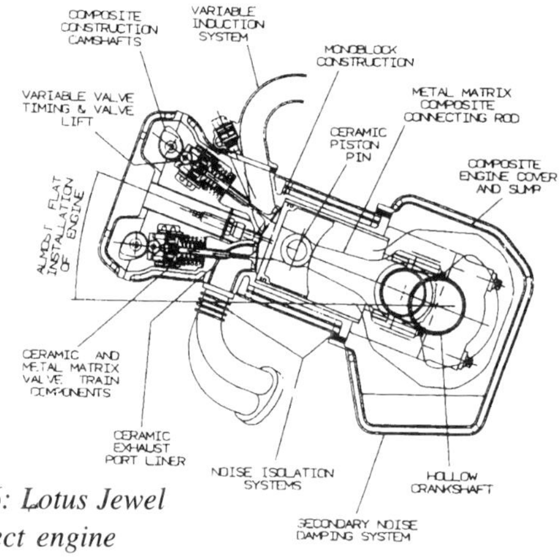

Fig 6: Lotus Jewel Project engine

Scania has become the first truck manufacturer to undertake series production of a turbo-compound diesel engine, Fig 7, with an efficiency of 46%, over 10% higher than the best gasoline engine figure. Exhaust energy is recovered in two stages; the first turbocharger (1) drives a compressor to force more air into the cylinders while the second is a power turbine (2) supplying additional power to the flywheel (4) through a hydraulic coupling and gear train (3). Power and torque are about 5% higher at 400 bhp and 1750 Nm. There is also a 6 gm/kWh saving in specific fuel consumption.

At the left of the figure the power turbine is seen downstream of the turbocharger turbine, the latter being small in relation to mass flow at high engine speed — where, pressure ratio across the power turbine being high, it still provides enough output to drive the compressor. At low speed when pressure ratio across it is reduced, the compressor's pressure ratio is maintained — partly because of the small turbine housing. Increased charge pressure at low speed thus makes it possible to inject more fuel and increase torque without air/fuel ratio becoming too low. Power turbine is connected to the crankshaft via a two stage spur gear set and hydraulic coupling. Mass inertia of the power turbine, converted to crankshaft speed, is about half that of a normal flywheel (gear ratio 1:30); the 'soft' coupling reduces the peak torque amplitudes. The set-up is believed by Scania engineers to be a more satisfactory route than a variable-geometry turbocharger, some of which contain more than 100 moving parts and are liable to reliability problems. However, it is pointed out that greatest advantage comes with engines operating under heavy and sustained loads.

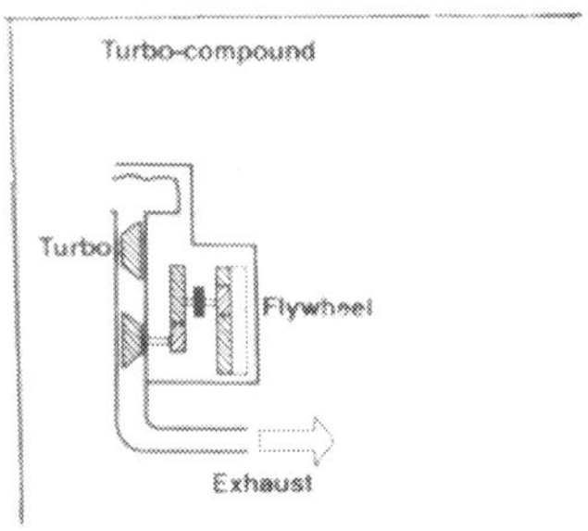

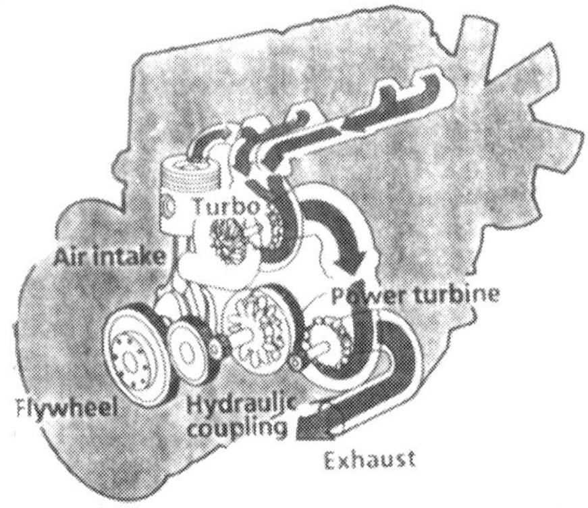

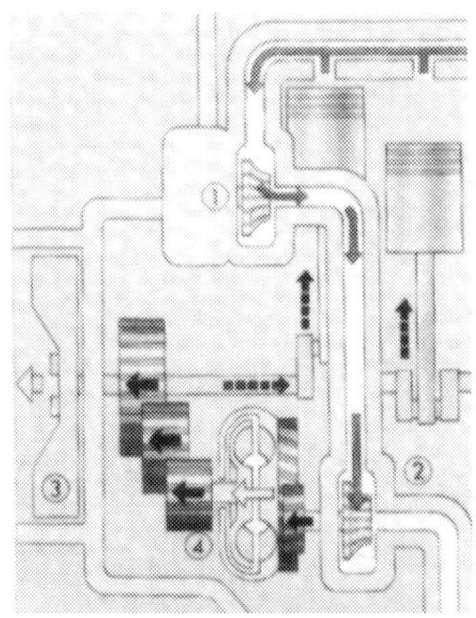

Fig 7: Scania turbo-compound diesel

Improved engine temperature management and differential expansion control

During the cooling cycle of an engine, after combustion, there can be an unduly long retention of heat by some of the components. This heat can flow back into the incoming air-fuel charge with resultant loss in engine efficiency. The design in Fig 8 thus aims to make combustion chamber and charge induction surfaces store and release heat selectively, to limit the piston surface temperature and control thermal expansion. The use of sprayed insulant coatings on the inside walls of engine combustion chambers is proposed here to manage the heat release from the combustion process. A low thermal expansion/high conductivity insert is also cast into the piston crown to inhibit differential expansion. An abradeable coating is also sprayed on the piston walls, above the top ring, in order to reduce clearance and therefore enhance heat transfer. Under crown oil-spray cooling is employed, too, together with a dual wall construction for inlet and exhaust manifolds, the net effect of which is to reduce the onset of knock and therefore allow the use of higher compression ratios to provide enhanced efficiency.

The problem of differential thermal expansion of engine cylinder head and block, in different materials, is said to be overcome in the design of Fig 9. An aluminium alloy head is secured to a cast-iron cylinder block by extra-long bolts which engage with threaded bosses deep down in the block. Above the threaded portions the bolts pass through oil galleries and are warmed sufficiently to cause the correct linear expansion to compensate for that of the cylinder head, between its support faces.

Improved turbocharger waste-gate control and move to constant-pressure cycle

A simple and reliable turbocharger pressure-controller, for providing enhanced operation of the waste-gate valve, is provided in the design of Fig 10 which permits standard exhaust bypass operation. The flow control element is positioned within the exhaust bypass conduit and the waste-gate actuator is responsive to bypass air pressure for actuating the control element. An air pressure control device has an intake port connected with the intake-air outlet and an outlet port providing bypass air pressure to the flow control actuator. When airflow paths between intake and outlet ports provides for continuous flow at least one other inhibits through airflow.

Provision of a sprung element in the connecting rod of a piston engine is achieved by a series of belleville washers in the design of Fig 11. They constrain a piston within a piston, the inner one of which moves on a steel liner in the aluminium body

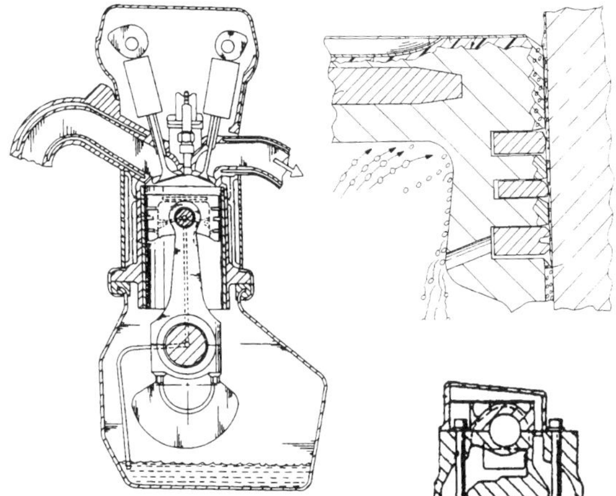

Fig 8: Patented temperature management system by Ford Motor Co (GB2293863A)

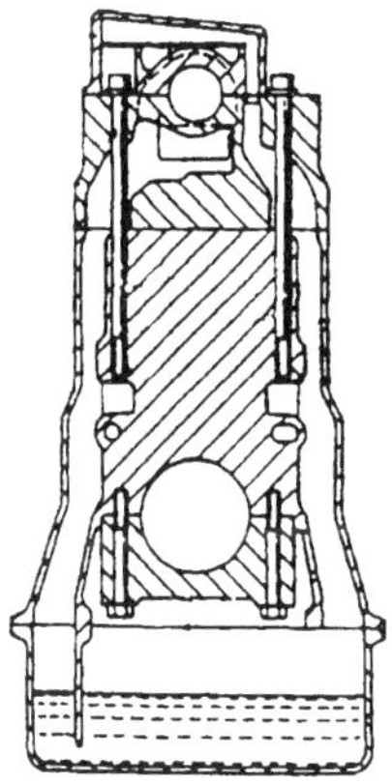

Fig 9: Differential expansion control by Rover Group (GB2312020A)

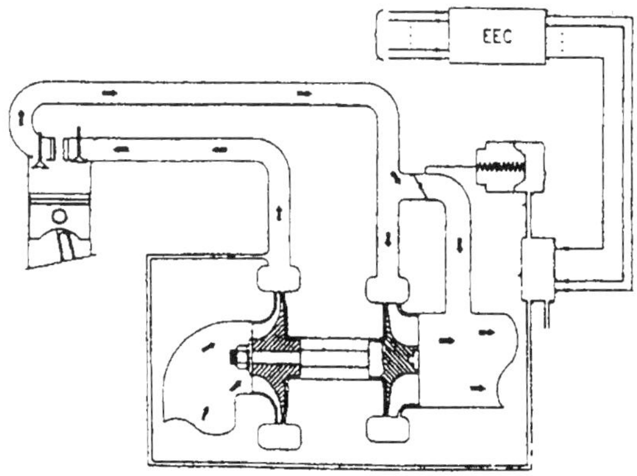

Fig 10: Turbocharger waste-gate control by Cummins Engine (GB2311556A)

of the outer piston. A retainer ring at the base of the liner restricts the movement of the inner piston. Because, after ignition, the first element of the 'stroke' is compression of the springs and thus an increase in clearance volume, there is a drop in cylinder pressure, and temperature, in the combustion chamber. As the energy stored in the springs is released after TDC, although ignition occurs before TDC, engine efficiency is increased because there is no longer any negative torque occurring before TDC, changing the theoretical air cycle of the engine to a constant-pressure one.

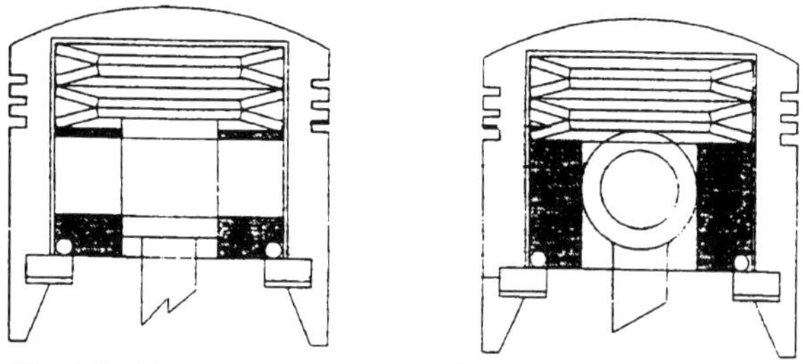

Fig 11: Constant-pressure cycle piston modification by GF Galvin (GB2318151A)

Engine induction systems

Low-cost supercharging and electromagnetic valve actuation

An inexpensive vehicle engine supercharger, having an integral by-pass mechanism, is proposed in the design of Fig 12. Of the rotary, positive-displacement type, it incorporates a duct within its casing to communicate with the suction and discharge ports at each end, for supplying naturally aspirated air to the engine. Rotors also compress the air entering from the inlet port on the right to supply the discharge port, to the engine, on the left. The by-pass passage is controlled by a valve and actuator on the extreme right of the mechanism. On the extreme left is the electromagnetic clutch which drives the supercharger but drive is broken when the by-pass valve senses a predetermined maximum boost pressure and opens the bypass port.

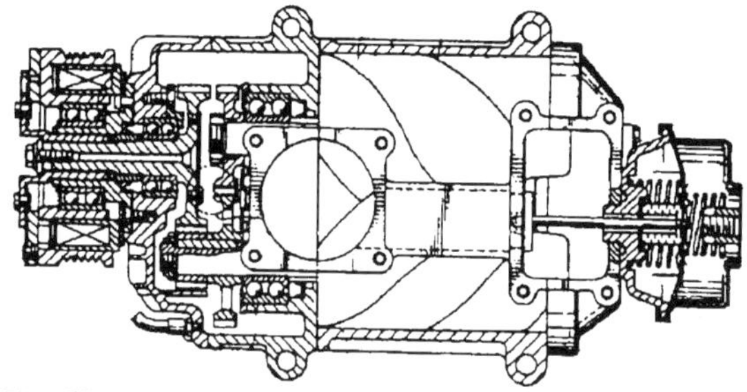

Fig 12: Low-cost supercharging by Tochigi Fugi Sangyo (GB2301149A)

Long valve opening time, and so adequate cylinder filling at high engine speeds, is among the advantages claimed for this electromagnetically actuated poppet valve, Fig 13. The armature of the 'solenoid' has switching magnets which hold the valve in open and closed positions. Compression springs are positioned between magnets and valve to avoid free-play in both the opening and closing modes.

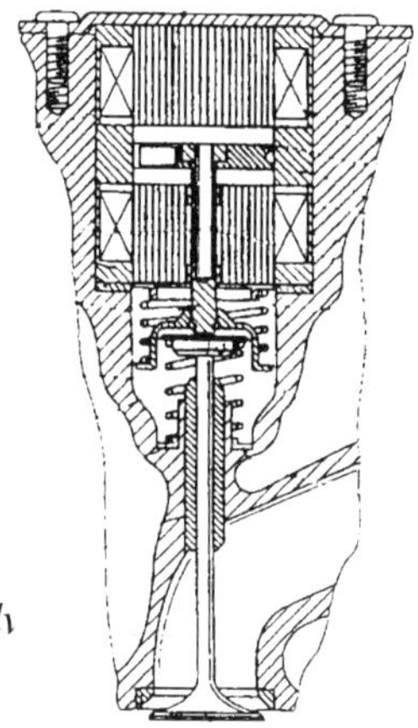

Fig 13: Electromagnetic valve actuation by Daimler-Benz (GB2312244A)

Better fuel-spray atomization, integrated air/fuel induction system, fuel injection rethought and storing swirl energy

The design in Fig 14 exploits the advantage that

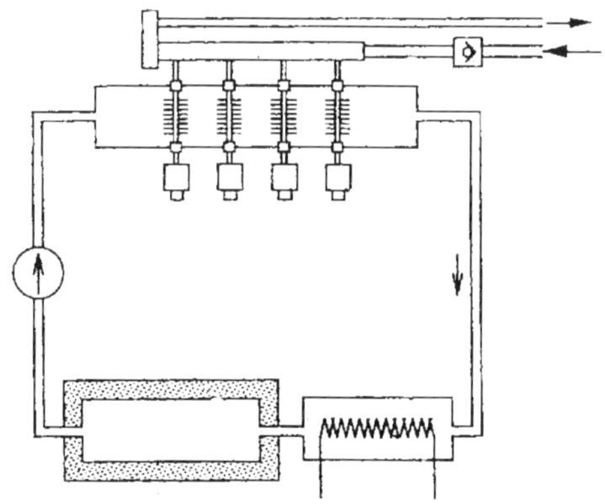

Fig 14: Better fuel spray atomisation by Ford (GB2312923A)

suddenly vaporized, or flashed, fuel effectively breaks up the spray from a fuel injection nozzle, when the fuel is subjected to a rapid pressure drop as it discharges into the inlet port or combustion chamber. The proposed engine incorporates a heat-exchanger for pre-heating the fuel, upstream of the injectors. Temperature is maintained at a constant level just below the fuel's boiling point and the heating fluid may be coolant water or exhaust gas. Heating can be assisted by an electrical coil and insulated reservoir.

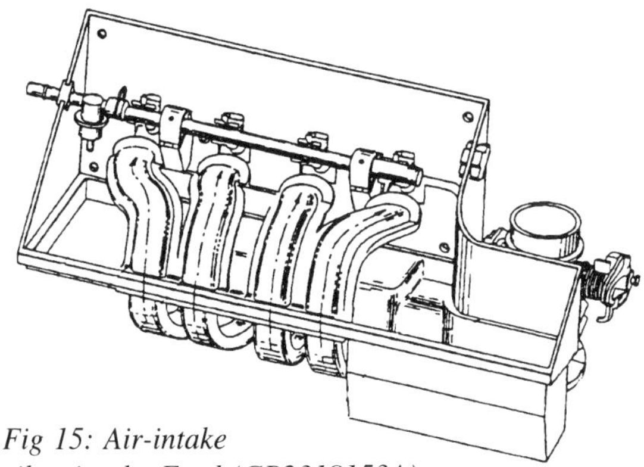

Fig 15: Air-intake silencing by Ford (GB2318153A)

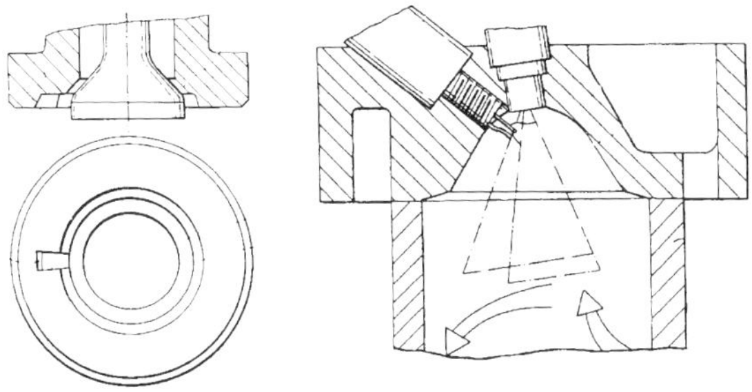

Fig 16: Fuel injection rethought by Robert Bosch (GB2295204A)

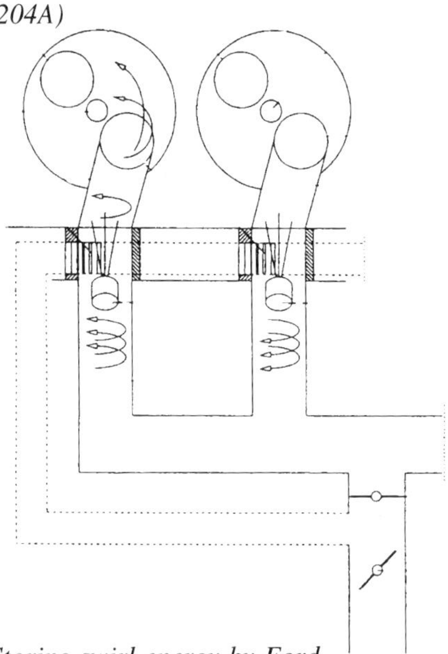

Fig 17: Storing swirl energy by Ford (GB2299373A)

Efficient engine air-intake silencing with minimum space usage around the engine is achieved in the design of Fig 15. The intake air-box and manifold ducts are integrated in one unit, the air-box volume surrounding the intake ducts. An air-plenum and throttle valve is also integrated into the assembly together with the fuel rail delivering to the injectors.

A fundamental problem has existed with direct fuel injection into engine combustion chambers. This is in the conflict between providing a fuel-rich mixture for ignition at the sparking plug yet providing a narrow enough spray cone to avoid wetting the cylinder walls with fuel which would fail to vaporize and result in unwanted hydrocarbon emissions in the exhaust. In the design of Fig 16, a poppet valve type injector provides a narrow conical spray but has a subsidiary conical arc sprayed towards the sparking plug. This is achieved by a recess in the annular wall of the injector downstream of the valve seat.

In an engine designed so that two intake port branches supply each inlet valve, the first branch being connected to a common plenum of an air supply manifold while the second is connected to a common plenum of a second manifold, this second manifold acts as a stratified charge storage arrangement as covered by an earlier patent by Ford. Now the company have extended this concept to lean burn engines so as to improve charge preparation and combustion quality, Fig 17. Air drawn from the first branch into the second swirls while in storage and retains kinetic energy for the next intake stroke. When the intake valve of the left hand cylinder is closed air is drawn through the lower throttle butterfly into the dotted line drawn duct under the action of the intake depression caused by the adjacent right-hand cylinder as its valve opens, there being a tangential swirl connection into the second plenum, downstream of which fuel is injected and upstream of which the plenum duct is shaped to retain swirl energy. The upper of the two butterfly valves may be opened at high engine loads.

Refinement and reduced emissions

More on the Sarich 2-stroke and a Rover K-series development

The promise of greater fuel economy using two rather than four stroke engines has been given a new interest by the Australian Sarich engine which has caught the attention of the volume builders. Sarich has done much to advance the cause of two-strokes by allaying fears of poor emissions performance by devising the acclaimed system of injecting fuel-air mixture into the cylinder. The metering and injection system is shown in Fig 18. Fuel is metered through chamber (a) via port (b) and compressed gas is introduced via an air-port (c) to displace the fuel and deliver it to the engine. Crucial phasing of the valves is controlled to ensure fuel inlet and outlet (d) valves are closed before the air one opens and that any residual gas is expelled before the next charge. Gas inlet to the chamber is controlled by valve (e) opening at a higher pressure than it requires to close the outlets.

A stretched version of the Rover K-series engine features variable valve timing in Fig 19. So-called 'damp-liner' technology has been used involving bore walls just 3 mm thick and a water cooling jacket of just 0.65 mm. An extra long crank throw then brings swept volume to 1.8 litres. The VVT mechanism continually varies the inlet cam period to optimize power. An eccentric disc controlled by the EMS alters the phasing of camshaft to crankshaft so as to lengthen the inlet valve opening period above 400 rpm. A maximum power of 143 bhp at 7000 rpm propels the 2359 lb vehicle (MG*F*) at its maximum speed of 130 mph.

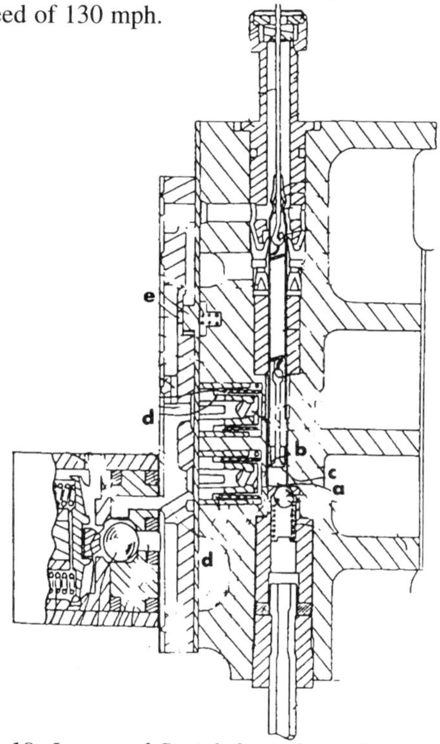

Fig 18: Improved Sarich 2-stroke, in Patent GB2210413A

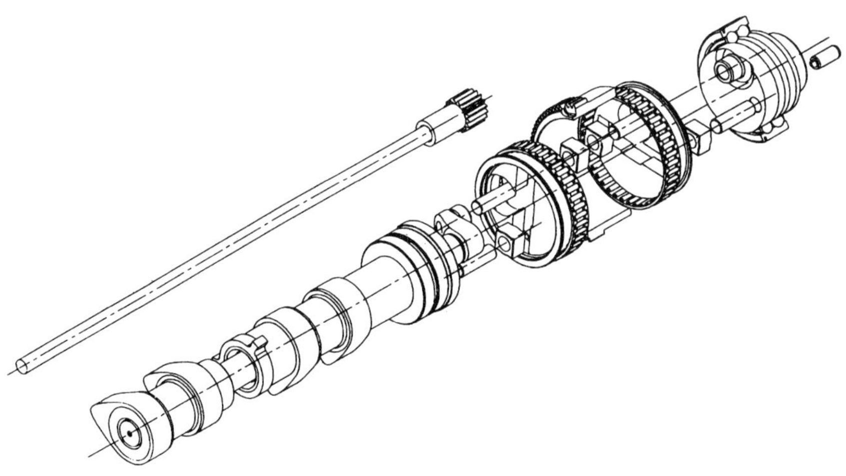

Fig 19: VVT for K-series

Variable valve timing made easy, enhanced fatigue life for valve gear and variable geometry engine intake

By using a lateral displacement device for the chain drive to an engine valve operating mechanism, a phase change can be obtained between the driver and driven sprockets in a system announced by Sachs Industries, Fig 20. Engine oil pressure is used to effect the displacement via a chain tensioner device. A check valve, shown in the sectional view of the tensioner, controls the operating pressure and a solenoid valve connected to the pressure cylinder causes the device to change from chain tensioning to displacement mode. The solenoid is switched by the engine management system and in the arrangement shown at A around 30 degrees of crankshaft angle displacement is involved in the phase-change. In the 3 or 4 position arrangement at B, both camshafts can be controlled and, by using a proportional solenoid valve, a possible future application may be continuously variable cam-phasing.

A valve drive system has been suggested which combines long fatigue life with maximum stiffness of all its constituent elements. In Fig 21 an IC engine camshaft is supported by a housing comprising pillars and beams which do not require complex machining of the cylinder head. A short bearing interval ensures avoidance of camshaft bending vibration and the beams joining the bearing support pillars are only interrupted by cut-outs for the extra short rockers and thus form an unusually secure mounting for the rocker shafts.

Obtaining induction ram effect at more than one 'resonant' speed is the object of the design in Fig 22. The carburettor or fuel-injection choke tube is connected to a plenum chamber in a barrel rotatable by a motor geared to engine speed. The effective lengths of the intake and outlets vary with barrel position by their formation, in part, by grooves in the barrel wall.

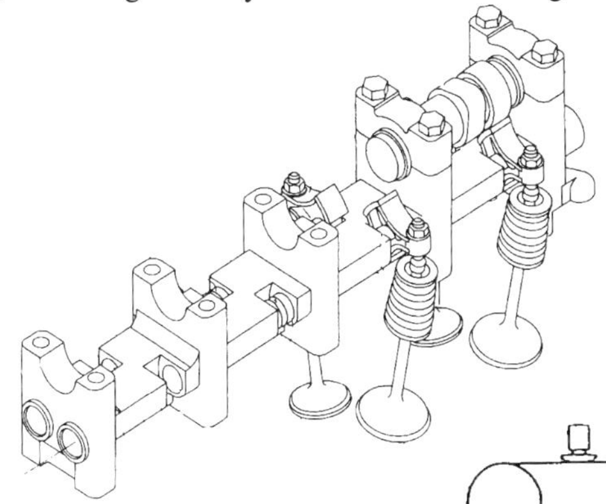

Fig 21: Enhanced fatigue life for valve gear by Mercedes-Benz (GB2297805A)

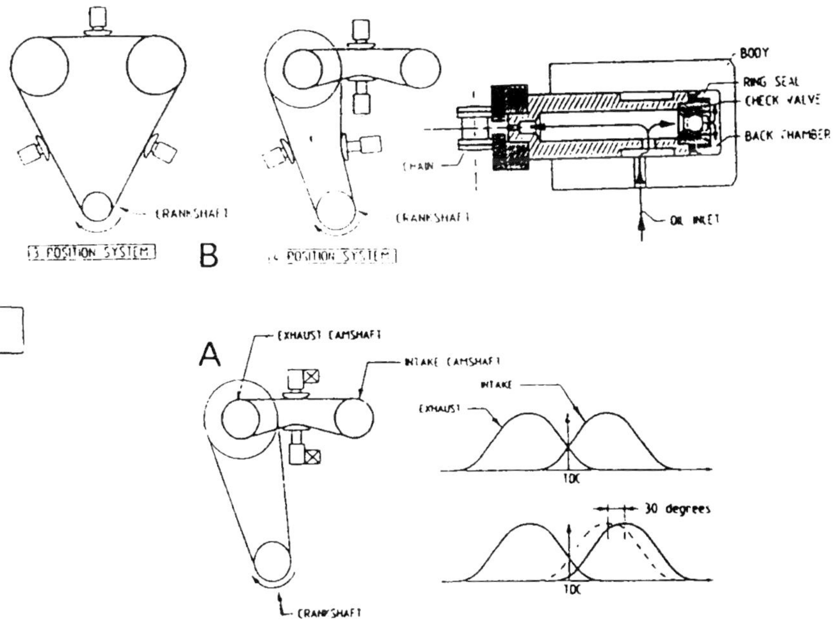

Fig 20: Sachs variable valve timing

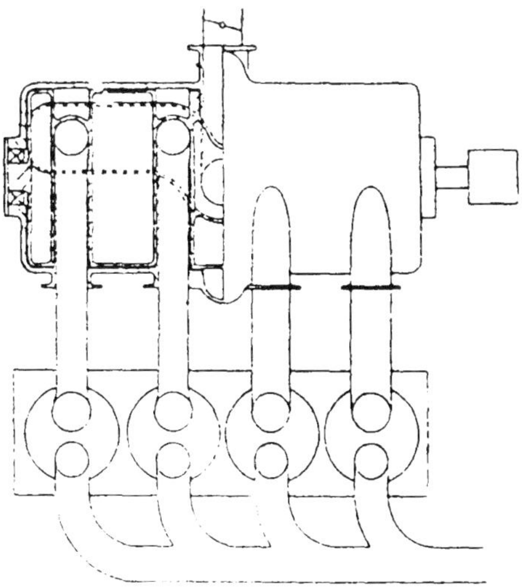

Fig 22: Variable geometry intake by Tickford, in Patent GB2205609A

Economical flywheel to crankshaft connection and secondary engine force balancer for motorcycle

A more cost-effective connection between engine flywheel and crankshaft is possible, without loss in reliability, with the design in Fig 23. It uses a collar, held by an expansion screw, in a concentric opening in the crankshaft. Dog teeth on shaft and flywheel form the connection, the joint being made firm by tensioning of an axial spring acting on an expansion screw where the two parts mate. The flywheel has a collar which extends into the crankshaft opening, the centre of the collar having a hole through which the expansion screw operates.

Normally there is difficulty incorporating a full Lanchester-type harmonic balancer into a motor cycle engine because of the high rotational speeds involved. In the design of Fig 24 , a compromise is reached in which a large, crankshaft-mounted, gear-wheel drives two smaller co-planar gears in opposite directions, one by a further countershaft gear.

Low emission CVs

Initiatives taken by MAN for diesel engine pollution reduction include the air-injection system shown in Fig 25 for dealing with visible smoke on initial take-off and subsequent acceleration before the turbo-charger achieves full compression. Normally fuel is not completely burnt under these conditions as the engine is not supplied with enough air — so an air reservoir is introduced which is fed from the compressor. The injection system comprises a pressure pipe with solenoid valve and non-return valve in the charge-air line at the inlet to the air distribution pipe. By means of an electronic controller, throttle pedal depression opens the solenoid valve to inject air for

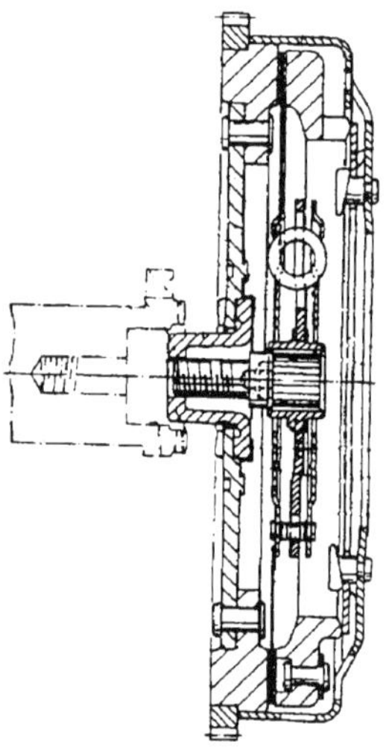

Fig 23: Economical flywheel to crankshaft connection by Fichtel & Sachs (GB2301166A)

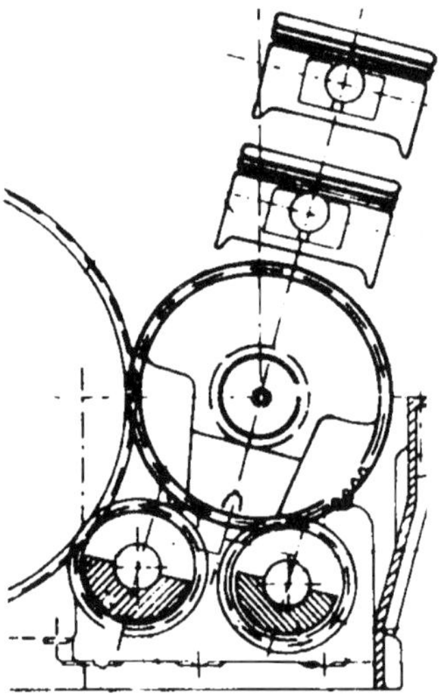

Fig 24: Balancer system by Triumph Motorcycles in Patent GB2186914A

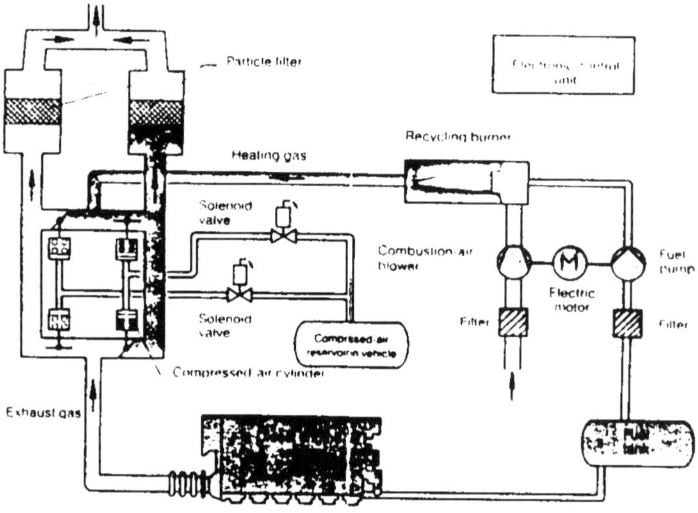

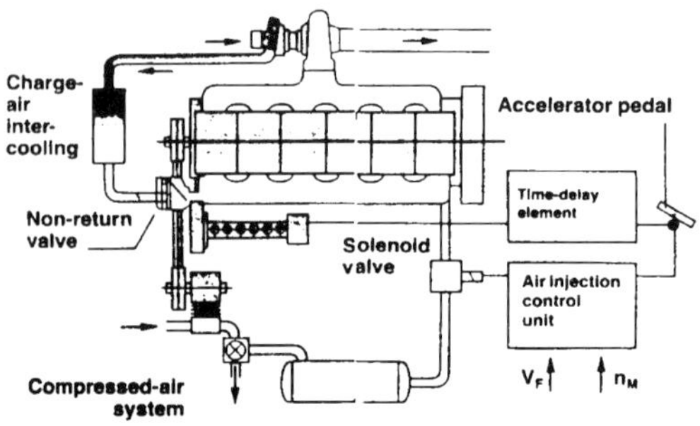

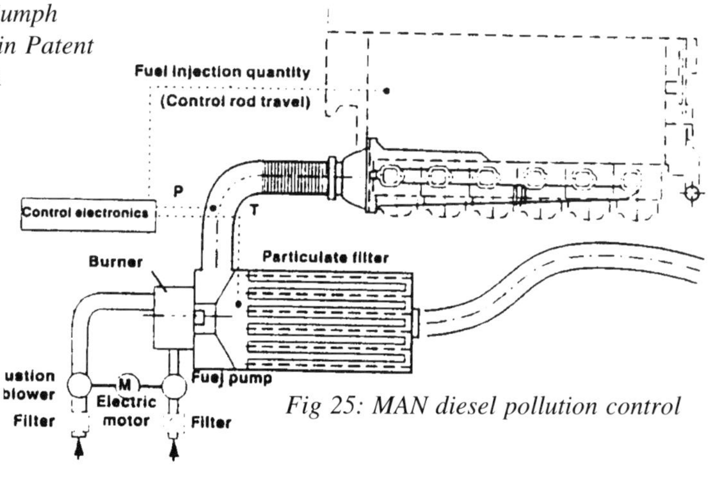

Fig 25: MAN diesel pollution control

a 0.7 second period. Closure of the non-return valve protects flow from the turbocharger while the ECU releases a matching quantity of fuel for injection. The system is used on the company's NL202 and NG272 low-floor buses. The company has also developed its particulate filter technology in which both duplex and full-flow arrangements are used to overcome the problems of filter clogging with deposits which can-

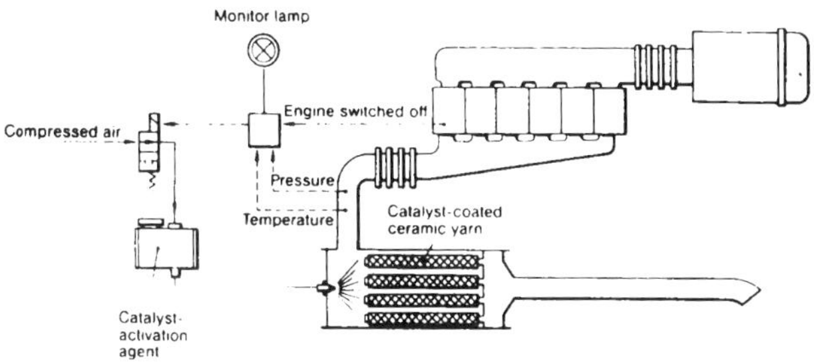

Fig 26: Filtering diesel particulates by Mercedes-Benz

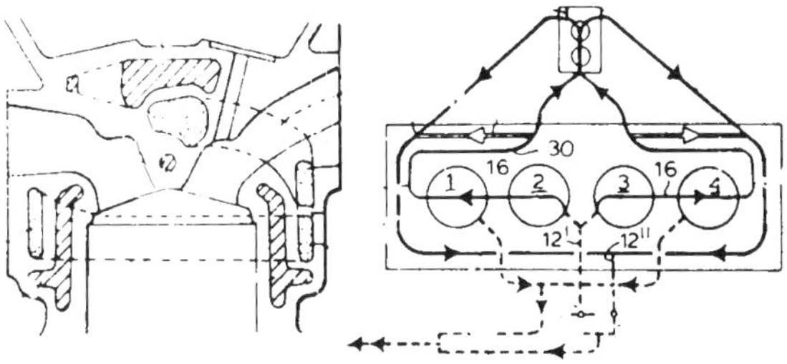

Fig 27: Speeding engine warm-up, by Ford, covered in Patent GB2199368A

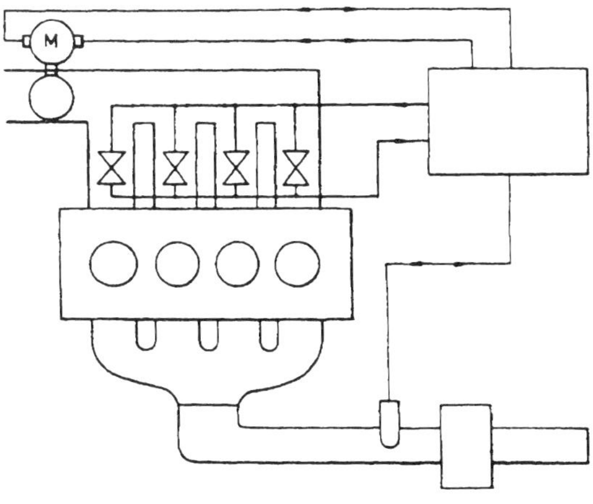

Fig 28: Quicker catalyst light-up time by Rover Group (GB2316338A)

not be burnt off in conventional layouts because exhaust gas temperature from low-consumption engines is too low. In addition, oxidiszing catalytic converters are being developed for exhaust gas post-treatment.

An exhaust filtering system for diesel particulates is now offered on Mercedes bus chassis. The so-called ceramic candle type filter features catalytic regeneration. The system, Fig 26, complies with emission limits for urban buses which came into force in the US during 1991. The company is reported to have rejected the philosophy of developing catalytic control systems originally engineered for spark-ignition engines, on the grounds that diesel emission control needs solutions with less flow resistance, proneness to blockage and better thermal and mechanical resistance in the harsher CV environment.

Speeding engine warm-up, quicker light-up time for denox converter and use of scotch-yoke crank

Exhaust ducting cast in the cylinder head is used in this design proposal, Fig 27. The ducts allow alternate paths for the engine exhaust according to position of two-way valves. As well as warming up the inlet manifold during initial running of the engine, exhaust gases also warm up the engine coolant and lubrication systems.

Faster catalyst warm-up time from engine start-up is the objective of the design of Fig 28 which involves providing excess fuel and air to enter the converter for part of the combustion cycle. A fuelling device controls the exhaust gas entering the catalytic converter such that a combustible mixture is introduced to enable the converter to more quickly reach light-off temperature or regain it if very cold operating conditions cause the catalyst to 'go out'.

The Collins CMC scotch-yoke engine, Fig 29, has attracted considerable interest. Exhibited recently by consultants Emdair who are R&D contractors to the Australian builders, the engine has inverted-tee cylinder formation in the 2-litre two-stroke version shown here. Compact size is thanks mainly to the scotch yoke mechanism joining the pistons to the crankshaft. It has been calculated that a vehicle designed to exploit the lower package size of the engine would achieve a 10% fuel saving over a conventional design.

Drive and steer systems

Transport of the future — guided lanes or electric/flywheel hybrid car propulsion?

Tribus, Fig 30, has been described as being to the trolleybus what light-rapid-transit is to trams. But it also applies to both goods and passenger vehicles — to urban and inter-urban traffic. A segmented bar pantograph collector overcomes the traditional limitations of trolley booms. This results in a multi-lane high speed equivalent to a trolleybus which could have attractions both for commuting and dense traffic motorway driving. The overhead wiring structure is vastly simplified from traditional trolleybus layouts and the ability to 'jump' from lane to lane, and through junctions, is made possible by reserve energy provided by on-vehicle flywheels and/or hybrid electric/petrol drive systems. On the three lane motorway installation shown, only the two inside lanes would typically be electrified. Inset is shown how two or more elements of the segmented pantograph bar are always in contact with overhead feed cables. The originators are Lawrence Designs, 40 Lower Broad Street, Ludlow, SY8 1PH.

The US company American Flywheel Systems is claiming ranges for electric cars of 300–600 miles against the 50–100 miles which is representative of existing battery-electric technology. The AFS 'mechanical battery', Fig 31, stores kinetic energy in vacuum-enclosed rotors running at high speeds on magnetic bearings. The charge current causes the rotors to accelerate up to speed and energy is drawn from the spinning windings by the unit acting as a generator. The technology is made possible by new materials for the fibre windings and the magnetic bearings. The vacuum housing is shown at (a), fibre windings at (b) and stationary shaft at (c); counter-rotating rim wheels are at (d), spoke tubes containing the magnets at (e) and magnetic bearings at (f) — (g) and (h) being the pick-up coils and end-bell respectively.

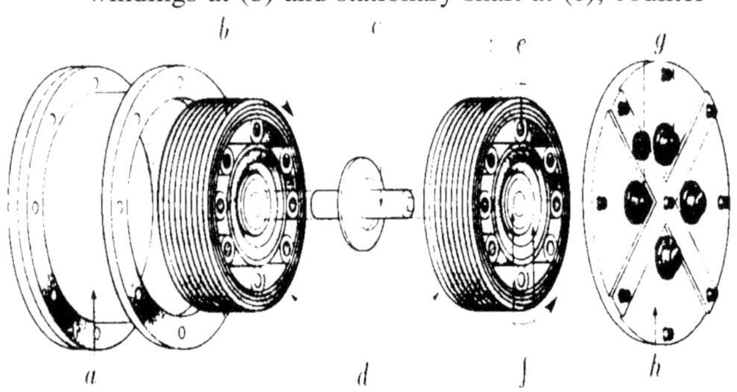

Fig 31: Flywheel battery

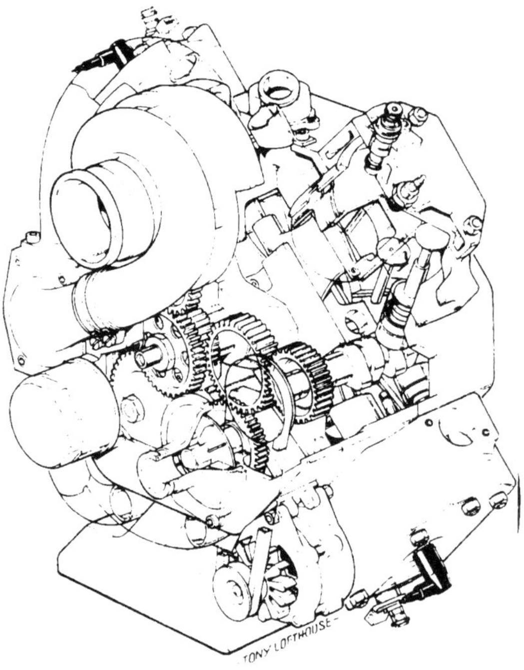

Fig 29: Collins scotch-yoke engine

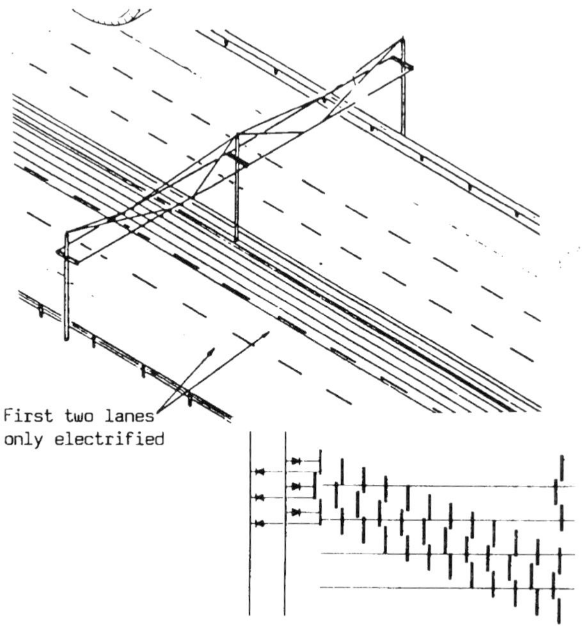

Fig 30: Guided lane system

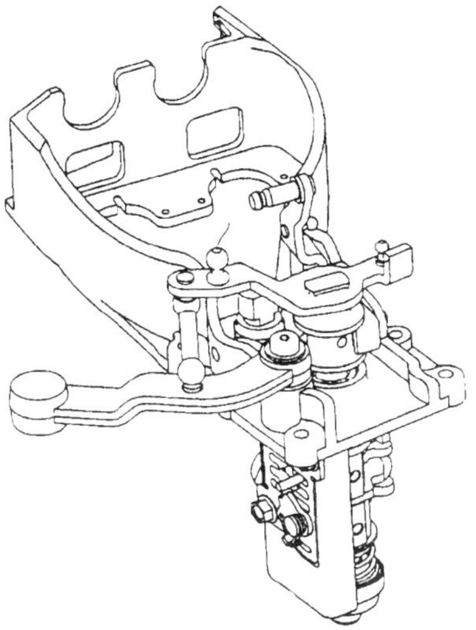

Fig 32: Sure-fire gear engagement by Mercedes-Benz (GB2308165A)

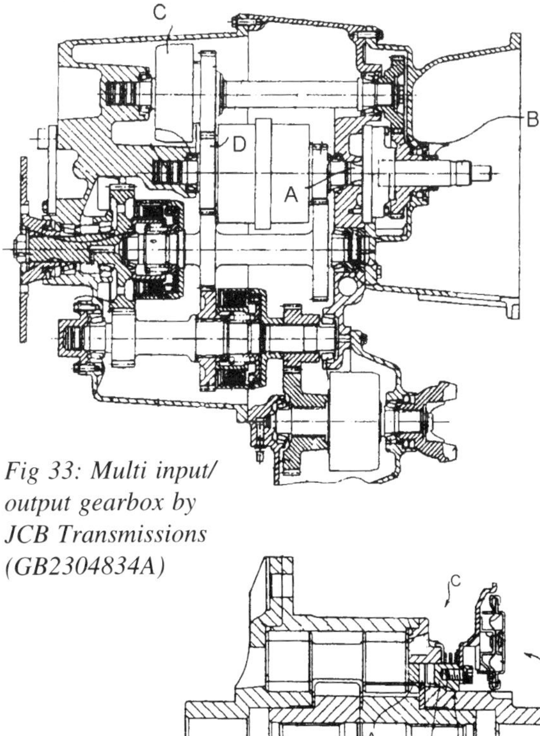

Fig 33: Multi input/ output gearbox by JCB Transmissions (GB2304834A)

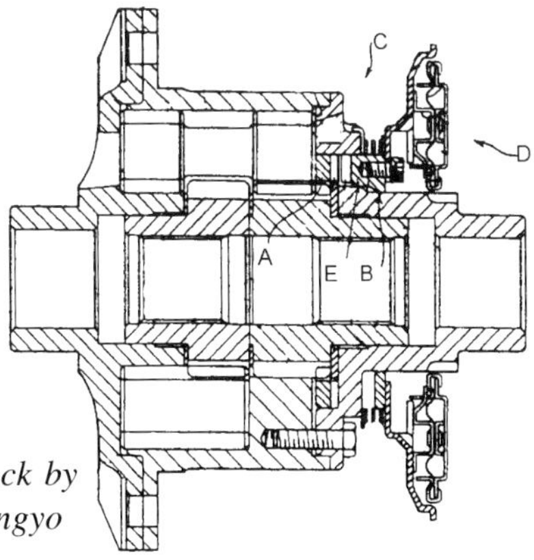

Fig 34: Rapid-actuation diff-lock by Tochigi Fuji Sangyo (GB2305224A)

Ensuring sure-fire gear engagement and multi input/output gearbox

Inertial effects which might impair proper gear-selection in a change-speed linkage, between shift-lever and selector forks, are negated in the design of Fig 32. An extrusion lever with an inertial mass at its end is linked by a ball-jointed pull-rod to the gear-change lever on the selector shaft. The geometry of the linkage is such that pivot arm of the ball-joint, on the extension-arm, to the selector shaft is larger than that between the ball-joint and the arm's own pivot-point. Also the pivot arm between the ball-joint and selector shaft axis on the change-gear lever is larger than that between the hinge-point which connects to the shift-lever, and that lever's pivot-point, on the selector shaft axis.

An additional transmission path, providing further gear ratio, is the objective of the design of Fig 33. The first input A has the conventional output while a second input B has a mechanism C associated with it, and a common gearwheel D, which allows incorporation of part of the change-speed train of the main gearbox to be used with the secondary transmission path which has output to a transfer drive at the base of the assembly.

Rapid actuation diff-lock, improved racing clutch and automatic drive selection

The differential housing in Fig 34 is turned by the final drive gear and contains a pair of auxiliary drive gears, engaging their respective pinions. These rotate in bearing apertures within the housing. A lockable part A is coupled to a differential part, turnable in one direction with respect to the housing. A locking element B also turns with the housing and a coupling mechanism C is located between locked and free elements to block rotation of the differential, operated by the adjustable member D. A cam mechanism E increases the coupling force.

Avoiding undue wear in the slipping of high performance car clutches is the objective of the design in Fig 35. The main, driver-disengageable, clutch is supplemented by a smaller, belleville-spring-loaded, clutch between the drive plate and the splined gearbox input shaft. The second clutch is arranged to slip at a higher torque than that of the main clutch, which is of a multi-plate carbon-carbon type. Because the high torques which cause slipping of the secondary clutch are only applied for brief periods, little heat is generated.

Power transmission to the front or rear axle of a vehicle can be made a function of the speed difference between the front and rear wheels but, thus far, the necessary mechanisms have been unwieldy. In the design of Fig 36, a compact drive splitting mechanism is devised offering considerable freedom of location on the underside of the vehicle. A trochoidal gear pump is mounted between a shaft and co-axially rotating cylinder, each connected respectively to front and rear axle prop-shafts. A piston slides in the cylinder in response to pressure generated by the pump to actuate a clutch which locks front and rear drive together.

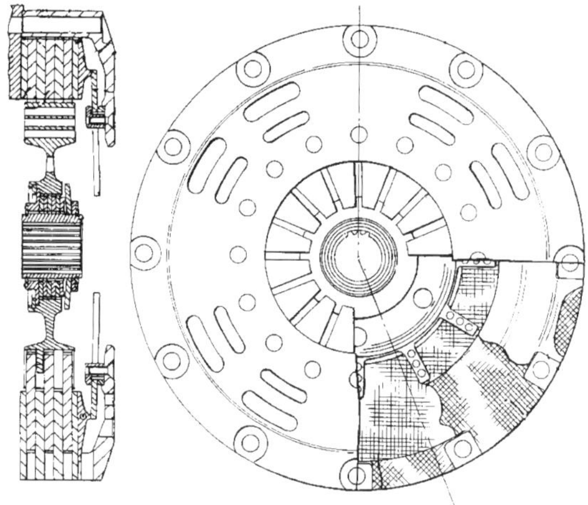

Fig 35: Improved racing clutch by Automotive Products (GB2294300A)

Easy-change CV-UJ assembly, Better performing plunge joint and better CV-joint packaging

Ease of replacement of constant-velocity universal joint assemblies is considerably improved by making the hub and joint housing members as separate units, Fig 37. The hub comprises a flange portion, to which the road-wheel is bolted, and an integral body supported by hub-bearings for which the body forms the outer race. Hub and joint housing each have interfacing conical surfaces as well as an interconnecting splined joint. A spigot on the joint housing has a projection on it which is deformed on assembly so as to lock on to the conical surface of the hub.

Attempts have been made to reduce the plunge resistance of tripode universal joints under high installed angles by the use of both inner and outer rollers mounted to the tripode trunnions, each having

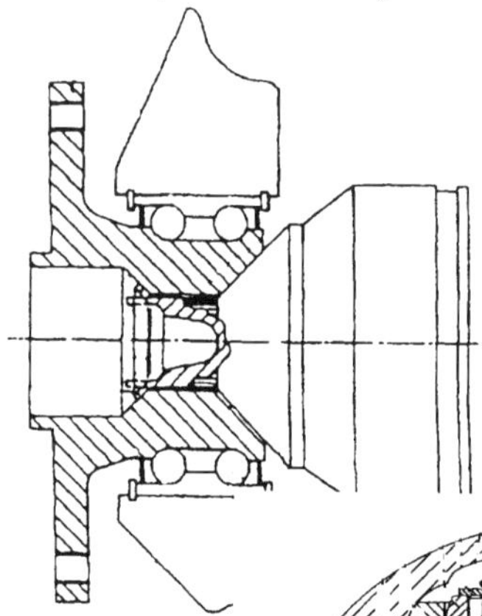

Fig 37: Easy-change CV-UJ assembly by GKN Automotive (GB2303428A)

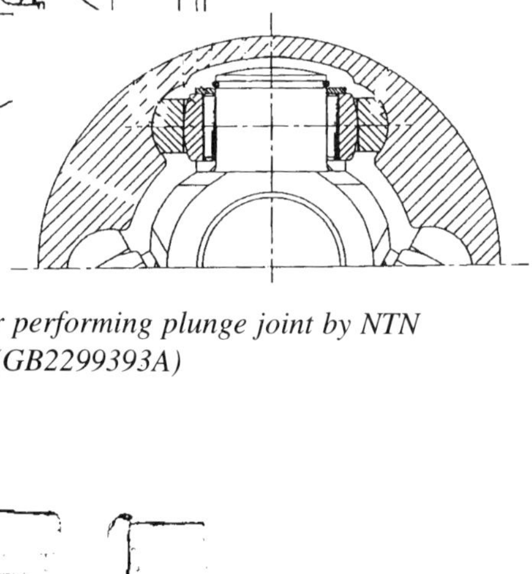

Fig 38: Better performing plunge joint by NTN Corporation (GB2299393A)

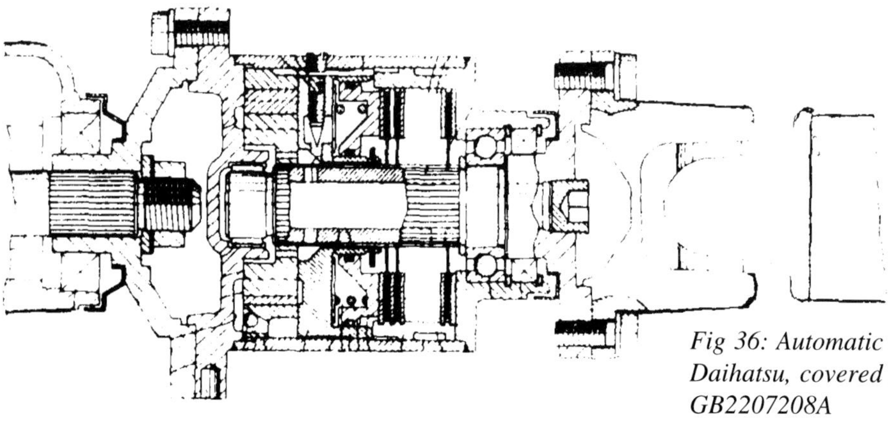

Fig 36: Automatic drive selection by Daihatsu, covered in Patent GB2207208A

spherical outer surfaces and cylindrical inner ones. Although induced thrust is lower than in earlier types further reduction is limited by the tendency of the outer rollers to follow the inner rollers because of friction between them with increase in both contact stresses and rolling resistance. In the arrangement of Fig 38, the inner surfaces of the outer rollers are such as to direct a load component parallel to the axis of each trunnion.

By offsetting the centre of the CV joint outwards of the centre plane of the front wheel bearing, several advantages have been achieved in this invention, Fig 39. A longer drive shaft, resulting, means less angularity on bump and rebound. Since the inner tracks of the wheel bearing lie axially adjacent to the outer running grooves in the outer member of the CV joint, bearing performance is unaffected by torque transmitted and the overall bearing diameter can be reduced. Finally the CV joint can be assembled from the outside into the hub member, which doubles as outer race of the CV joint and inner race of the wheel bearing.

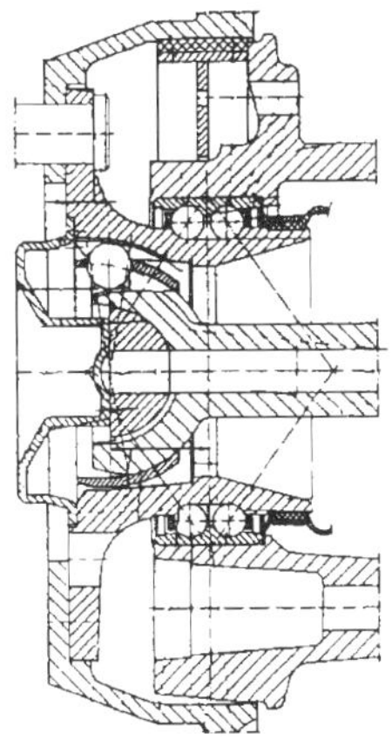

Fig 39: Better CV-joint packaging by Lohr and Bromkamp (GB2295365)

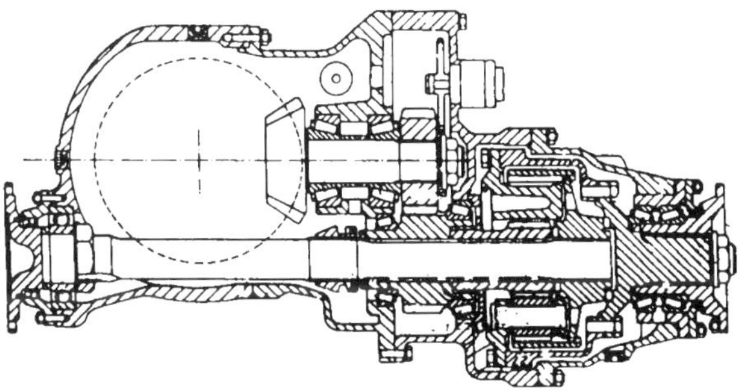

Fig 40: Balanced torque triaxle drive

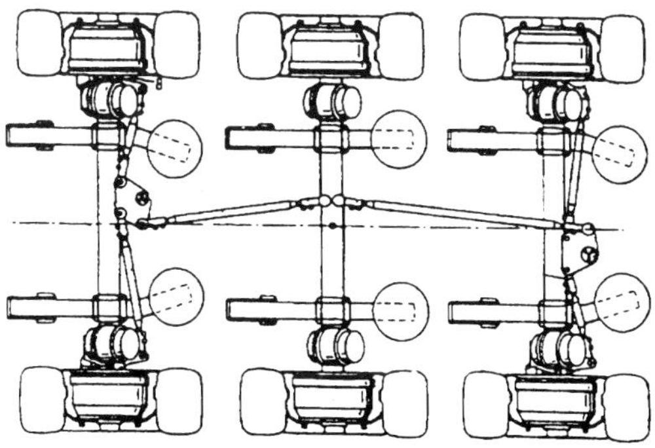

Fig 41: Sophisticated auto-steer for trailer by GW Mitchell (GB2315471A)

Balanced torque triaxle drive, sophisticated auto-steer for trailer

In order to overcome the problems of uneven torque splitting between the three axles of a triaxle drive, the design in Fig 40 by GKN Axles proposes the use of a spur gear epicyclic differential unit at the point of torque input to the bogie. This is designed to split torque unevenly, passing some 67% of the input torque down a through shaft inside another tubular shaft which conveys the remaining 33% to the first axle inter-wheel differential unit. The initial 67% can then be split evenly between second and third axles in the normal manner.

Automatic steering of the leading and trailing axles of a three axle trailer bogie is possible, in the design of Fig 41, in response to the relative movement of the main chassis with the bogie suspension sub-frame, a change in direction of the steering being effected with just one movement of the steering mounting plate. The movement of the steering mounting plate, relative to the main chassis between its outermost positions, is such that the line of action of the joining-ends of the primary link crosses the axis about which the main chassis rotates. The leading and trailing axles are pivotable with respect to the triaxle subframe, a linkage mechanism coupling each of the steerable wheels to the steering mounting plate on the main chassis. The mounting plate rotates between two positions which govern the direction of steer-angle.

Wheeled tanks challenge tracklayers and more efficient engine-coolant airflow

A challenge to tracklaying armoured fighting vehicles was made by Timoney Technology of Ireland in a presentation to Autotech, Fig 42. Arguing against the assertion that wheeled vehicles are more vulnerable than tracked ones in a combat situation, the authors argued that an 8 X 8 off-roader could often keep going with one wheel station out of action. A notable feature is overall height of just 1.3 metres to the turret ring, some 500 mm lower than other vehicles of this category. Axles are fitted with inboard hydraulic disc brakes and have differential lock facility. Patented wheel hubs have a one-piece stub-axle casting with lugs to carry the wishbone ball-joint bearings. The planetary gear system is also inboard of the stub axle and the ring gear is attached to and rotates with the hub housing — importantly permitting a small diameter rotary air seal for the centralized tyre inflation system. Suspension is of the double-arm type with two springs mounted on the lower wishbone, resulting in low overall profile and minimum hull intrusion. Suspension movement of 400 mm is supplemented by high deflection of the large tyres. Maximum speed for the 24–31 tonne combat weight vehicle is over 100 km/h.

Reduction in back-flow of air drawn through a vehicle engine coolant radiator, by the fan, is the objective of the design in Fig 43. The cowl which extends backwards from the radiator, to the fan, incorporates a C-shaped scoop facing the clearance around the fan. Back-flow is thus inhibited by turning it back on itself to create eddies within the fan-blade tip vortex region.

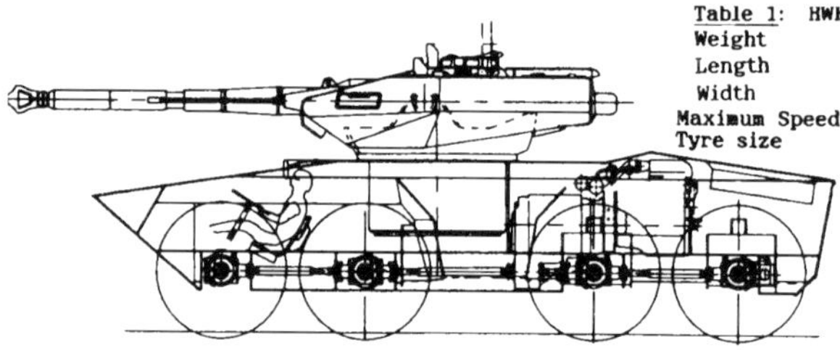

Table 1: HWFV Specification

Weight	26 t/31 t
Length	6542 mm
Width	2950 mm
Maximum Speed	112 km/hr
Tyre size	24R20.5

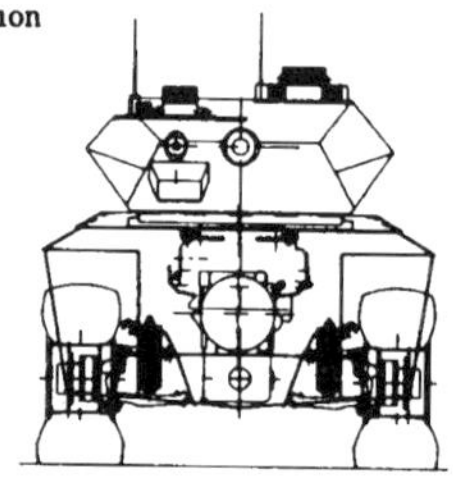

Fig 42: Wheeled tanks challenge tracklayers

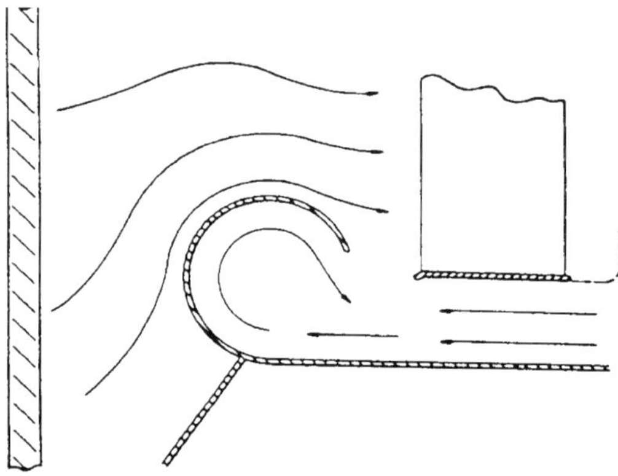

Fig 43: More efficient engine-coolant airflow by Rover Group (GB2311562A)

Suspension development

Transverse spring composite suspension and optimizing unsprung mass with wheel travel

A transverse single leaf spring made from polymer composites forms the lower control arms for the wheels and hubs in the suspension system design of Fig 44. Where the spring attaches to the central mounting it is in the form of an inverted dihedral and remains clamped as such even after the normal load of the vehicle has deflected the outer arms of the spring into an upwardly inclined attitude.

In the suspension configuration for driven wheels of Fig 45, part of the mass of the transmission unit is supported as sprung mass by the body. At the same time, vertical movement of the transmission is allowed so that the wheels can also travel vertically over an increased distance before the universal joints reach their angular limits. The suspension arm for each driven wheel is connected to both sprung and unsprung parts of the vehicle while drive to the wheels is through articulating shafts. During motion of the suspension, vertical motion of either of the wheels will be transmitted through the respective arm to produce a smaller vertical movement of the transmission unit.

Advanced suspension linkage

Active rear axle kinematics is the name given by BMW for its rear suspension design on the 850i coupe, Fig 46. As well as providing anti-squat and anti-dive, wheel-camber is adapted to reduce tyre strain at high speed and roll-steer is built-in which achieves load reversal reactions. As well as four wheel control links, a fifth joins two of the others to modify their relative motions. Two lower and one upper transverse link form the 'wishbones' while a trailing link provides fore-aft location. This link has a secondary connection via an intermediate link to the upper transverse link. Coil-spring and damper are

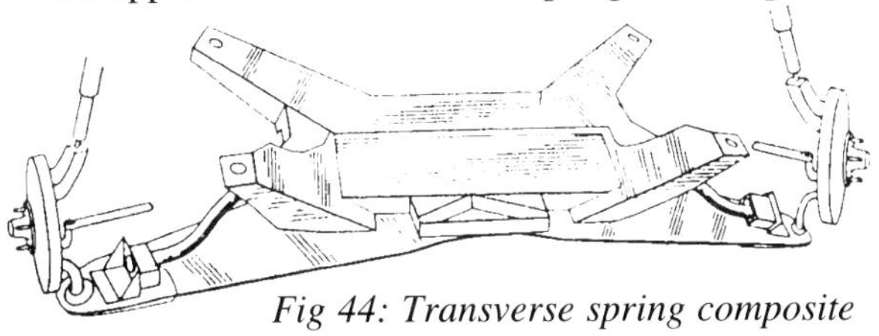

Fig 44: Transverse spring composite suspension by BV Booher in the USA, covered in Patent GB2202498A

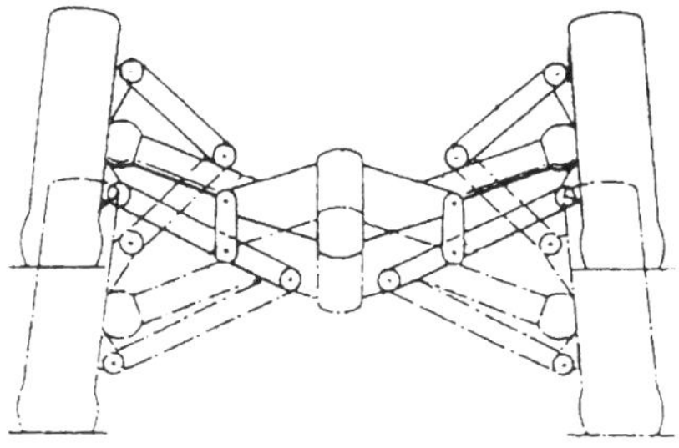

Fig 45 : Optimizing unsprung mass with wheel travel by Rover Group (GB2302066A)

mounted adjacently on the rear wishbone. The wheel-carrier pivot is above and ahead of the wheel-axis, to achieve anti-dive/squat. The intermediate (fifth) link is a strut which transfers braking torque on the wheel carrier, through the upper 'wishbone' to the trailing link. Because the strut lower pivot has less velocity than the wheel-carrier pivot on the trailing arm, during vertical travel of the wheel, this velocity is also seen at the upper wishbone. This causes the wheel carrier to rotate more slowly than the trailing link about the wheel-axis, and twisting of the wheel carrier on braking is suppressed without losing the horizontal compliance in the anchorage bushes of the diff-carrier at the root of the transverse links. The link also compensates for horizontal forces on the wheel either side of the vehicle so no swivelling of the whole suspension, with the diff-carrier, takes place. The elasto-kinetic function arises from the interaction of the anchorage bushes with the linkage geometry. In braking, for example, the two links forming the lower wishbone are subject to differential tension and compression. At the same time the anchorage bush of the trailing arm yields. Transverse link motions are such as to ensure that resulting motion of the wheel carrier is purely horizontal, without causing steer effects as the wheels move vertically. With the short wheelbase of the coupe compared with the saloon, and the comparatively high driving torques, smooth behaviour under load reversal conditions has to be ensured. This is helped again by the design of the trailing arm anchorage. At high lateral accelerations, wishbone links are distinctly skew to the body of the tilting vehicle. With load reversal caused by sudden removal of driving torque, the diff-carrier swivels due to the resilience of its mounts. Simultaneously it puts the pivot point of the upper wishbone link within it, horizontally, and the front/rear lower wishbone links are displaced by (fv) and (fh) respectively. Tilt of the wishbone also causes lateral displacements of the pivot points (rv) and (rh), causing understeer in both wheels. This effect is only

Fig 46: BMW 850i suspension

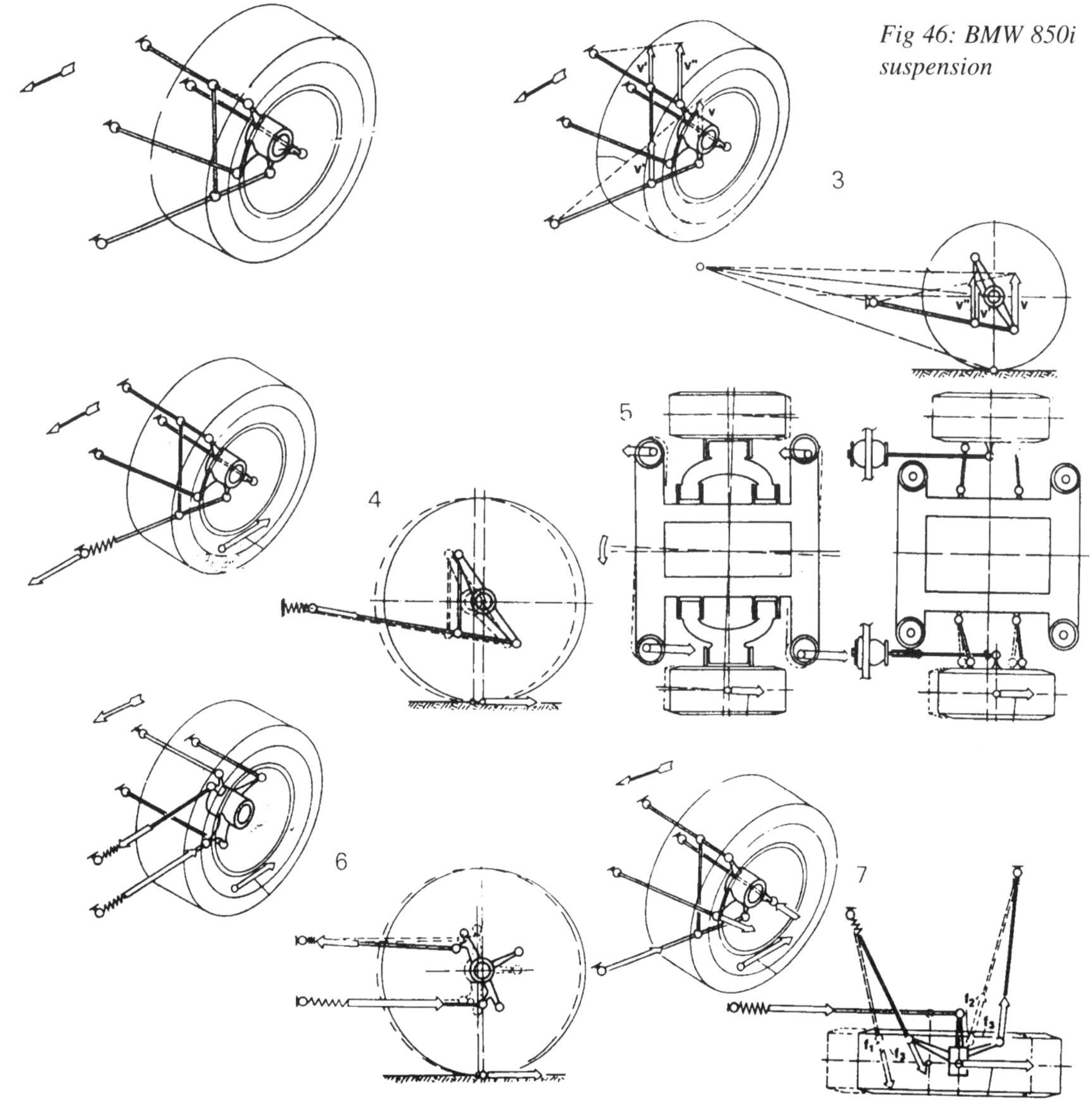

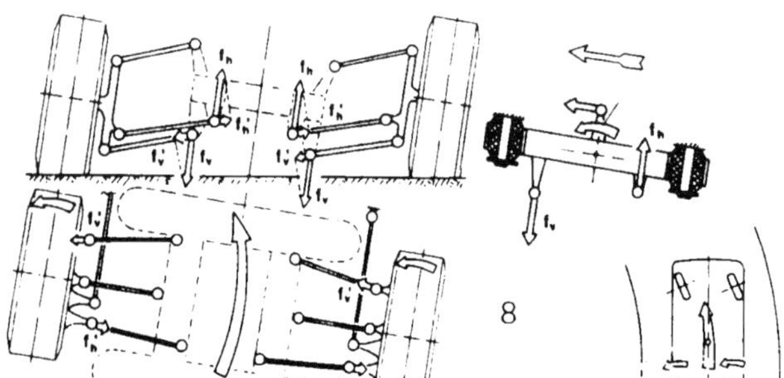

(1) Axle construction schematic

(2) Schematic of five-arm wheel suspension

(3) Function of anti-squat and anti-dive control

(4) Effect of longitudinal elasticity in the BMW integral rear axle suspension

(5) Effect of braking moment on conventional wheel suspensions

(6) Effect of one-sided longitudinal forces on different wheel suspensions

(7) Distribution of forces on wheel-controlling parts of the BMW integral axle when the brakes are applied

(8) Processes occurring in the BMW integral axle when the load is suddenly reversed

achieved because the suspension of the trailing link is independent of that of the diff-carrier from which the transverse links pivot.

Improved suspension geometry control, avoiding suspension compliance conflict/steer fight in braking/accelerating and smoother action for suspension strut

The desire to change from rigid beam axle to independent suspension on vehicle categories such as 4X4 sports/utility vehicles may be restricted by concerns for loss of accurate suspension geometry, particularly when an off-centre differential is required to clear an engine sump and unequal length drive-shafts are involved. In Fig 47 a compact form of IFS is proposed. Tubular members which surround the drive-shafts are pivotally connected to the hub carrier units and chassis-mounted differential casing in a 'wishbone' arrangement. The wishbone length is maximized, by the position of the pivots, to give better geometric compromise.

A torsion bar front suspension system for a vehicle which is simple in construction, and yet does not result in bending of the torsion bar when the vehicle is subjected to lateral loading, has been achieved in the design of Fig 48. By ensuring the torsion bar axis is at a small angle to the axis of the lower suspension arm fulcrum and laterally displaced from it, appropriate values of La, Lb, Lc, S and θ can be chosen so that sufficient suspension compliance is available in the fulcrum bushes without any bending moment being applied to the torsion bar. The spring plate connecting the bar to the arm is slender enough to be flexible in the fore-aft direction.

Reduction of the steer effect in braking is claimed for this suspension configuration, Fig 49, for the steered wheels of vehicles. Compliant bushes, B and B', normally permit the suspension arm to rotate about an axis A_m as the road wheel rises and falls but this can change to either A or A' under appropriate loading. During braking or acceleration the angle θ and the compliance of the bushes allows the suspension arm to turn in the horizontal plane and compensate for steering displacement of the road-wheel hub.

Reduction in ride harshness is achieved in this independent wheel suspension linkage configuration, Fig 50. The centre portion of the spring-damper strut is used as an anchorage for the anti-roll bar, a link from which is angled to the axis of the suspension strut in such a way that transverse reaction loads from the ground-level cornering forces are balanced by loads induced in the anti-roll bar so that sideways deflection of the damper tube-and-rod is minimized.

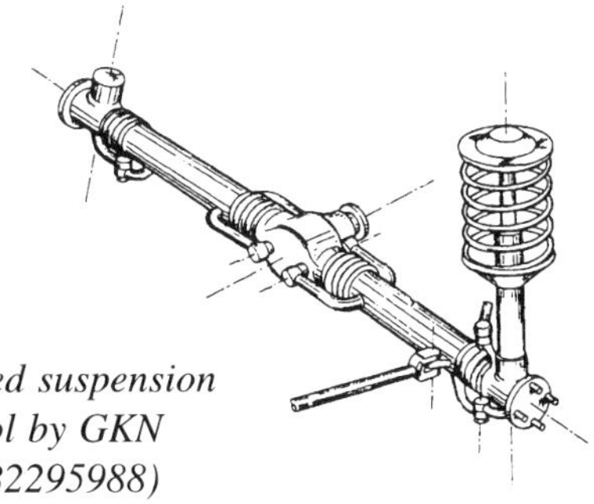

Fig 47: Improved suspension geometry control by GKN Automotive (GB2295988)

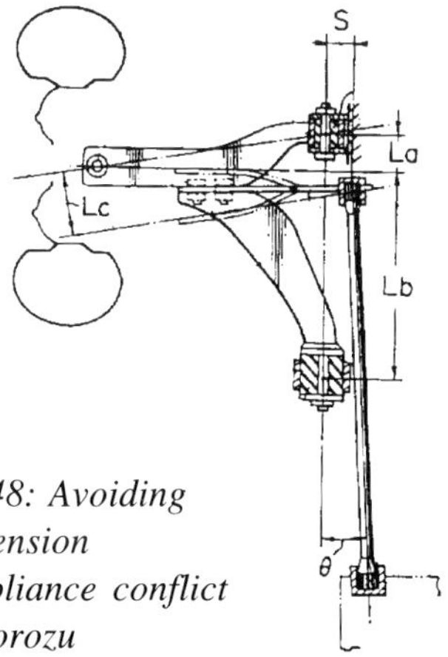

Fig 48: Avoiding suspension compliance conflict by Yorozu Corporation (GB2294668A)

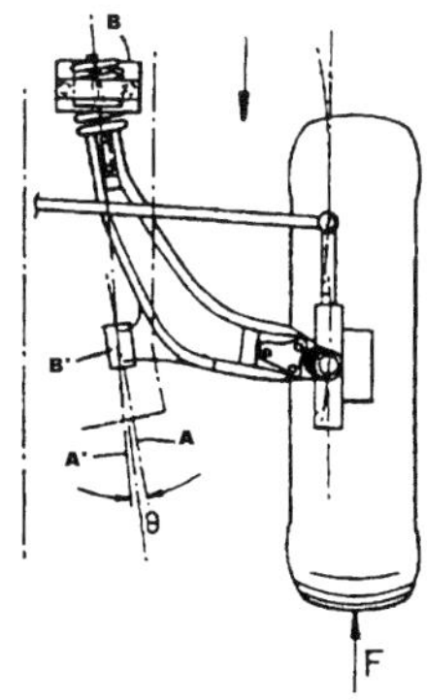

Fig 49: Avoiding steer fight in braking/accelerating by Rover Group (GB2311499A)

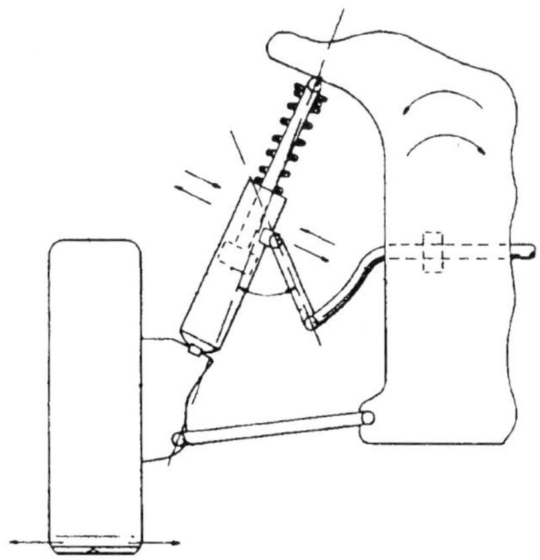

Fig 50: Smoother action for suspension strut by Rover Group (GB2306411A)

Fluid-actuated anti-roll control, non-stick front forks and freer moving bogie suspension

Lower weight and greater compactness is said to arise from the use of this fluid-actuated anti-roll system, Fig 51, which claims to achieve a high degree of lateral stabilization. Hydraulic actuators act on the upper ends of telescopic suspension struts; the actuator pistons are hydraulically linked in a way that motion in one is mirrored in the other. Excessive load on one suspension unit is thus transferred to the opposite side unit varying according to vertical suspension travel.

Overcoming the problem of high friction in seal systems used on the front forks of motor cycles is the subject of the design in Fig 52. The usual telescopic fork arrangement gives way to a twin-wishbone suspension arrangement. Universal joints at the outer ends of the wishbones allow steering motions as well as vertical suspension travel. A pivoting triangular framework is also used to transfer steering torque from the handlebars to the forks which support the front wheel.

Less chance of bearing seizure at opposite ends of the walking beams of a tandem bogie suspension results from this design, Fig 53. Forked axle brackets connect to the beam ends via exposed bolts with bushes and convenient stroke mechanism — which also lowers the vehicle to the ground on electric jacks. Steering involves the use of a highly flexible steering wheel including a fork or ball used to give the user a stable support. A voice command system is used for a number of the secondary controls.

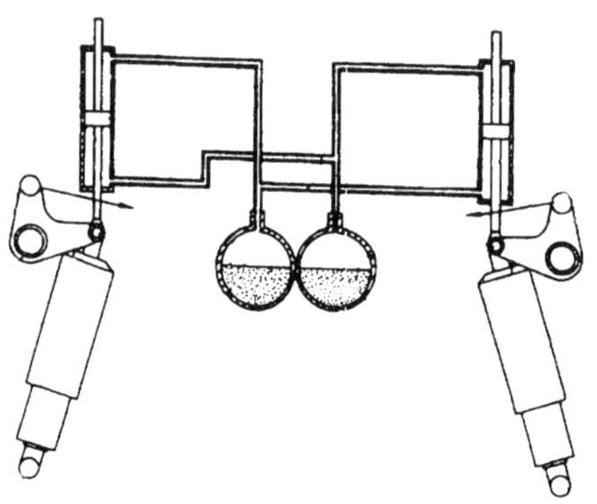

Fig 51: Fluid-actuated anti-roll control by AMG Motorenbau & Entwicklung (G B2315716A)

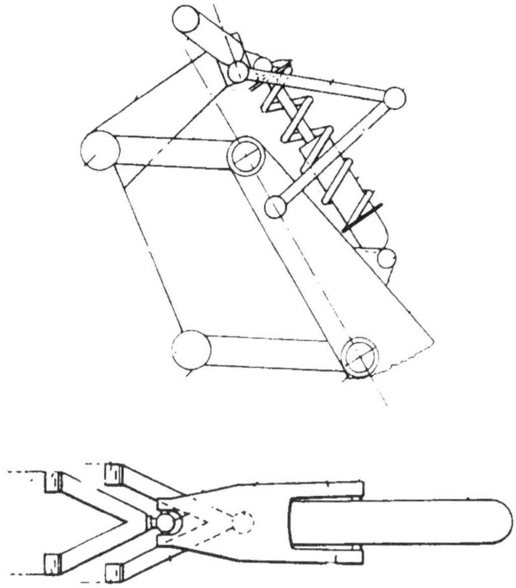

Fig 52: Nonstick front forks by NH Hossack (GB2207645A)

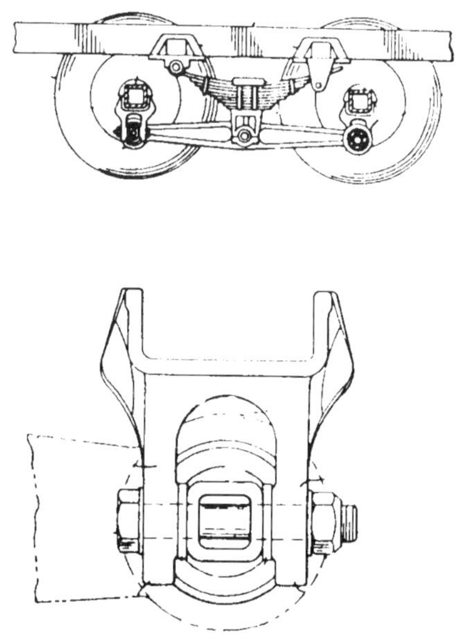

Fig 53: Freer moving bogie suspension by the US Boler Company in Patent GB2192596A

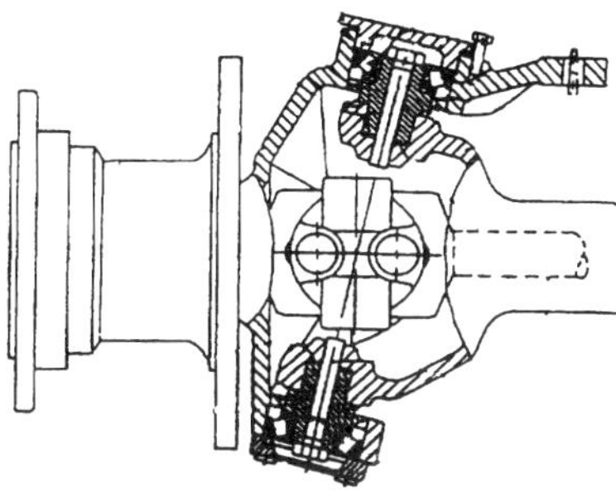

Fig 54: Easy prestressing for swivel pin bearings by Claas KGAA (GB2316050A)

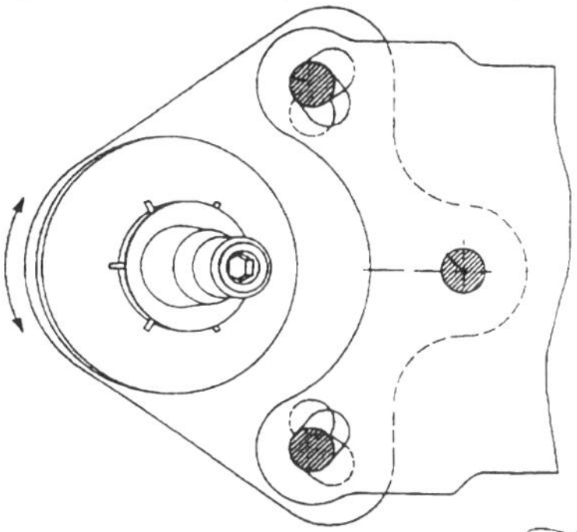

Fig 55a: Uncomplicatec steering-castor adjustm by Daimler-Benz (GB2315469A)

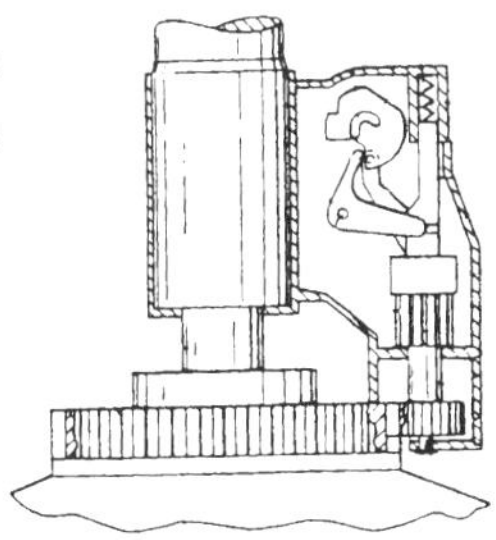

Fig 55b: Steering column immobiliser by Marshall Wolverhampton (GB2303108A)

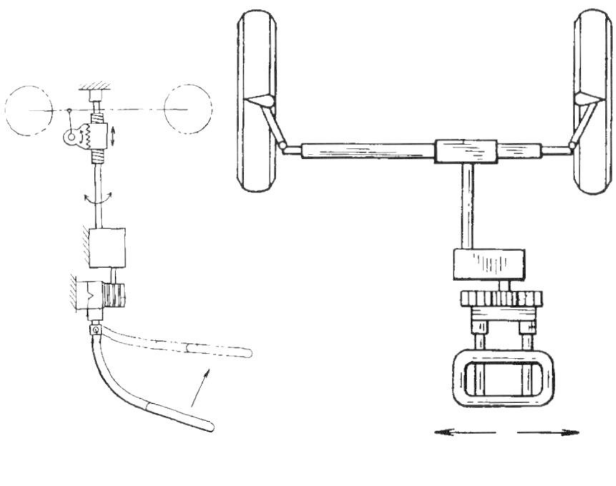

Fig 56: Tiller steering by Kazunaga Morimoto in GB2293143A

Easy prestressing for swivel pin bearings, uncomplicated steering-castor adjustment and a steering column immobilizer

An unusual arrangement of steering swivel pins on this steer-drive axle for industrial vehicles, Fig 54, allows prestressing of the taper roller bearings to be carried out without difficulty. Outwardly facing swivel pins are used in a design which is said to be less expensive than normal arrangements of comparable performance. Forked ends of the axle body, containing the pins, engage with forked ends of the hub carrier. Each swivel is in the form of a circular pin provided with central collar and constrained at one end in a bore of the axle body fork. One of the corresponding bores in the hub carrier fork has an internal thread into which a cover is screwed with an inner collar bearing against the outer race of the taper roller bearing.

The problem of providing degrees of castor adjustment without weakening the integrity of wheel location, or involving heavy expense, is alleviated in the design of Fig 55. A steering ball joint incorporates a wide flange which overlaps the transverse link to which it is connected and there is a connecting dowel which pivots one part with the other. The parts are clamped together through cleverly profiled holes for the securing bolts. These allow three different angular positions between the joint and the link so as to provide stepped adjustment of castor with secure fit when the clamping bolts are tightened. A comparatively modest torque can be used to lock a steering wheel/column against rotation by using a high ratio gear pair between the column and the clutched locking device, Fig 55b. A member which has coded input response to a key or other immobiliszer controls the action of the clutch, which is based on teeth cut into the shaft of the pinion of the gear pair.

Back to tiller steering and an anti-dive motor cycle suspension?

As the concern for driver safety increases, the very concept of a circular steering wheel attached to a rigid column will be questioned. This is the view of an invention, Fig 56, which replaces the wheel with a laterally moving handle shaped to optimize protection of the thorax region, on impact. The handle is connected to a rack-and-pinion by deformable arms, and an hydraulic valve, which transfer the side-to-side steering motion into swivelling of the road wheels.

Two fore-and-aft suspension arms form a four-bar linkage which gives anti-dive suspension geometry to the front wheel of a motor cycle when braking. In the arrangement of Fig 57, geometry is such as to allow positioning of a steering arm which articulates without steer fight on vertical travel of the suspension.

Control strategy for 4-wheel steer, self-steering in forward and reverse and a necked hub member

Lotus have released more information about the active rear steer system they have under development — which they say is an extension to their active suspension work using closed loop control strategies. This allows steering to respond to both driver inputs and external disturbances. Vehicles thus equipped would therefore respond to crosswinds or split braking in a controlled manner. There is also less sensitivity to vehicle load and tyre performance — unlike open-loop and predictive-type strategies which can only be optimized for specific operating conditions. The system, Fig 58, is said to operate at normal power-steering system pressures and work is under way on production-intent actuators and controllers.

To overcome the limitations of existing self-steering axles, this design by an Italian inventor, Fig 59, comprises a tubular member which slides over the axle while road wheels pivot directly on the frame of the vehicle. Two tie-rods positioned on either side of the axle have their ends linked eccentrically to the wheel hubs. Thus, while reversing or moving forward, actuators are caused to restore the tie rods and therefore the wheels to their neutral position.

Improved heat insulation for the tyre and greater space for the brake actuator cam arise in this hub design, Fig 60. The tapered hub section and necking between the bearings produce the desired configuration.

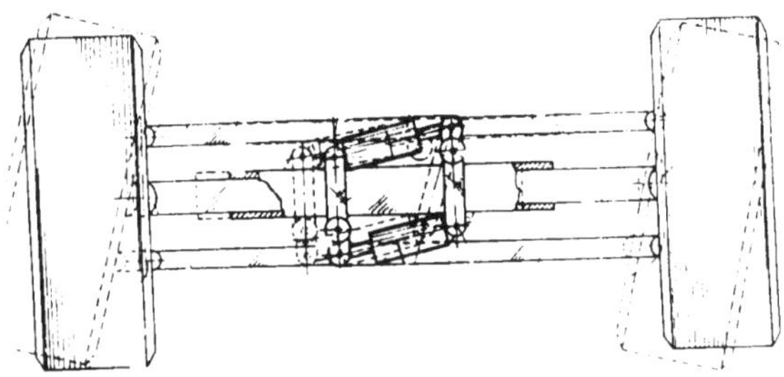

Fig 59: Self-steering system by Sansavini Bertozzi, covered in Patent GB2202811A

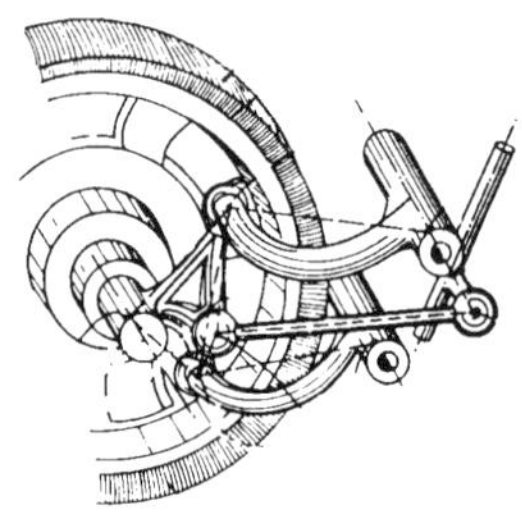

Fig 57: Anti-dive suspension by Elf France, in Patent GB2201642A

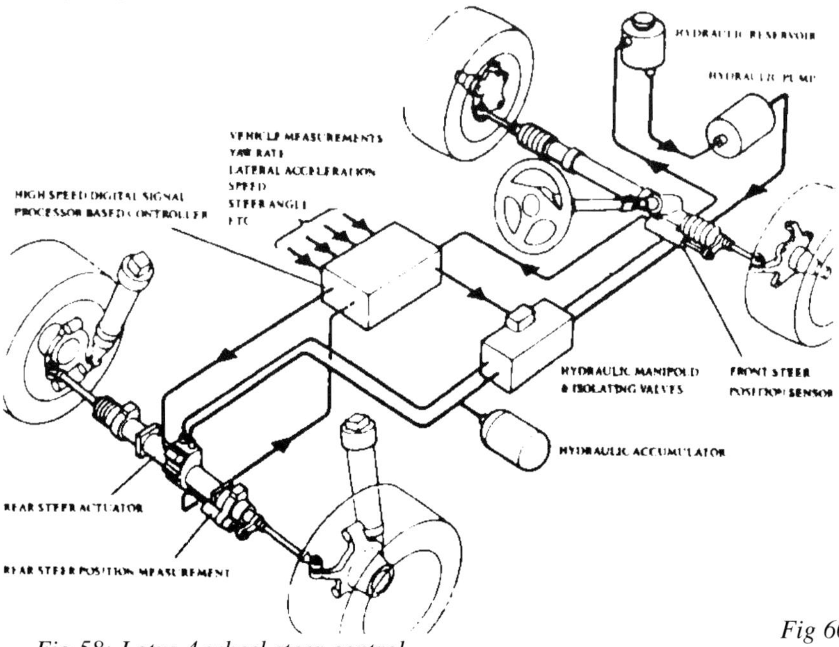

Fig 58: Lotus 4-wheel steer control

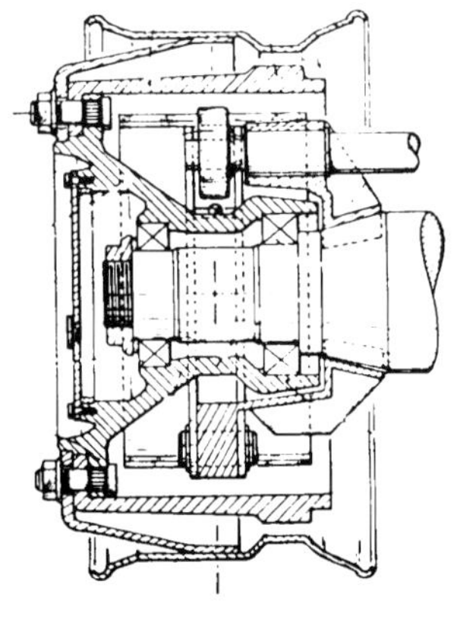

Fig 60: Necked hub by Trailor in Patent GB2192959A

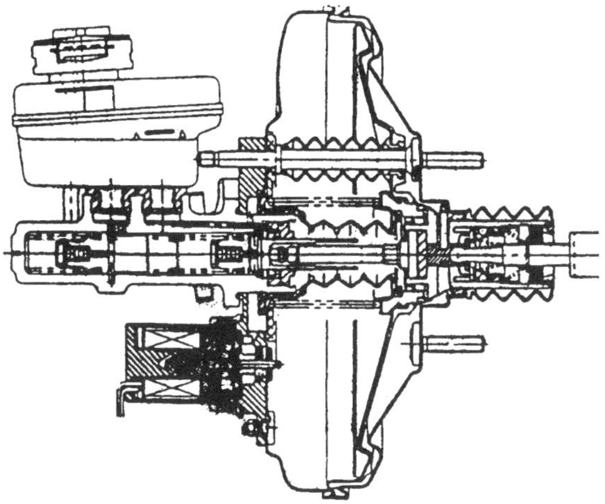

Fig 61: First electronic booster

E-Booster	Modlfied Basic Unit LSC 80 (10")	
Master Cylinder	22.2 CV/CV	
Function I	*Automatic Braking*	
Function II	*Automatic Booster Operation*	
	Input Rod Force	-
	Input Rod Speed	-
	Kneepoint	80 bar
	Response Time	200 ms
	Release Time	180 ms
Function III	*Variable Ratio - Pedal operated*	
	Input Rod Force	1400 N
	Input Rod Speed	0 - I00 mm/s
	Kneepoint	125 bar
	Response Time	170 ms
	Release Time	180 ms

Fig 62: Initial booster + proportional valve specification

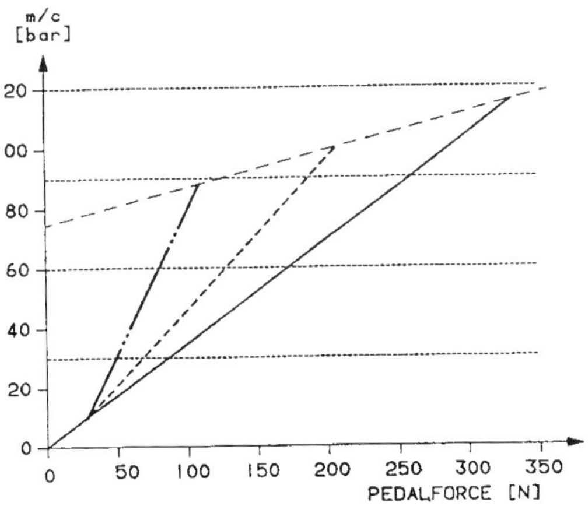

Fig 63: Variable ratio characteristic (full-line base ratio; dashed line medium ratio; chain line high ratio)

Braking systems

Electronic brake actuation system

Lucas Varity have determined that the typical driver is usually inhibited against applying optimum pressure to effect an emergency stop. The company are therefore advocating the use of an electronic actuation system (EAS) based on the electronic booster seen in Fig 61 that has been replacing the fast-vacuum booster, fitted to many cars, and which was limited to providing assistance only at a fixed ratio and often reacted too slowly in emergency.

A proportional control valve has been developed such that the output stroke is proportional to the input control current signalled through the brake pedal. A control loop is partially closed via an ECU which measures output/input force ratio and compares it with a vehicle-specific algorithm. An initial specification was as shown in the table of Fig 62 and involved electronic booster control superimposed on a mechanical booster. Performance curves for a series of possible ratios are in Fig 63.

A parallel development of the long-stroke (LS) booster meant that the benefits of electronic control could be applied to heavier cars and replace tandem booster arrangements which tend to be restricted to either high-performance cars or heavier van derivatives. The company is concentrating on the development of the LSC 115 T tandem power unit, Fig 64, which has an integrated electronically proportioning solenoid control valve. Shown enlarged in Fig 65, this valve is seen to be an electrical analogue of the proven servo control valve, but with input/output ratios compared electronically rather than by rubber reaction disk.

The company believe the full potential of such systems can only be realized when legal insistence on mechanical back-up is dropped and 100% reliability of the electronics can be guaranteed. Money saved could be used to incorporate other electronic features, Fig 66, such as ABS and engine management. The ECU would be able to compare actual vehicle deceleration with the theoretical value based on a given booster output force, and thus forewarn of brake-fade. The ECU could also relate engine torque to achieved acceleration for assessing loading on the vehicle so that the system is informed of prevailing road gradient. In integrating ABS, the booster would provide the energy source and the combined system

ECUs could help to create the next stage of automatic braking, in controlled traffic conditions, involving intelligent cruise control. Functions such as traction control, hill-hold and variable pressure boost for ABS could be incorporated.

The first stage of the project will be seen on some Mercedes cars early in 1996 as 'brake assistant'. For this application the vacuum chamber of a conventional vacuum booster is equipped with the Mercedes position sensor. This tells the ECU that the booster piston has travelled a given distance in less than a pre-defined time and switches on the solenoid valve. Atmospheric air then enters the booster working chamber to amplify driver effort with maximum available servo power. Without this device full decelerative advantage with ABS is lost. In an active stability control feature, the ECU helps to control side-slipping out of a corner by selectively applying braking force to one of the outside wheels, generating braking force from the booster and modulating it by the ABS. In stage two, hill-hold will be added such that brake pressure will automatically be applied when a vehicle stops on a hill and automatically released when it pulls away. This will involve the proportional solenoid valve and the booster generating the necessary braking pressure by metering the air intake into its working chamber. Such metering is also essential for automatic braking and adaptive cruise control, ACC. Current cruise control, it is argued, can suffer a lack of sufficient engine deceleration on steep gradients such that the vehicle gains speed; EAS can supply precisely metered braking pressure to prevent this. Beyond this, EAS can generate pressure for ACC so that a vehicle is kept at a safe distance from the one ahead, without having to rely on engine-braking which might prove inadequate. For stage three the problem of varying brake pressure requirement with degree of vehicle payload is tackled. Here, not only is brake booster pressure metered but also the pressure level boosted proportionally to the pedal force (measured by a sensor on the piston rod) applied, so that constant deceleration is obtained with a constant pedal force irrespective of payload.

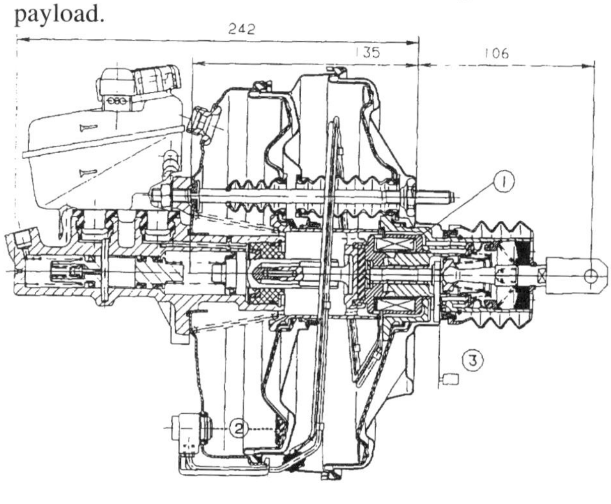

1 Coil for Solenoid
2 Travel Sensor
3 Functional Switch

Fig 64: LSC 115 T with integral solenoid valve

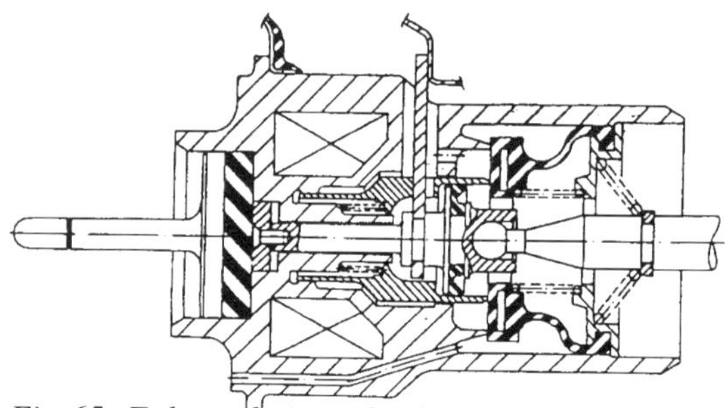

Fig 65: Enlarged view of valve

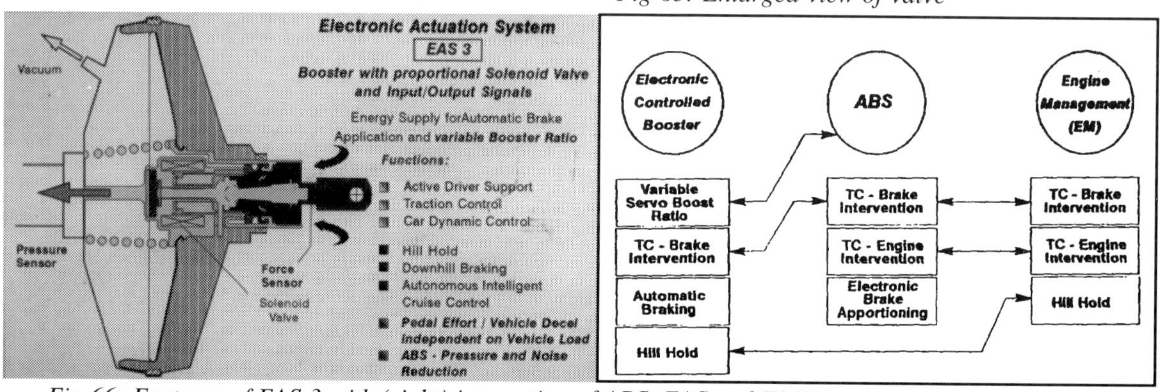

Fig 66: Features of EAS 3 with (right) interaction of ABS, EAS and EMS

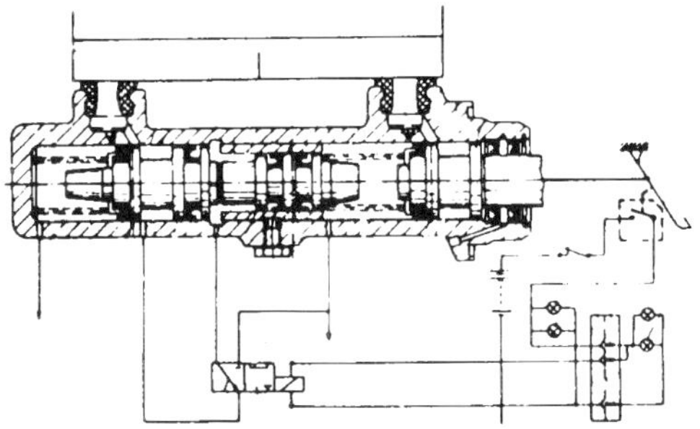

Fig 67: Brake apportioning for solo or coupled mode by Daimler-Benz from Patent GB2146720A

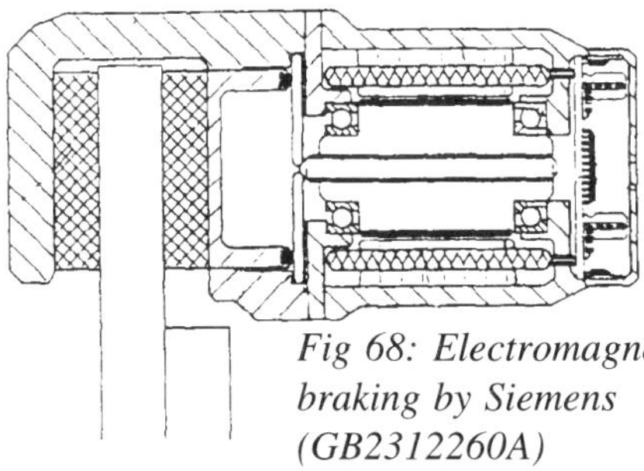

Fig 68: Electromagnetic braking by Siemens (GB2312260A)

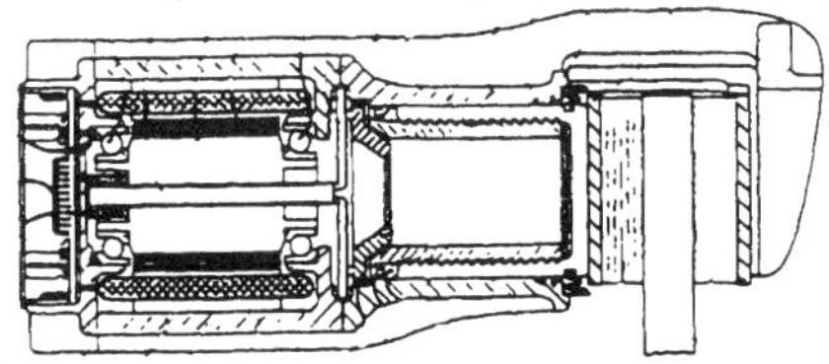

Fig 69: Auto-adjusting electro-brake by Mercedes-Benz (GB2315527A)

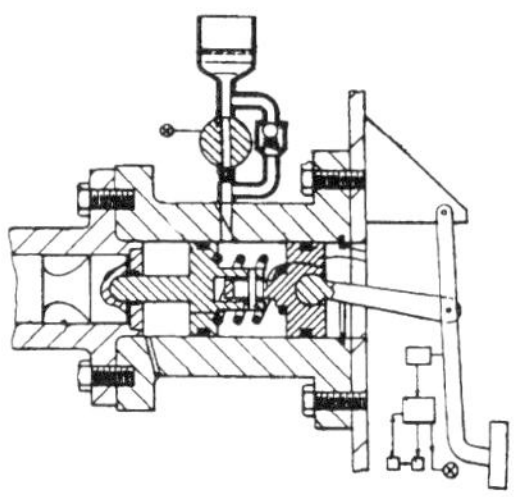

Fig 70: Fail-safe electric brake actuator by Delphi Automotive (GB2305478A)

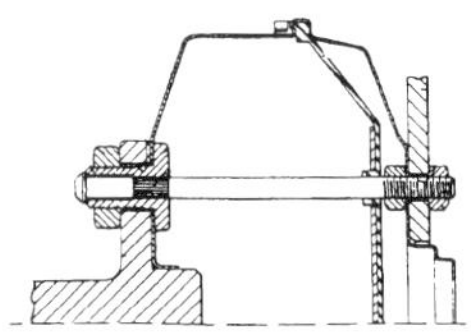

Fig 72: Brake booster by Delphi France (GB2312718A)

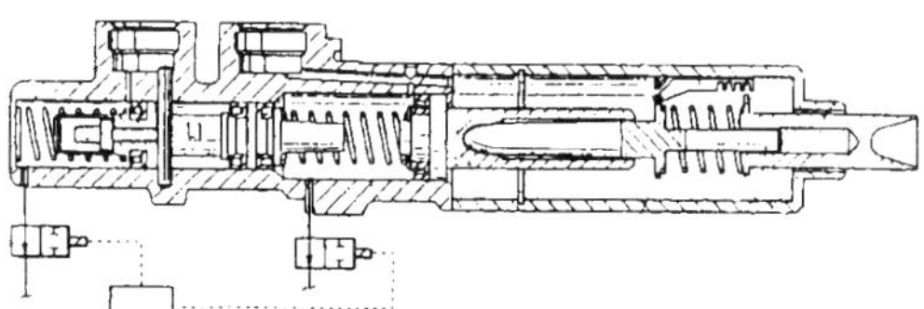

Fig 71: Fail-safe brake actuator by Mercedes-Benz (GB2307529A)

Brake apportioning for solo or coupled mode, electromagnetic braking, auto-adjusting electro-brake and fail-safe electric brake actuators

A car braking master-cylinder is 'tandem-like' in configuration and the pressures in front/rear wheel brake chambers apportioned according to vehicle load. This effect is altered according to whether or not a trailer is coupled, in the design of Fig 67, by means of a 3/2 way solenoid valve.

A means of applying braking selectively to individual wheels is provided in the design of Fig 68. The brake actuator is rigidly mounted on the disc-brake calliper and uses electromagnetic operation. The actuator comprises an electric motor, the armature spindle of which is threaded so as to produce translatory motion/force against the pad pack-plate for rotary motion of the armature.

An electrical brake actuation system is covered in the design of Fig 69 which has the advantage that braking fluid does not have to be relied upon to balance out brake-pad wear. The electromagnetic actuator comprises a spindle axially driven by an electric motor via a threaded roller drive with the axial force boosted by a lever mechanism in transferring it to a pressure plate acting on a brake piston. A mechanical correcting device is incorporated for pad-wear balance.

An indication of the applied braking force is supplied to the driver in this electric, or electro-hydraulic, brake actuator which is backed-up by hydraulic brake application upon detection of failure. A sensor behind the brake pedal detects any electrical faults and triggers the adjacent hydraulic system into actuation, while providing a reaction force at the pedal. The hydraulic unit has a through bore, to the master cylinder, containing two pistons which can slide relative to one another, Fig 70.

Only a relatively small movement of the brake pedal is necessary in the design of Fig 71, in the event of hydraulic pressure failure, before mechanical coupling of piston rod and piston takes place within the master cylinder. Mechanical coupling is achieved between two co-axially sliding members when the length of the piston stroke exceeds a threshold value. A locking catch, at A, is spring-loaded in the open position normally but is closed by the action of a cam when the stroke exceeds the threshold.

Easy-assemble brake booster, compact integration of clutch servo actuator parts and set-speed down-hill braking

In the Fig 72 design a booster housing having front and rear walls is held together by through tie-bolts which also form the mountings to a vehicle bulkhead and a means of securing the brake master cylinder to the housing. At the master-cylinder side of the booster, the outer wall of the housing is assembled against a splined collar which also gives the booster a measure of controlled collapse on impact.

A compact clutch actuator, which can be easily installed, is the essence of the design of Fig 73. Pressurized fluid is used to actuate the clutch and the device incorporates a pump and drive, to the left of the diagram, feeding a pressure reservoir within the fluid container, to the right of the diagram. There is a valve, shown centre, between it and a servo cylinder. Between reservoir and container there is a preset passage-constriction formed, by their integration, to restrict and control fluid flow.

Problems of automatic down-hill 'active' braking have been identified as due to uneven axle loadings due to the shift of vehicle centre-of-gravity on the incline. This causes wheel-locking in the lightly laden axle and consequent undue use of the ABS. In the design of Fig 74 the braking effort distribution is such that only the highly laden axle is braked in normal cases. Only when deviation from the set speed or slip at the braked wheels exceeds a certain value will the brakes be applied on the lightly laden axle.

Easy height-adjust for trailer coupling and auto-adjusting towbar

Existing adjustable-height trailer couplings are unwieldy to adjust because of the heavy knuckle forgings, incorporating the hirth gears, being awkward to remove and replace for resetting the engagement of the gears. In Fig 75 a parallelogram linkage is formed out of the front and rear supports of the coupling. The hirth gears are at the lower end of the front support and a gas spring damps the movement of the mechanism during adjustment. The geometry is such that up to half of the total towing force is taken up by the rear member.

Loss of payload carrying area on a drawbar trailer truck combination arises from the need to have an inordinately long drawbar so that prime-mover and trailer do not foul one another on sharp cornering manoeuvres at low speed. In the design of Fig 76, the drawbar is made to increase in length on acute angled corners and in normal driving can be set much closer than is usually the case, to the advantage of available load space within a maximum overall length of combination. A tractor to trailer angular measuring device on the tow bar has an arm that engages a stop displaced from the axis of the towing hook. This signals a control unit which extends a hydraulic cylinder in the telescopic drawbar.

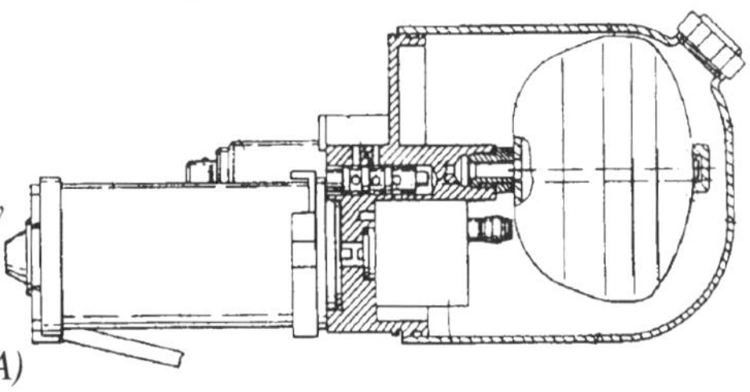

Fig 73: Clutch actuator by Fichtel & Sachs (GB2301404A)

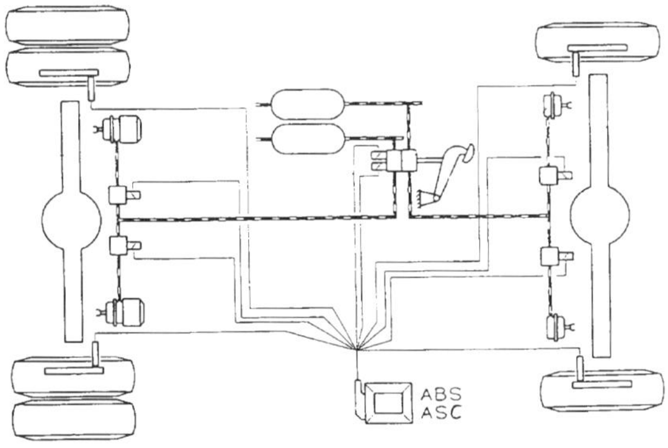

Fig 74: Set-speed braking by Wabco GmbH (GB2317930A)

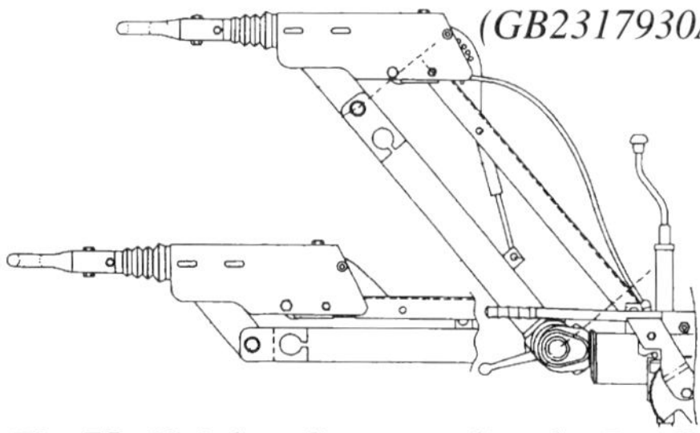

Fig 75: Height-adjust coupling by Bradley Doubledock (GB2318107A)

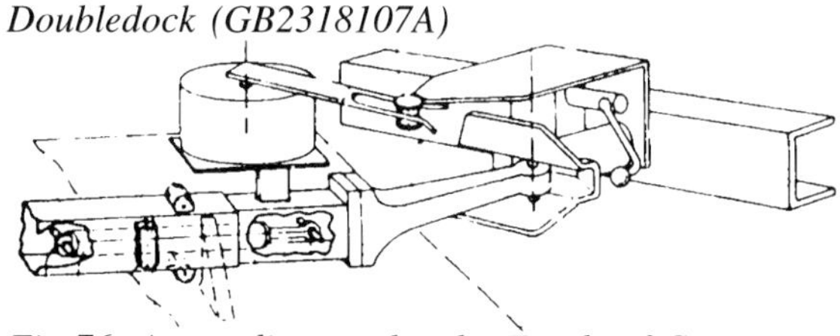

Fig 76: Auto-adjust towbar by Fruehauf Corporation

Chapter 7:
System developments: body structure/systems

A decade of patents specifications and company releases provides an insight into automotive product development up to the period of new designs covered in the first five chapters. In the area of body structures and systems, the trends to reducing vehicle weight and increasing occupant safety are clear in the examples which follow.

Structuralsystems

Controlled collapse

Safety and weight reduction continue to be the main drivers of structural developments. Better control of frontal collapse is receiving particular attention. Avoidance of bending collapse of the front longitudinal members of a vehicle body in impact is the objective of the design in Fig 1. Because such members usually involve a curved section to provide clearance with mechanical systems it is difficult to prevent the onset of bending collapse, under end load, prior to the desired controlled longitudinal collapse of the box sections. While vertical beads are formed into the vertical walls of the box-members to induce longitudinal buckling, the claimants have found that inclining these at an angle is successful in cancelling the bending moment induced by the front end-load. Asymmetric side beads have also been used in vehicle body front-end longitudinals to control crush characteristics in frontal impact. The beads act as fold initiators but are not fully effective as the corners of the members take greater load than the sides. In Fig 2, asymmetric corner divots are introduced as triggers and the arrangement particularly suits larger and heavier vehicles which have thick gauge longitudinals. Because the divots help to reduce the bending moment of inertia at the cross sections where the beads are, this proposed design relies only on specially configured corner divots — with long and short axes arranged to pair with one another as seen here. Two-stage energy absorption is another approach, avoiding premature firing of the occupant-protection airbags . This is the object of the bumper configuration in Fig 3. Two-stage impact energy absorption is obtained by a soft outer bumper, supported on relatively weak collapse-tubes, deforming under minor impact without firing the airbag. Heavier impact causes the more rigid inner bumper to collapse on its support tubes, with sufficient decelerative force to fire the airbag. Structurally efficient longitudinals are also needed behind the front-end. An improved structural efficiency for the platform underbody of an integral-bodied vehicle results from the use of a novel design of side-rail in Fig 4. The side-rail is a built-up beam having channel section flanges and a form of corrugated web in between. Preferably a pair of rails are sandwiched between upper and lower floor panels with the whole structure supporting a pillar-arch and cowl.

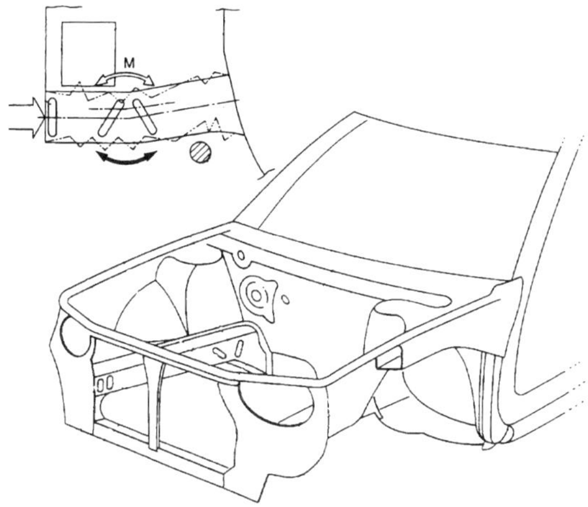

Fig 1: Patented front-end by Fuji Jukogyo (GB2318552A)

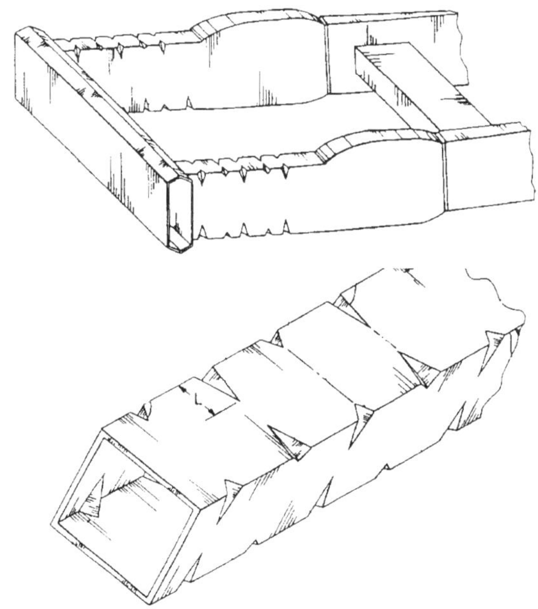

Fig 2: Patented controlled collapse system by Ford (GB2295358)

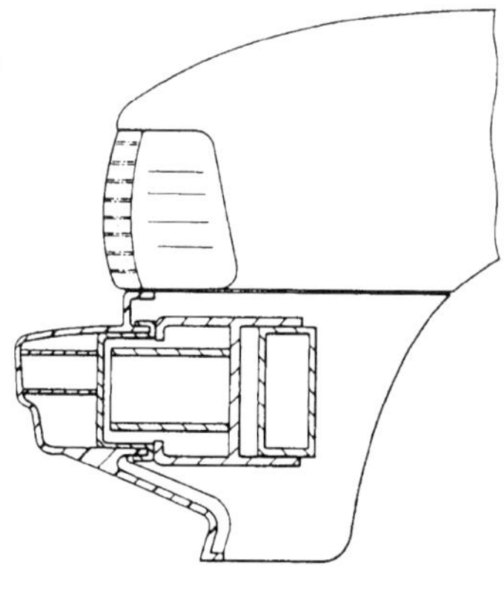

Fig 3: Two-stage energy absorption by Rover Group (GB2308100A)

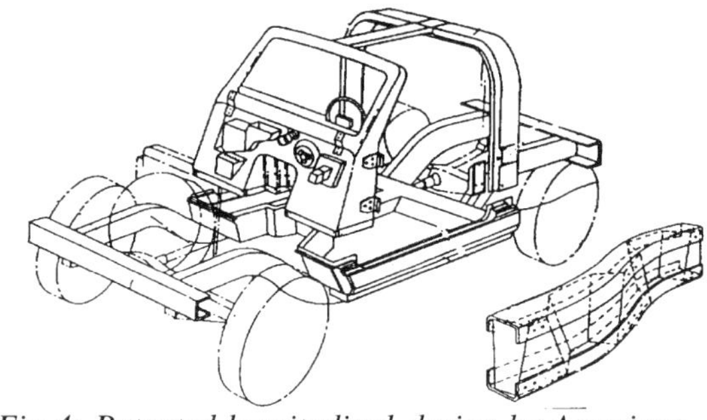

Fig 4: Patented longitudinal design by American Motors (GB 2187686A)

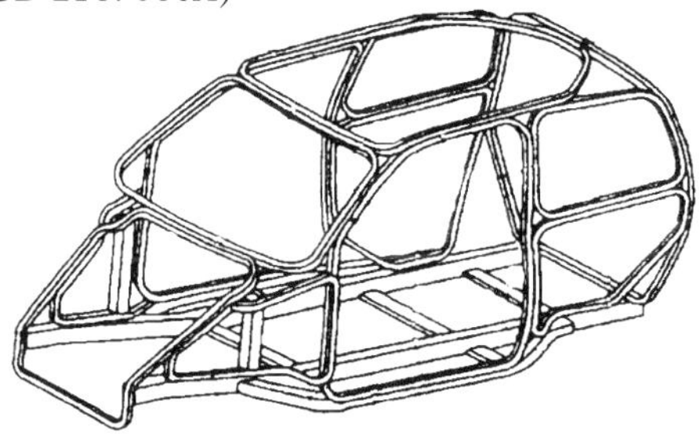

Fig 5: Patented body shell construction by Honda (GB2305639A)

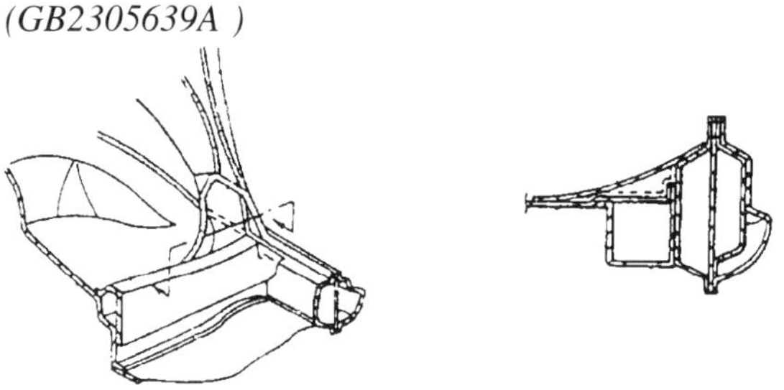

Figs 6/7: Patented side impact protection structure (and section) by Fuji Jukogyo (GB2306922A)

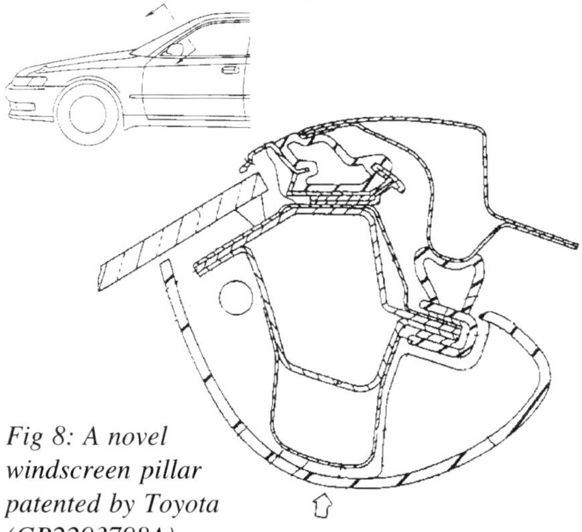

Fig 8: A novel windscreen pillar patented by Toyota (GB2293798A)

Body shell integrity

In giving structural integrity to the whole body shell frame a number of new approaches have been tried. In this body-shell made from rings, a light alloy integral framework for a car body is made possible. In the design of Fig 5, which affords minimum material wastage in the form of off-cuts. Almost the entire shell framework is made from extruded strip formed into polygonal looped rings. The rings are interconnected over the lengths of their side to obtain good structural integrity.

There is much effort to concentrated on improving structural integrity in side-impact. Preventing intrusion into the passenger survival space of the sill and rear quarter panel, during side impact, is the object of the design in Fig 6/7. At the intersection of sill and rear quarter panel, shown in the cross-section view, an additional longitudinal crash tube is introduced and the joint is braced across the vehicle by a U-sectioned transverse member which forms a box-tube when welded to the floor panel.

Combining rigidity with energy-absorbency is a further requirement of the structural engineer when he considers the vehicle interior. The construction of vehicle body pillars and rails is generally such as to provide the highest possible strength and rigidity for a given weight of material and cost of fabrication. This conflicts with the requirement for a degree of energy absorption when these structural members are exposed to the vehicle occupants. So a better compromise is achieved by using the elements shown in this typical section, of a windscreen pillar, Fig 8. The conventional pillar section has an extra member sandwiched between its outer and inner 'panels' so that panel thicknesses can be reduced accordingly. Then the inner panel has an extra, U-shaped section spot-welded to its walls so as to provide an additional energy-absorbent panel of yet thinner gauge. The usual deformable plastic garnish panel then separates the absorber from direct contact with the occupant in an impact situation. Also shown in the sectional view are the door pillar and weatherseals which define the shape of the basic windscreen pillar section.

Exploiting wall forces to grip occupants is another interesting approach. An unusually high degree of occupant protection in side impacts is afforded by the seat and airbag layout of Fig 9. The upper portions of the seat backrests are extended outwards to almost touch one another and the adjacent areas of door trim. Platforms thus formed contain airbags which fire forwards but exert side pressure on the occupants by reacting against the door or sidewall trim.

Extra protection from B-pillar impact can also be designed in. Means to permit the vehicle occupant to swing laterally in a pendulum effect, with pivot at the upper side rail, is proposed in the design of Fig 10. A yieldable portion of the B-pillar to sill joint collapses, with an adjacent section of the floor-pan, under side impact. The patent also covers proposals for controlled collapse of the B-pillar based on separate deformable portions.

Taking the injury out of side impact is also possible with the arrangement seen in Fig 11. Normally the B-pillar bends under side-impact at the discontinuity between the upper, window, and the lower, door, portions of the pillar, at occupant chest level. Here the centre-pillar has a designed-in strength discontinuity portion at seat level. In order to ensure this failure mode takes place, top and bottom anchorage supports for the pillar also must be reinforced. This is achieved by using a box-sectioned pillar rooted to box-sectioned transverse members at roof, X, and floor level, Y, which make up a ring frame around the vehicle and include reinforcement of the floor centre tunnel, Z.

An associated proposal, Fig 12/13, to that in Fig 6/7 comes from Toyota and covers the provision of an energy absorber within the trim of the door to further increase side-impact protection. Conventional cushioning devices added to the trim protrude into the passenger survival space. Here the cushions are provided within the thickness of the door and its associated trim case and configured to optimize the energy absorbency and allow the occupant to take the impact at hip level.

Chassis/body shell elements

Slotted construction for chassis frames is an interesting way of reducing the weight of vehicle structural elements. The possibility to make a lightweight vehicle chassis frame in aluminium alloy, without the need to form a complex one-piece tapered sidemember, is clear in the design of Fig 14. Relatively short lengths of constant section extruded sidemember portions are joined together by strong light interconnection members of hollow ribbed construction which can have such items as spring hanger

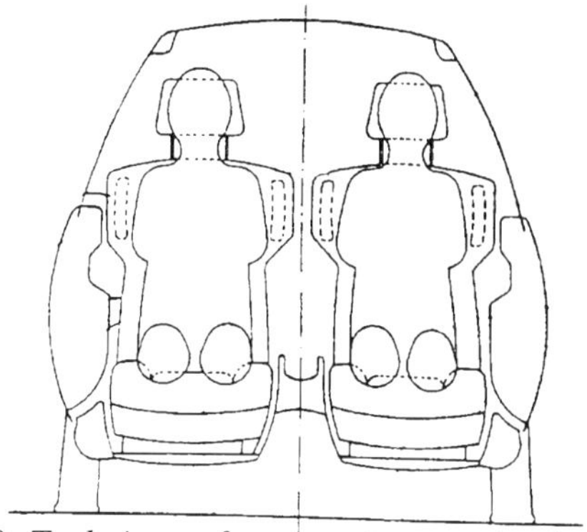

Fig 9: Technique of occupant protection in side-impact by Mercedes-Benz (GB2309440A)

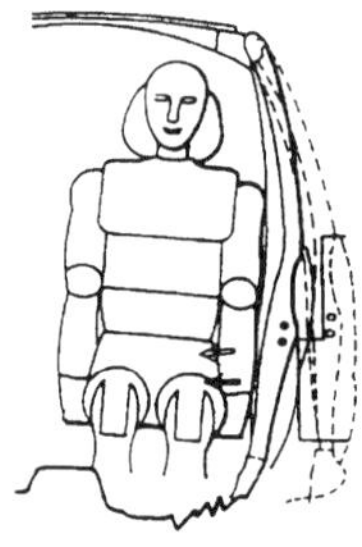

Fig 10: Design for protection from the B-pillar by Fuji Jukogyo KK (GB2311259A)

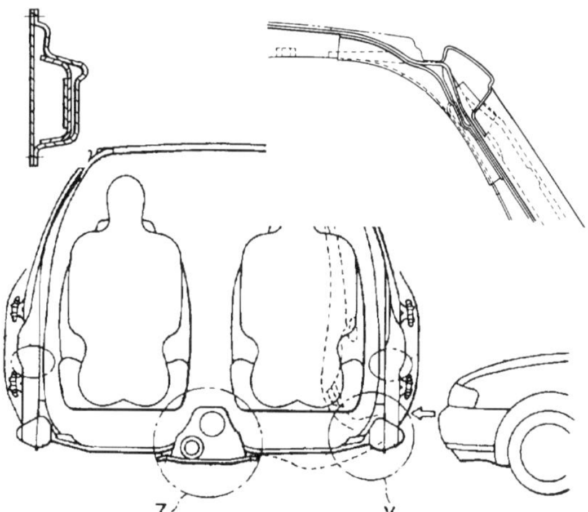

Fig 11: An improved survival space under side-impact arises from this arrangement proposed by Fuji Jukogyo in Patent GB2292716.

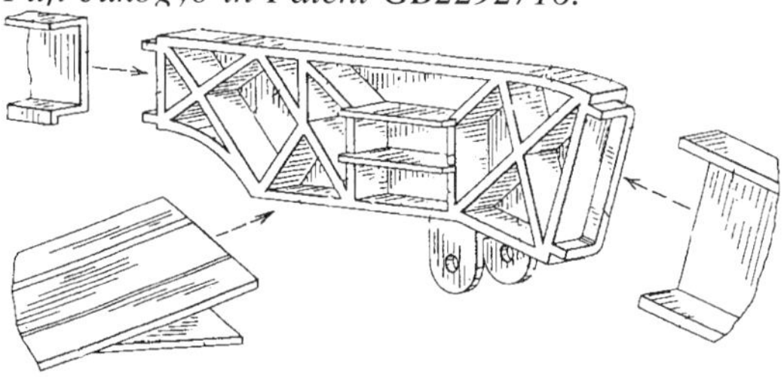

Fig 14: Patented sidemember construction by Dana in GB2313577A

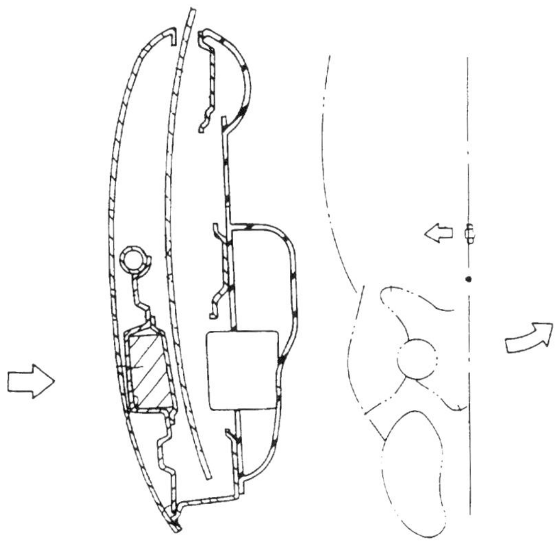

Fig 12/13: Associated proposal in GB2292913A,

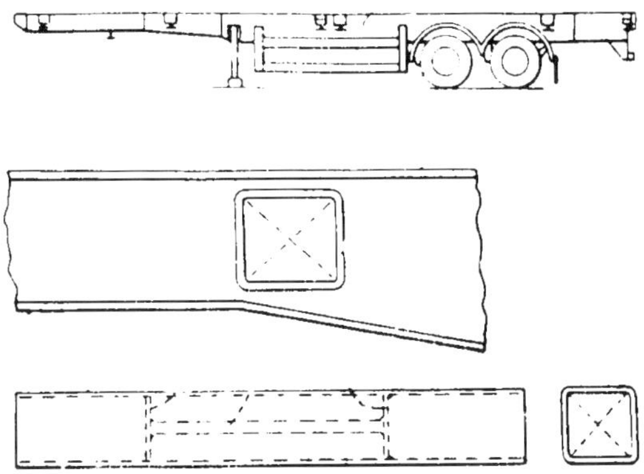

Fig 15: Technique for maintaining shear strength, patented in GB2203393A

brackets integrated with them.

Protecting chassis sidemember shear strength is a particular problem which occurs on skeletal semi-trailers, for container transporting, because of the comparatively shallow depth of the main longitudinal beams. This means that transverse, bolster, crossmembers of rectangular section cannot easily be inserted through them and welded in position. The cut-outs which pierce the sidemember webs can be too large to allow any residual shear strength in the sidemembers. In this arrangement by Crane Fruehauf, Fig 15, a pair of plates and connecting tube are inserted inside the crossmember so that the plates correspond to the cutouts in the sidemember. After welding up they replace the lost web area in the sidemember.

Even a dual purpose oddments-compartment can be used to enhance occupant protection in impact. The proposal for a centre-console locker in Fig 16 includes provision for a measure of side-impact support and the possibility of having its contents warmed by heater ducting. The box structure has two cross bulkheads, two side ones, stiffened by folded flanges, and a floor bulkhead. The sides also form the outer walls of fore/aft running heater ducts.

Stiffening up a convertible body is often necessary when the roof panel is removed from a car body shell. Increasing the torsional and bending rigidity of the rear-end of an open convertible car structure is the object of the design in Fig 17. A vertical structural wall forms the front-end of the hood stowage compartment at the very rear of the vehicle. At its outer

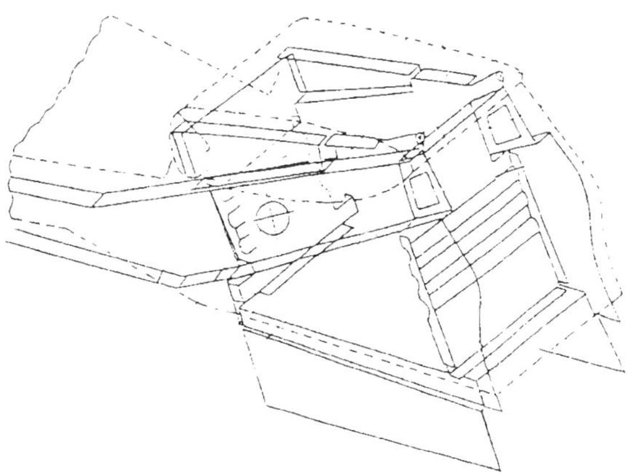

Fig 16: Oddments box doubling as structural member in patent by Daimler-Benz (GB2312190A)

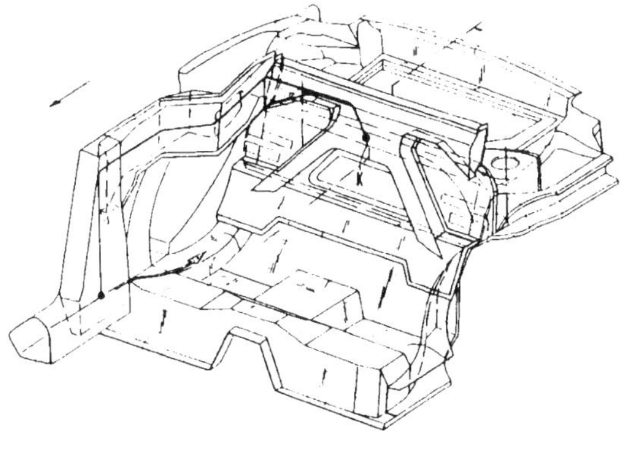

Fig 17: Patented method of strengthening convertible body, by Porsche (GB2308104A)

edges, and in a cruciform across its leading side, rail structures react loads imposed on the body structure by the rear part of the chassis, the upper crossmember directing the loads to the flux point K. The forces reach this bulkhead via cranked box-section members, and a tapered pillar of substantial section, which connect to the sills and main chassis.

A suspension-location structure can be used to supplement body shell structural rigidity. This applies to Porsche in the chassis construction of the 911 Carrera. Of particular interest is the rear suspension arrangement which involves a sophisticated subframe structure, Fig 18, which is elastically mounted to the body. The LSA lightweight strut axle comprises a five-link wheel location system, the links being arranged in two planes, two forming the upper 'wishbone' and three, including a tie-rod, forming the lower one. As the wishbones are effectively each a pair of tension/compression members, these are largely relieved of twisting or bending moments. Combined with its brakes and drive-shafts the whole is a pre-assembled module. Each suspension strut is pivot-mounted at either end, to wheel-carrier and sub-frame respectively. The suspension 'struts' are die-cast by the Acural process and are of C-section, with webs reinforced by diagonal ribs. These are in aluminium alloy AlSI7Mg with 240 N/mm^2 yield strength. The sub-frame members are of similar construction but use a H- rather than C-section. Total weight of the assembly is 17 kg.

Many of the smaller body-builders consider the way ahead in structural efficiency lies with composite bodies. Three dimensional weaving of glass-fibre preforms by the technique announced recently by Cambridge Consultants is likely to open up new opportunities for composite structures. By computeriszing the complex instructions to a Jacquard loom, which works by passing longitudinal fibres through loops in vertical wires, moved vertically as the shuttle carrying the thread moves between them, automated construction of the preform is possible. A box beam having longitudinally oriented fibres in the flanges and at 45 degrees on the webs can readily be programmed. Vehicle seats, bumpers and side beams are seen as other applications as the technique allows curved forms and flanged edges to be achieved without difficulty. The main limitation of the process is the 50–55% fibre content limit in GRP needed for the resin to penetrate between the fibres in resin transfer moulding. A resin injection technology such as RTM or network injection moulding is seen as a key part of the automating process, for which production rates of between 10,000 and 100,000 per year are seen as optimal by CCL. Fig 19 shows straight, curved and tapered parts which it is possible to construct, making the truck chassis frame to be a potential application as well as body side panels. It is even possible to envisage economical integral construction of monocoque van bodies provided a large-enough loom could be found to handle the size of the preform.

Structural/suspension innovations have also reached production in a Lotus model. Torsional stiffness of 6600 lbs ft / degree has been achieved for the open body structure of the Elan, with a steel and

Fig 18: Porshe suspension sub-frame

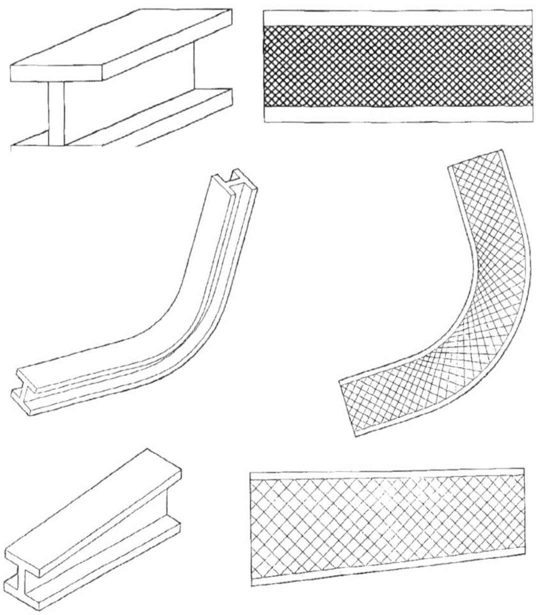

Fig 19: Parts made from the CCL process

polymer composites combination structure, Fig 20. Body panels are produced by the VARI process. A constant curing temperature and newly modified resin systems have since improved the production cure time. Also a new preform system is used to produce the panels which enables sharper corners to be achieved. Body shape gives 0.34 drag coefficient and front/rear lift coefficients of 0.30 and 0.065 respectively. The chassis structure has departed from the traditional Lotus pure backbone with the integration of the composite platform which is a single-piece 3 mm thick moulding, including inner sill, toeboard, heelboard, 'A' and 'B' posts. The resin is an isopthalic polyester reinforced by continuous filament glass fibre with additional reinforcement at the mating points with the backbone outriggers. The latter are made from 18 gauge steel which is E-coated and wax injected. Body panels are joined by elastomeric polyurethane adhesives. There is also a steel subframe reinforcing 'A' and 'B' posts, sills and rear suspension anchorages. The relatively large number of separate 2 mm thick skin panels makes for quite easy future facelifting, according to the company. Bumpers are of RRIM and doors, unusually, swing in an arc inside rather than outside the front A-panel.

A patented 'interactive wishbone' suspension assembly is mounted on a separate raft of heat-treated aluminium alloy. The raft allows very stiff fulcrum bushes to be employed but without passing road noise; it also permits abnormally low caster angle and accurately controls suspension geometry under all conditions. Torque steer effects are also reduced while allowing longitudinal wheel compliance without changes in handling. There is not the normal excessive loss of caster during braking, caused by deflection of the fulcrum bushes, and braking stability has actually been enhanced by offsetting the hub spindle rearwards relative to the steering axis. Tracking stability is also optimized by commonizing camber change geometry, with bump travel, on both front and rear suspensions. Roll centre heights (30 mm,

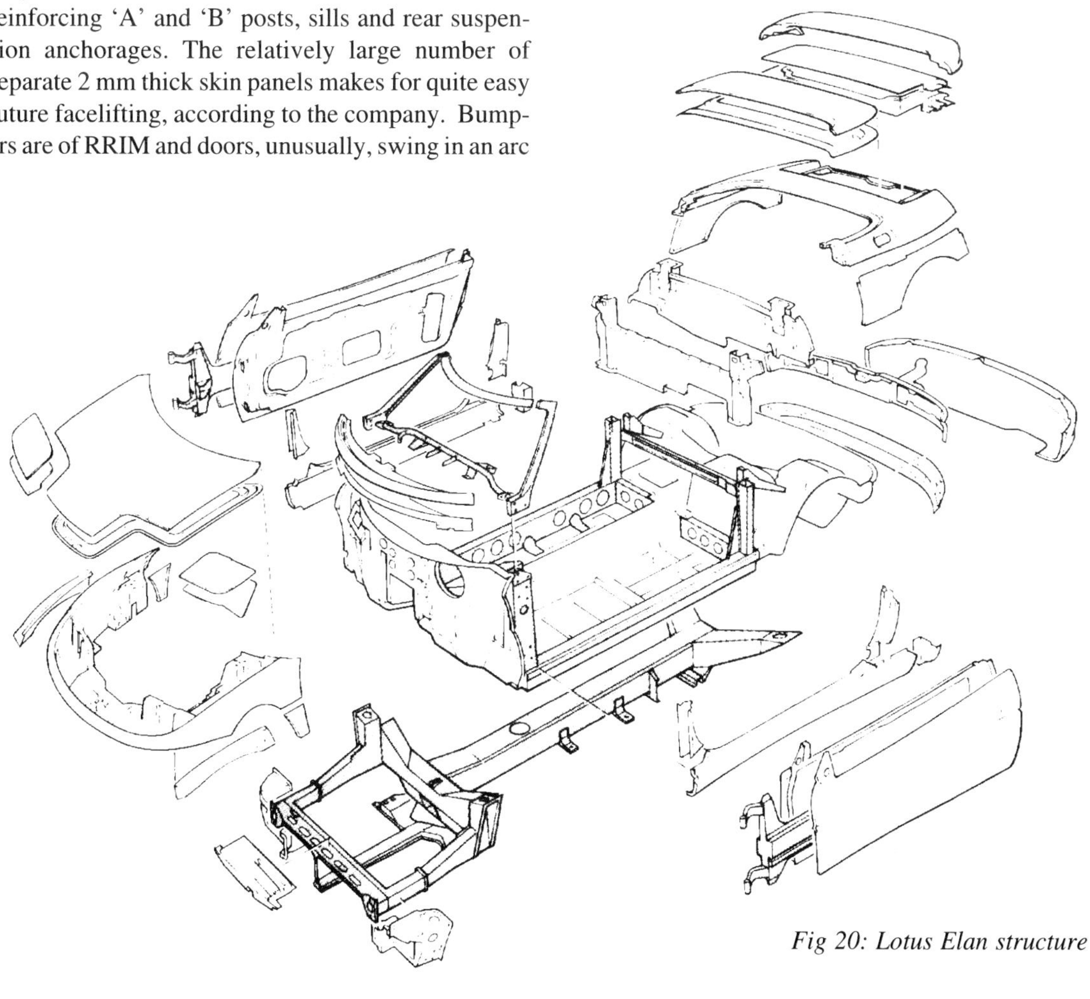

Fig 20: Lotus Elan structure

front, 60 mm, rear) remain virtually constant for normal roll angles and steering geometry of 60 per cent Ackermann has been chosen.

Many believe the achievement of optimum structural rigidity in car bodies will require the use of a punt-type structure with sandwich floor panels. A sandwich structure for a roadster chassis is seen in Fig 21/22. Mazda engineers[1] released details of this sandwich panel sports-car project in SAE papers, in which the admirably simple 'punt' structure is described. This approach is felt to be justified in open-top cars which otherwise demand costly and weighty understructure modifications when derived from standard saloons. Structural efficiency was measured by calculating the strain-energy involvement of each panel by finite-element techniques. Aluminium alloy face and honeycomb core are used to obtain body bending rigidity of 1.0 X 10^3 kNm and torsional rigidity of 0.7 kNm/rad from the 100 kg structure whose natural frequency in bending and torsion was above 40 Hz. The latter figure is well above 'shake' prone adapted saloon open-tops. Main floor panel is 100 mm thick and attaches to the structural side sills as shown inset.

Car body systems

Occupant restraint

Occupant protection has also been considerably affected by developments in vehicle seating and surroundings.

Preventing submarining or similar motion of a child in its safety seat is proposed in this Volkswagen patent, Fig 23. The shell of the seat is a carrier body for airbags which inhibit motion of the child during vehicle impact. Inflation pressure is also controlled so as not to over stress the child during this event. A further airbag is mounted in a play-table which is an integral part of the seat. A weight sensor and infra-red sensor determine the mass and size of the child and a preferred arrangement also includes inflatable shoulder and lap belts to increase the restraint effect. The airbag and gas generator assemblies are preferably of the plug-in type, the base of the sockets being electrically wired to the main harness of the vehicle, as shown.

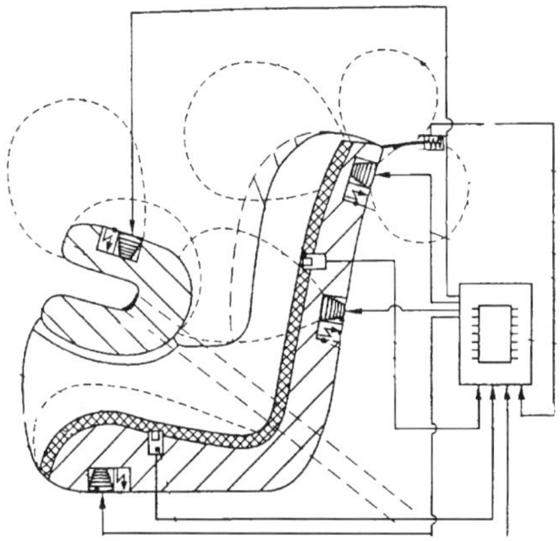

Fig 23: Child safety seat by VW in GB2290505A

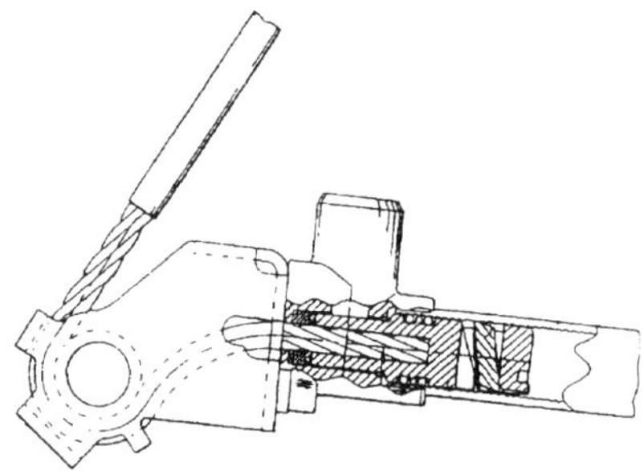

Fig 24: Patented pre-tensioner by Allied-Signal (GB2304027A)

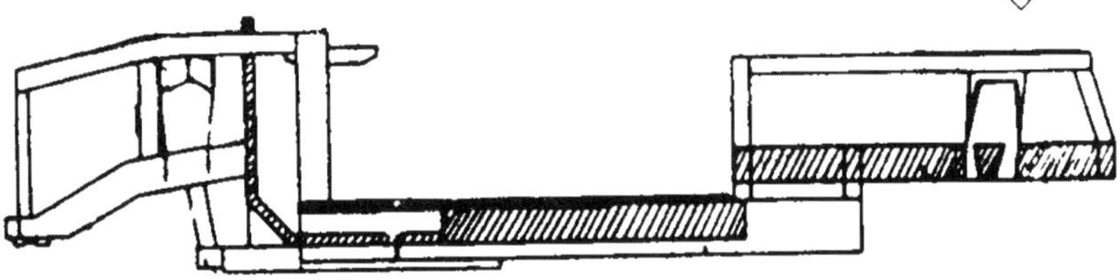

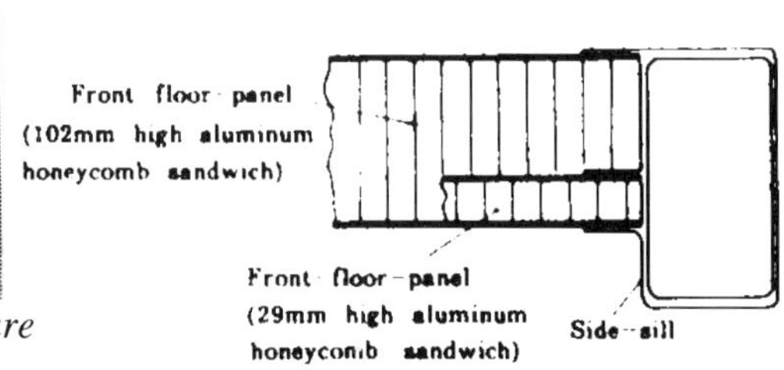

Fig 21/22: Mazda punt structure

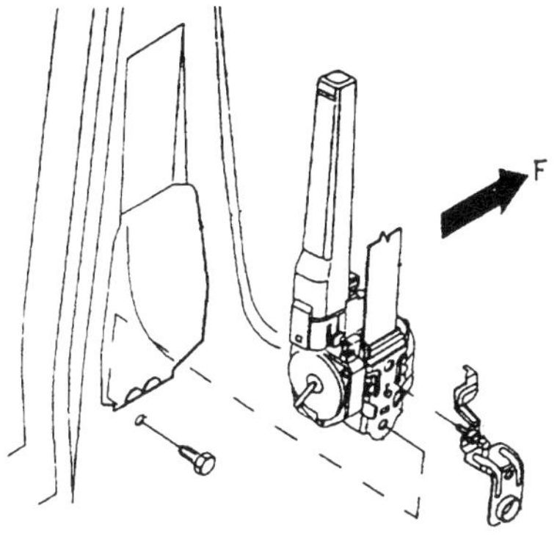

Fig 25: Safe install pretensioner by Allied Signal SpA GB2315983A

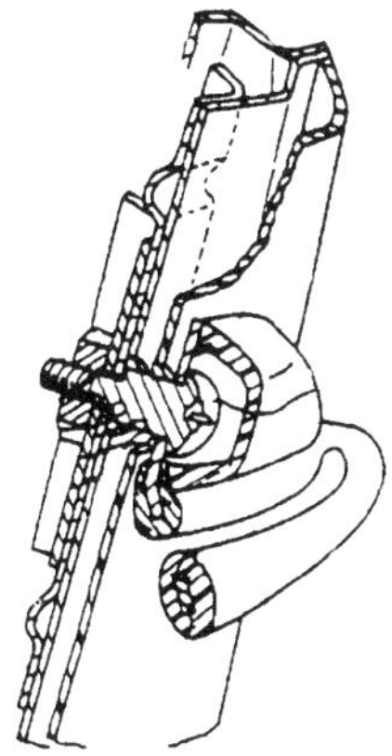

Fig 26: Energy-absorbing anchorage by Tokai Rika Denki Seisakusho (GB2307168A)

Fig 27: Side airbag within seat back by Allied Signal (GB2305638A)

Pretensioners for seat belts are another safety system receiving attention. In Fig 24, a relatively easily assembled and installed seat belt pretensioner mechanism has a strong piston locking element which, while allowing free movement of the piston in the webbing tensioning direction, locks the piston in the web-loosening direction. The cylinder is internally threaded at one end for setting the pretension and at the other has a toothed profile into which the locking element engages as well as two O-ring seals spaced at one tooth pitch.

The hazard of handling seat-belt pretensioners, while armed is overcome in the design of Fig 25, which also makes it difficult for unauthorized dealers to fit pre-used units and provides a visual indication to safety inspectors when this happens. The action of fitting the unit to the vehicle causes the lever of the arming bracket to be actuated and the bracket incorporates a flange for limiting access to the fixing screws. The arming bracket is secured to the retractor frame by a twin headed arming screw, the heads being separated by a weak portion of the stem. On tightening, the stem shears to detach the outer head and leaves the inner one protected by an enveloping sealing ring.

Energy-absorbing belt anchorages are another approach to lessening impact injury. In Fig 26, an upper seat-belt anchorage is provided which incorporates an additional measure of energy absorption. The body of the anchorage which supports the webbing has a specially shaped support-pin passing through it, and the adjacent body centre-pillar, which engages an exterior reinforcement plate in a special way. The energy absorber is positioned around the circumference of the outer end of the pin and is arranged to expand outwards under crash loading on the interior belt mounting.

Out-of-view side-impact airbags protect passengers from impacts in other directions. In the design of Fig 27, no obtruding into cabin space, or detraction from the interior styling, results from a modular side-impact protection airbag which fits within the seat back. A specially constructed extruded seat-back support housing has a fold over part, behind which the bag (plus inflator) is stowed, which allows the bag to deploy forwards, between the occupant and the door.

In another ingenious design the frontal airbag module is the horn push, Fig 28. Provision of tactile indication of when sufficient pressure has been applied to sound the horn is inherent in this steering wheel configuration which also prevents unwanted application of the airbag. The airbag module is movable with respect to the rim of the wheel and is mounted on three resilient elements which are also electrical contacts of the horn switch. The force/deflection characteristics of these elements are such that the resistive force rises to a maximum during initial movement and then falls to a minimum and rises once more to a second maximum with continued movement.

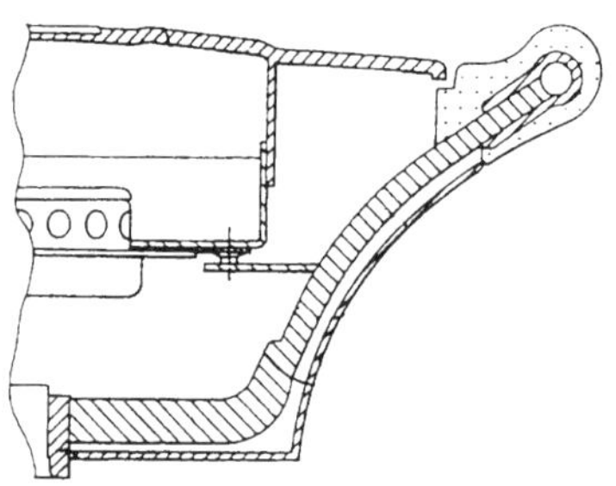

Fig 28: Airbag module as horn push by Autoliv (GB2309123A)

Side protection is also possible with an open window in the design of Fig 29. Protection of vehicle occupants during side impact, when a side-window is either broken or open, is the objective. A strap that threads through the deflated airbag is anchored to structural pillars either side of the window. When the bag inflates, the strap is tensioned by the disposition of inflating cells around the strap such that the occupant is constrained inside the window opening.

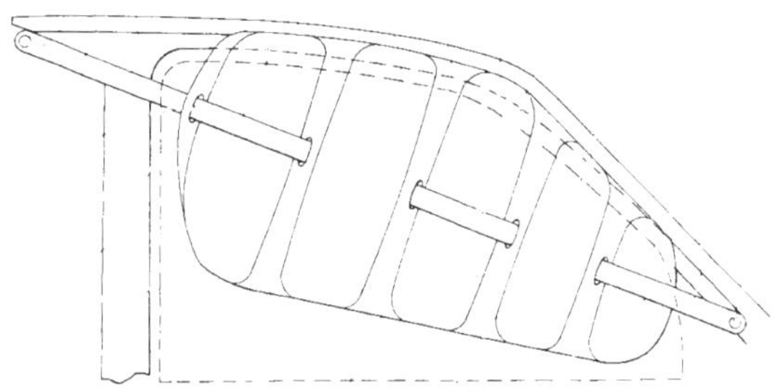

Fig 29: Protection from side-window shattering by Autoliv Development (GB2312877A)

One school of thought believes that occupant protection is best served by having seat belts anchored into the seat, particularly in open roadsters. Seats on the M-B 300 SL, Fig 30, have ten way electric adjustment with memory pre-set and a structure which includes anchorage for the shoulder belt inertia reel. It was designed with the aid of the material supplier, Norsk Hydro. Possibility to adjust the seat-belt harness to almost any occupant size is an important contributor to safety in an open car which otherwise has to rely on safety-reel anchorage by the half-height C-pillar. Since the crash loads on the seat from the safety harness are very considerable the integrity of the structure and its slide mechanism is crucial and required a 5800 element FE analysis in the SL design

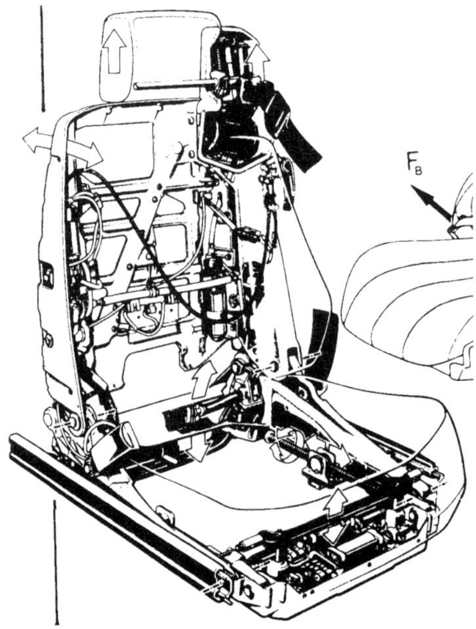

Fig 30: Seat of Mercedes 300 SL

A particularly compact seat-belt pretensioner has been patented by an Asian manufacturer. Avoidance of a long cylinder and rack to effect pretensioning of a seat-belt retractor is one of the objectives of the compact design in Fig 31. The retractor uses gas pressure to move the rack and rotate the pinion gear. A clutch mechanism transmits motion between the member attached to the pinion gear and a belt take-up spindle, the clutch mechanism being driven through step-up epicyclic gearing.

More effective seat-belt locking is claimed for another ingenious design. In particular, a faster, more

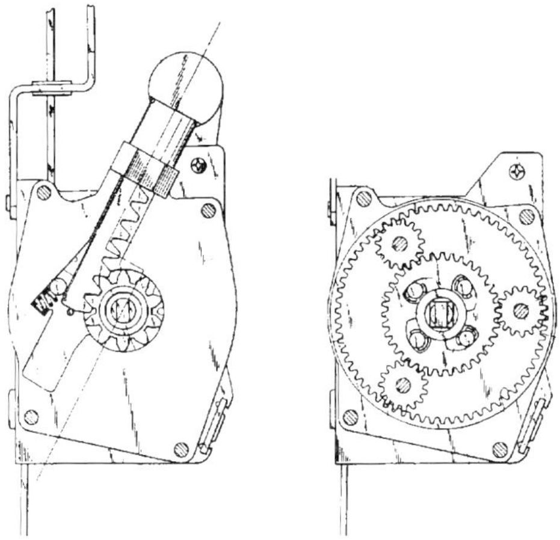

Fig 31: Compact pretensioner by NSK (GB2292304A)

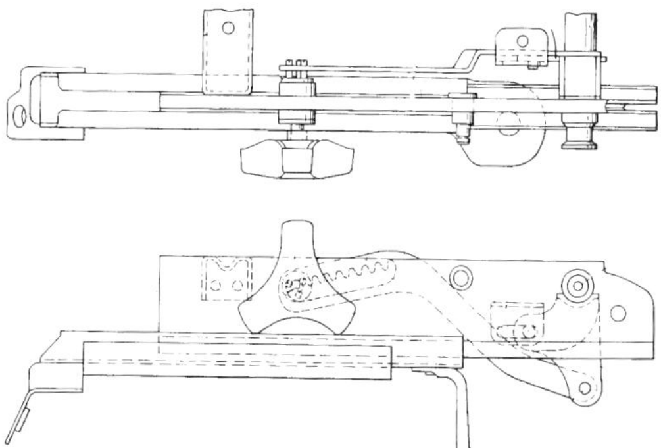

Fig 33: Adjustment mechanism by Dunlop Cox, GB2293972A

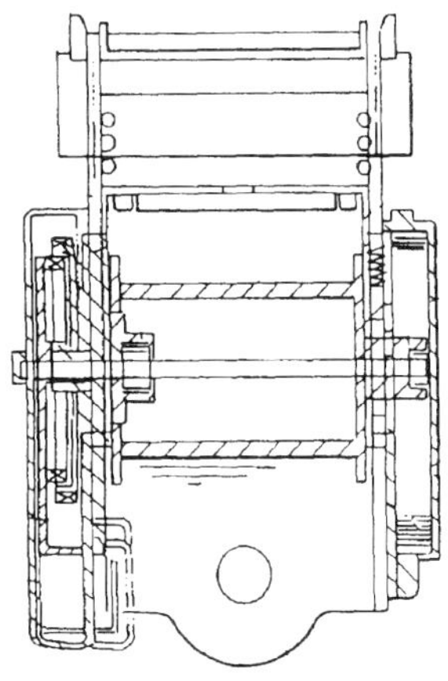

Fig 32: Improved belt locking by Allied-Signal (GB2300797A)

effective, release of the clamping wedge and spool assembly of a seat-belt retractor is achieved, in the design of Fig 32, when the loading on the belt drops below the spring bias force. The spool assembly is mounted so that linear movement with respect to the wedge housing pushes one or more clamping wedges vertically to lock the seat belt. Ideally, the spool is coupled to the wedges by coil springs and it is supported in a slider which is also the spool-assembly frame. The slider moves linearly with the spool to activate the wedges.

An improved manual height/tilt adjuster for a seat has solved a long-standing problem. Seat adjustment mechanisms using double pinned discs to engage notched racks have previously been impractical because of the difficulty of turning a coaxial handwheel through 180 degrees. For the design in Fig 33, a 3-pinned disc allows a more manageable angle for hand-turning and incorporation of such an actuator in the mechanism shown, and when fitted either end of a vehicle seat slide, allows a convenient way to provide front and rear height adjustment.

A simplified headrest adjuster is also a welcome advance. In many small cars the headrest has little use, apart from the protection afforded against whiplash injury, as different sized drivers cannot adjust the fore-aft position even to make contact with the rest. In the design of Fig 34, lighter weight construction and improved comfort follow from a locking mechanism design for a seat head support which allows an adjustable position for the rest. Key to the mechanism is an actuating lever within the retaining yoke of the headrest. At its top end is a release button and at the bottom end the locking element.

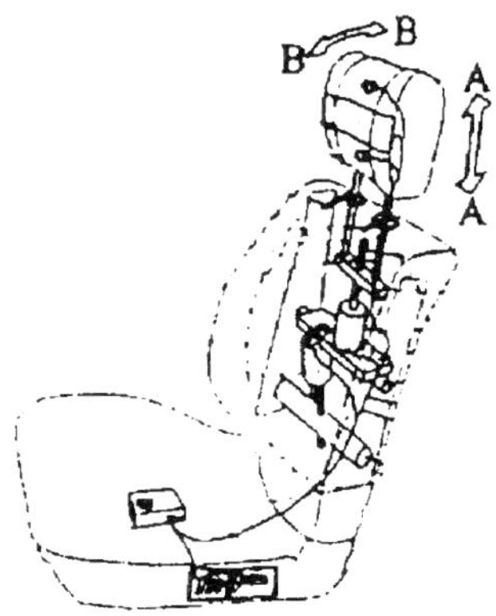

Fig 34: Simplified head rest adjuster by Volkswagen (GB2302706A)

Controlling headrest position is also improved by a mechanism for helping to prevent whiplash injury. The design in Fig 35 incorporates an anticipatory crash sensor which signals when an external object at the rear of the vehicle has greater than a threshold closing speed. Another sensor determines head location with respect to the headrest. A processor works out the correct headrest repositioning to minimize injury and actuates the mechanism accordingly.

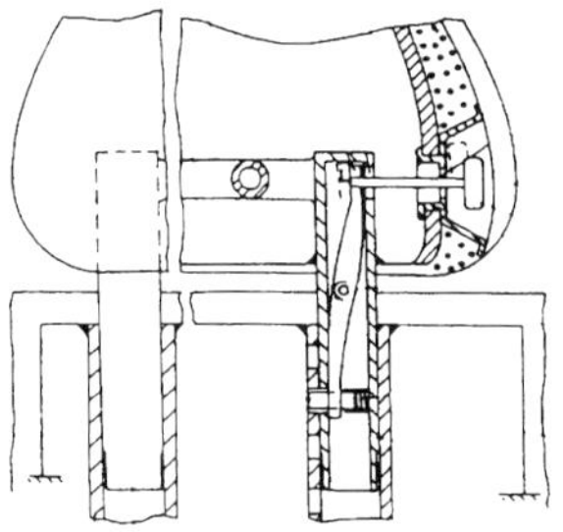

Fig 35: Controlling headrest position by Automotive Technologies (GB2301906A)

A simpler seat slide control that would seem to favour powered drive is seen in Fig 36. Substantial simplification of a seat slide and adjuster mechanism follows from the adoption of a system based on belts and pulleys rather than a serrated rack and pawl. Four seat-mounted pulleys are involved plus a simple brake mechanism, at the leading edge of the seat, which clamps the belt in any position and therefore provides finer adjustment than is possible with a toothed channel.

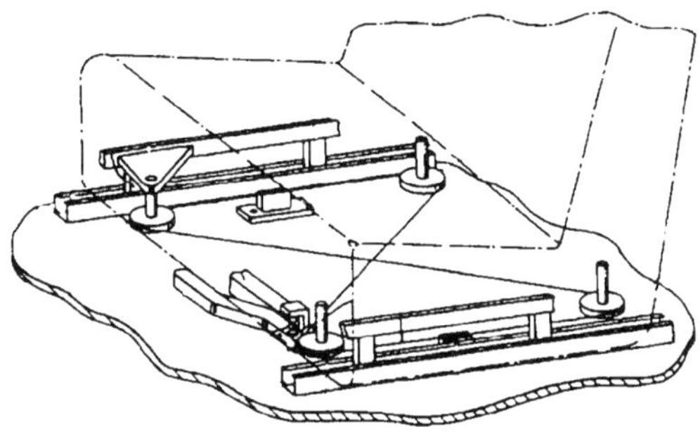

Fig 36: Patented design for a power-slide by Rover Group (GB2306880A)

A quieter belt-retractor for a seat belt is seen in the interesting design of Fig 37. It enables the ratcheting noise to be eliminated, of the pawl dragging over the locking teeth of a seat belt retractor mechanism, on rewind. A spool with a ratchet is used for carrying belt webbing while a rewinder is used for rotating the spool. An upturned cup-shaped member acts as a sensor for detecting acceleration of the vehicle over a given threshold, causing a lever to move upwards and move a pivotally mounted pawl to lock the spool. A blocking arm, frictionally coupled to the spool, disables the pawl when the spool is rewinding, its movement being limited by arcuately spaced abutments.

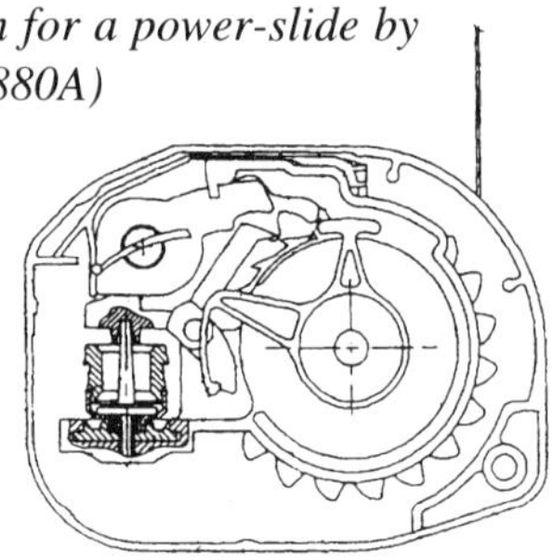

Fig 37: Quieter belt-retractor by Allied-Signal (GB2303044A)

Doors, windows and panels

An interlocked split-tailgate is a way of overcoming difficulties that have arisen with two-piece tailgates on cross-country vehicles and estate cars — of independent securing of the upper (window) portion and the lower (flap) portion when long protruding loads are required to be carried. In the design of Fig 38, a gas strut is used to automatically open the window portion when the latch is released. However, if the window is open when the flap is open, it cannot be closed unless the flap is closed first, whether the flap is hinged horizontally or vertically. The flap can also be independently locked to the body via the D-pillar. In the closed position the window rests against a seal

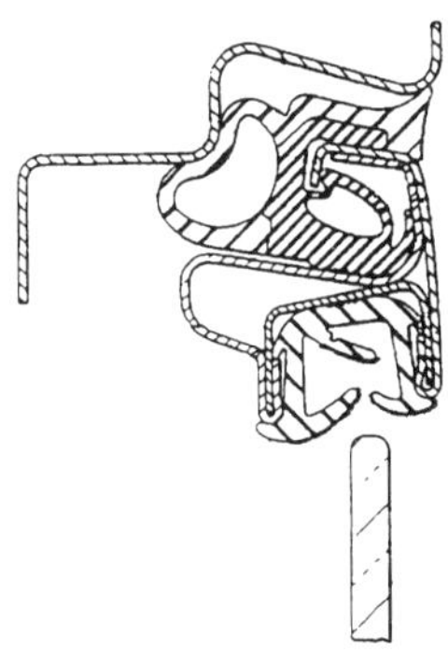

Fig 39: Door seal by Draftex (GB230994A)

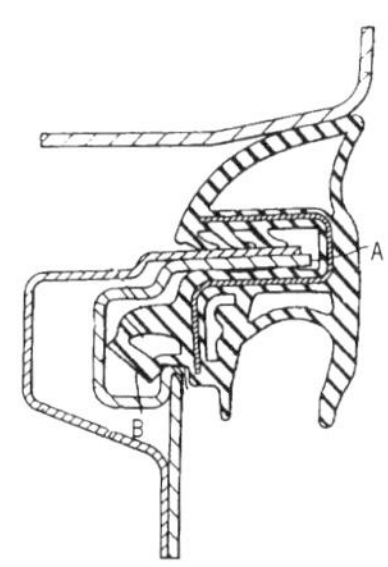

Fig 40: Easy-assemble door strip by Draftex (GB2293191A)

(a) which extends around to the roof, via the side pillars. In its lower part the window rests against seal (b) on the flap and the window is locked to the flap with a mechanism that prevents the flap being opened while the window is closed. The window raising linkage ensures no damage can be done to it when the flap is opened and closed. Reference to the inset enlargement of the window hinge linkage shows the window to have a horizontally hinged tilting lever (c), with a roller attached to it, which has a second lever (d) pivoted to it, the angular motion of which is limited by a stop (e). If the unlocking mechanism is actuated when the window is closed, the window moves from the closed position into the partially open position (f). Further automatic opening of the window can then only occur after a manual push because the 'power' lever arm is then too short. Closing of the window under manual pressure is arranged so that the window pivots a given distance until the roller rests again on its track. Under further pressure, the tilting lever pivots about axis (g) and the roller moves on its track until the shackle catches behind the latch at the bottom of the window.

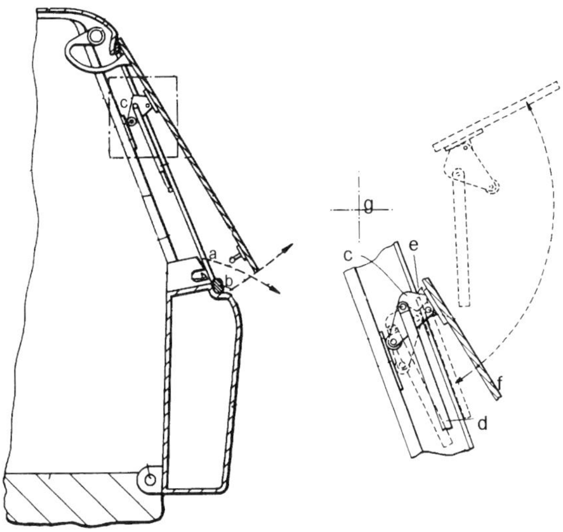

Fig 38: Opel split-tailgate proposal covered by Patent GB 2290060A

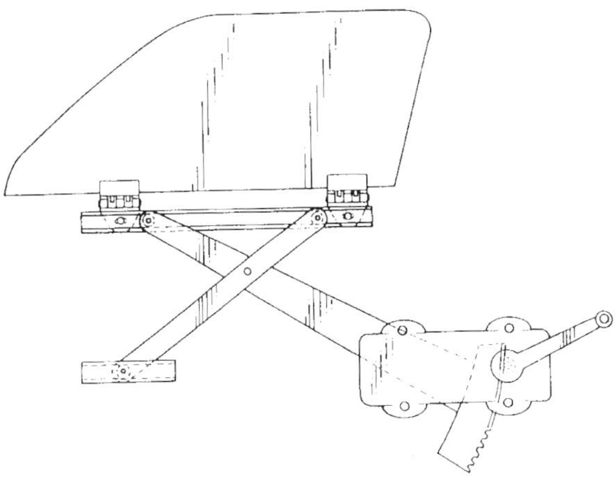

Fig 41: Winder mechanism by Nifco Inc (GB2313873A)

Improvements in door sealing are the subject of many patent applications. In the design shown in Fig 39 the upper rail of the door frame is made with a lipped channel section whose open-end faces inwards to receive the specially designed weatherstrip section. The lip of the channel holds the weatherstrip in place after its integral locking contour has snapped into position. The door periphery is then sealed, on closure, in two places by direct compression of an inbuilt 'balloon' and by shear deformation of the outer edge of the seal.

Weatherstripping without tools is possible with the design in Fig 40. The ability to assemble and remove a door weatherstrip without special tooling follows from the metal reinforced elastomeric strip fitting over flange A in the usual way but being locked into position by a snap-over section of the resilient strip B which can be withdrawn with modest force.

A successful yet inexpensive window winder for curved glass taxed designers for many years. Unless a complex compensating mechanism is introduced into a window regulator used for curved windows, there is a tendency for a wrenching force on the glass-holding clamp. This is because the clamp rises in a linear path while the glass rises in a curved one. In the design of Fig 41, the glass clamp is rotatably mounted to the raising linkage so no unwanted torque is induced.

A long-life window-winder drive is promised by the design of Fig 42. Deterioration in operation of window regulators can occur in conventional chain driven systems due to stretch of the associated wire rope — and the inability of the spring-tensioner to compensate for the stretch. Here an integrally formed chain has interior teeth and a flat outer side with a wire rope wrapped between inner and outer sides. An electric drive is used with the endless chain, which engages driver and driven gears enclosed by a chain protector. The chain-runs lie in flexible tubes.

Easy access for door hardware assembly is the object of the design in Fig 43. A further improvement in modular door construction comes in this design in which hardware mechanisms such as door latch/lock and window regulator are built into a central framework. It also forms the structural integrity of the door and supports items such as rear-view wing mirrors and hinges. Inner and outer non-structural skin and trim panels are snap fitted to the central framework after it has been fully assembled with the necessary hardware, without access hindrance.

An extending door handle to ease operation is seen in the design of Fig 44. The problem encountered by relatively small stature people, in driving two-door cars, of being able to reach the door handle to close the door from its fully-open position, is addressed by this design. A strap-type door handle is made extensible by means of a linkage between the door and its frame-surround. As the door is pivoted open, one end of the resiliently flexible handle is moved towards the other end causing it to bow outwards and be easier to grasp by the vehicle occupant.

The current sophistication in door latching and locking systems is exemplified by the Bosch integrated door latch, lock and auto-closing system. The company have revealed details of their electronically controlled automatic door closing and security system in a recent paper[2]. Only one actuator motor is involved and the functions of the device are integrated into one system. In an effort to minimize the number of mechanical parts, for closing and locking, only motor, gearbox, driving/stop plates, guide mechanism and ratchet are required; door slider and lock slider are located on one bracket. Door position is seen by the displacement of the drop latch, monitored by two proximity switches, while that of the driving plate is seen by two Hall-effect sensors. Shown at Fig 45(a) are the door in open and the first ratchet positions; the auto-closing system is inoperative when the closing pin is introduced in the first ratchet position as this actuates the controller via a proximity switch. The ratchet then holds the drop latch and prevents the door from springing open. The driving plate is connected to the gearbox output which turns the drop-latch into the closed position via a sprung

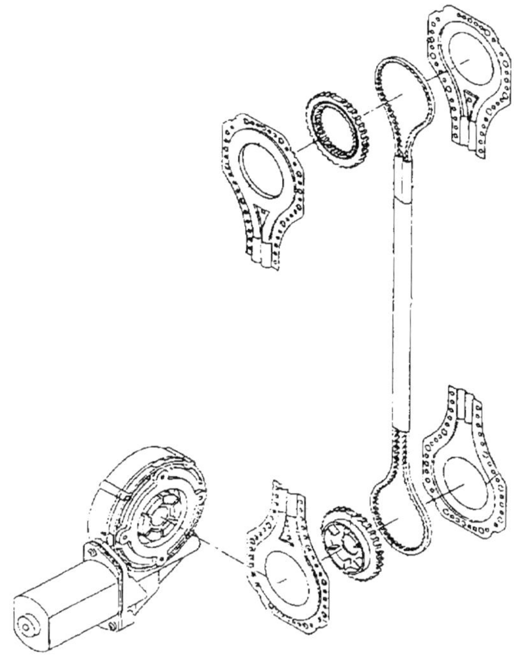

Fig 42: Long-life winder by Chin-Yun Huang (GB2313407A)

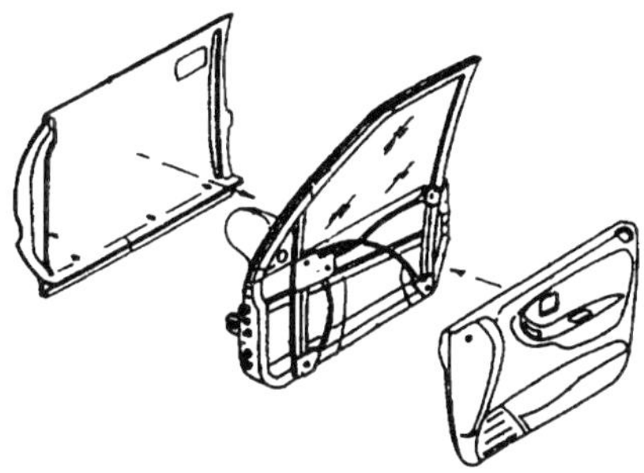

Fig 43: Easy access to door hardware by Honda (GB2315513A)

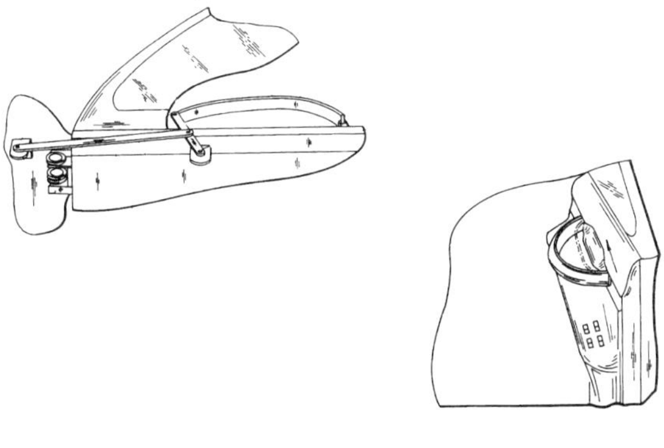

Fig 44: Extending handle by Prince Corporation (GB2318385A)

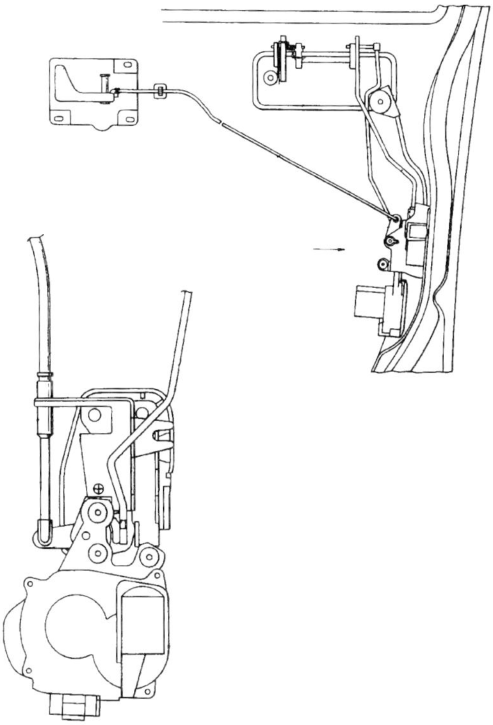

Fig 46: Rack and pinion mechanism by Rover Group, covered in Patent GB2291109A

stop-plate. The drive plate then completes one revolution and the door can then be mechanically opened by the exterior/interior handles. A console switch signals the actuator, via the CAN bus, to enter the central locking mode. The driving plate is then motored to the door-locked position (b). The driving pin on the stop plate prevents the drop latch from being handle-opened from the outside — but interior handles will turn the stop plate against its spring with the slider. Turning the locking key signals the actuator to enter the anti-theft mode in which the driving plate takes up the position monitored by the sensor; both inner and outer handles are then disconnected as in (c).

Another way of foiling the thief is with a vehicle locking system which overcomes the grave limitation of many existing systems, that once the thief has prised the lock from the body the latches can be operated by the pull rods linking them to the lock. In this arrangement, Fig 46 , the lock actuator is a key-driven rack and pinion, the rack of which is connected to a Bowden cable that links with the latch. The lock, rack and pinion, and the end of the Bowden cable are all located in a casing such that the locking function cannot be operated without the key to the lock, even if the lock is prised out of the door.

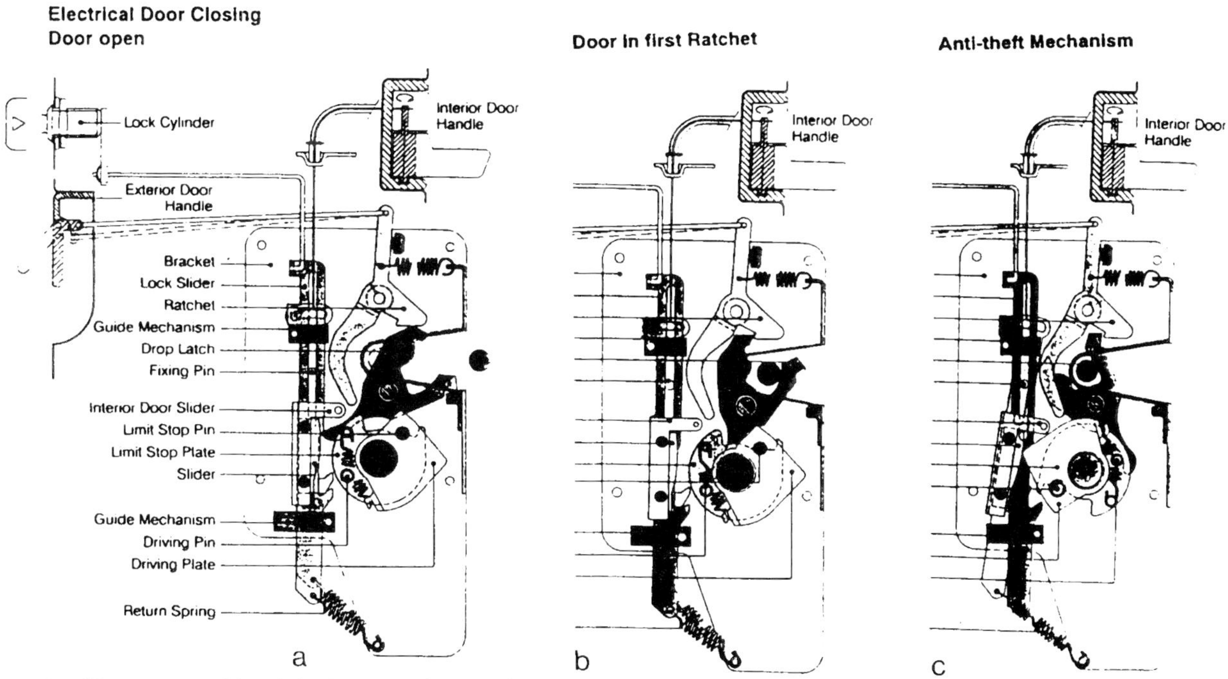

Fig 45: Integrated latch/lock system by Bosch

A way of minimizing operating forces on a central-locking system is the design of Fig 47 in which the electric motor only needs to be sized according to the operating forces of the locking lever system. Manual operation can also be carried out only against the resistance of these forces, in the event of motor power failure. Thus a rotary latch, pawl and release lever have an operating lever which comprises at least one operating lever plus a locking lever system . This involves an interior locking lever, central locking drive and locking element connected to it. The drive is a reversible motor with power take off having an eccentric control pin. The pin is controlled in a way that it rotates in either direction to move the central locking element into 'locked' or 'unlocked' position. This element has a forked receiver with control surfaces on its sides which cause the interior locking lever to be rigidly connected to it. Part of the control pin arc of rotation is outside the forked receiver and the central locking element has a stop face on both sides, for the control pin. Control movements of the pin are restricted by its striking one of these faces and the motor being switched off as this happens.

A particularly compact power door latch is obtained from the design in Fig 48. Existing powered-closing devices for vehicle doors cover a large projected area of the door. This is because the cancelling lever normally projects beyond the output member. In this design a compact layout is achieved by the cancelling lever overlapping the output member, as well as the member connecting the output member to the intermediate lever being shorter. The output member is rotated by a motor, moving the latch from a half-latched to a full-latched position. The cancelling cam surface is brought into contact with the connecting member to disconnect the output member, and the intermediate lever, when the opening handle is turned.

Another body mechanism which receives considerable design attention is that for convertible hoods. A compact folding-roof arrangement for boot-stowage is covered in the design of Fig 49, which maximiszes the space within the stowage cavity. To pivot the boot-lid frame upwards a hydraulic jack is extended and a slot in the lower end of the jack slides down over a fixed pivot. This initial motion of the slot causes the foot of the jack to pull on a cable, to unlock the boot lid. Further jack-extension raises the boot

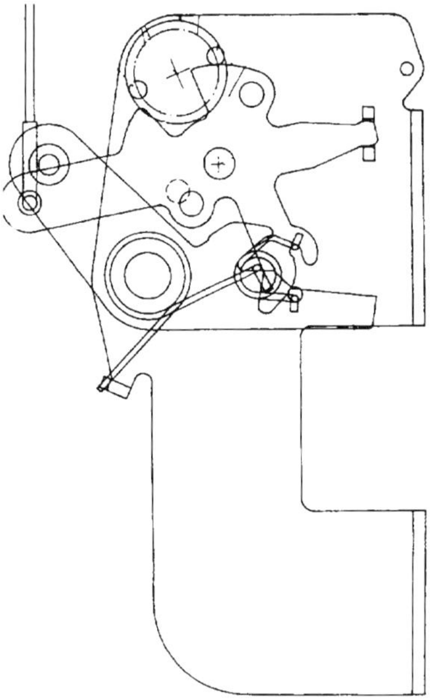

Fig 47: Reduced locking force system by Kiekert AG (GB2292768A)

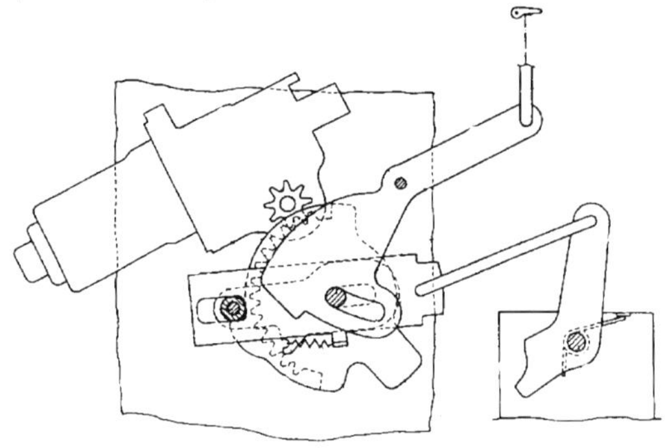

Fig 48: Compact door latch by Mitsui KKK Kaisha (GB2313408A)

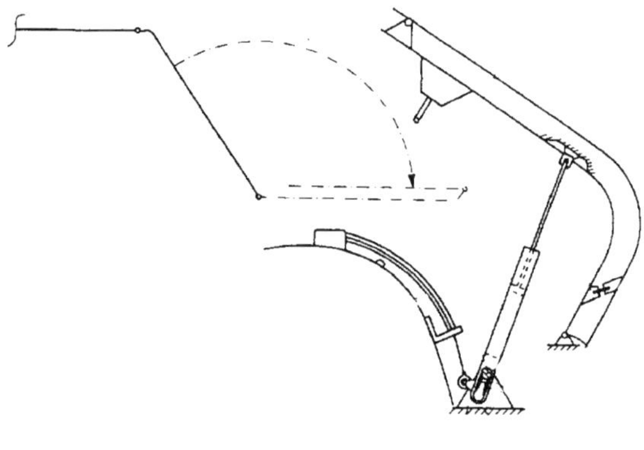

Fig 49: Boot-stowage of folding roof by Mercedes-Benz, covered in Patent GB2300671A

lid; jacks are positioned at each side of the boot, behind the wheel arches. Leaf springs are positioned at the side of the slots to stabilize the pivot positions.

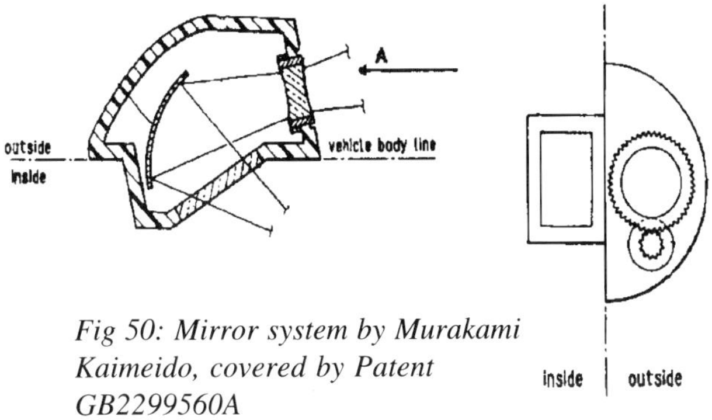

Fig 50: Mirror system by Murakami Kaimeido, covered by Patent GB2299560A

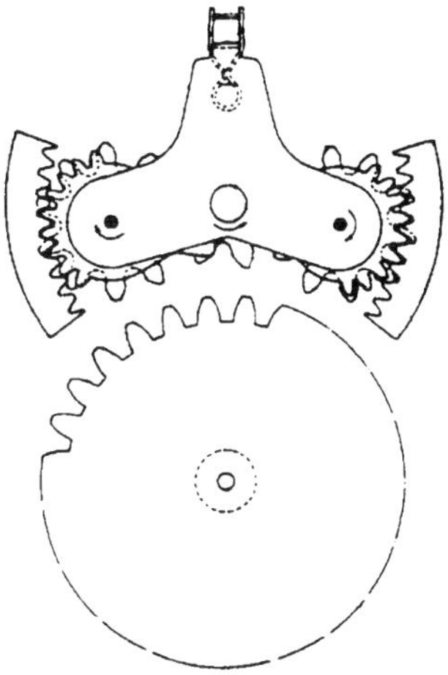

Fig 51: Sliding door drive by Mitsui (GB2305228A)

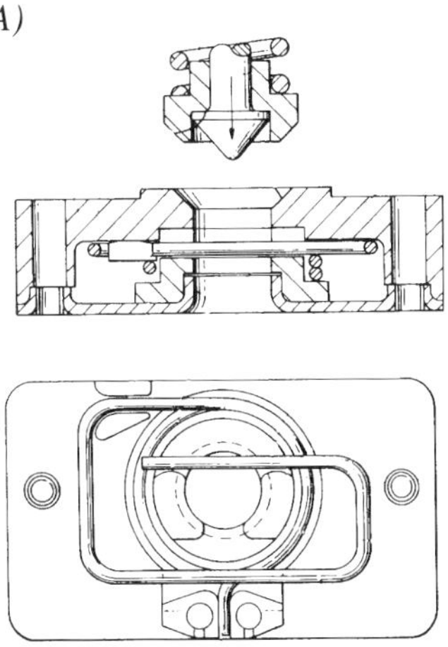

Fig 52: Reliable bonnet latch by Bloxwich Engineering (GB2312921A)

Trim and fittings

A mirror system without rain-blur is seen in the design of Fig 50. Substitutes for conventional rear-view mirrors have been proposed which avoid the considerable protrusion of the mirror from the sidewall of the vehicle. However, a drawback of such systems is the blurred image obtained in wet weather when the image is distorted by rain droplets. In this patented design, a housing on the vehicle side has a concave mirror inside it to reflect light collected by a circular concave lens at the rear of the housing. The lens is coated with a water repellent and is mounted in bearings so that it can be spun by gear teeth, on its periphery, driven by a motor pinion.

An improved sliding-door drive which does not require exclusive use of power is suggested in Fig 51. A clutch mechanism, for transmitting power from motor to movable member, has racks fixed to stationary members and engages with gears to drive the movable member. The clutch unit is supported on a swinging member, spring-biassed to a central neutral position. Gear A is engaged with the output gear B, and the stationary rack, while gear C is frictionally coupled to gear A. On energizing the motor the assembly swings away from the neutral position to engage with the movable-door drive gear.

A more positive and reliable bonnet release mechanism is proposed in the design of Fig 52. This improves on the conventional type comprising a headed striker pin mounted on the bonnet and adapted, as the bonnet is closed, to pass through an opening in a catch plate attached to the grille surround and become engaged by a catch lever. Here a latch for retaining a striker pin does so by the action of a latching member in the latch on the head portion of the pin. The latch has a resilient member, which is chordal to the bore, to retain the striker pin in the latch. A release ring rotates about the axis of the bore and has an abutment which engages with the resilient member to spring it from the bore and release the strike pin.

Cushioning a bonnet top impact is the object of the design in Fig 53 for a pedestrian protection device. In the cases of frontal impact with the vehicle, uses an external airbag triggered by a deceleration sensor. Protection is given particularly against violent head impact from the whiplash effect of the body folding after leg impact from a low bonnet. The airbag is stored in a protective cover which ruptures on inflation of the bag, the cover being contained in a nudge-bar extension of the normal bumper bar, which is itself covered with soft material to minimise leg injury.

Protection of the occupants is the object of this controlled-collapse trim system in Fig 54. Providing a crash impact waveform with a number of progressive deceleration peaks is an objective of a trim finisher for fitting to the interior of the windscreen pillar. Earlier attempts to do this have usually resulted in undue bulk of the trim member with consequent intrusion into the passenger space and/or obstruction of the front doorway. Here the generally channel-section plastic trim member has reinforcing ribs at right angles to its surface and arranged with different clearance dimensions from the screen pillar. Fastening of the trim to the pillar is via integral bosses parallel to the ribs which engage with a sprung plate with lugs engaging spring-clip fasteners fitted in holes in the pillar.

A simplified level-sensing system for lamp adjustment is seen in Fig 55. Undue complexity, and a large number of individual parts, in existing vehicle level-detection devices have resulted in costly solutions for automatic on-vehicle headlamp adjustment and suspension levelling. In this proposal a sensor is used having just two angularly moveable parts, respectively coupled to the vehicle body and suspension. The 'fixed' part is attached to a body-structure member in a way that allows the sensor to be coaxial with the fulcrum of a wheel suspension control arm, an attaching leaf-spring from which holds the 'moving' part of the sensor. The rubber bush of the suspension control arm compensates for any motion of the arm at right angles to the axis of rotation of the sensor moving-part.

A system of fibre-optic light division for headlamps is seen in the design of Fig 56. This headlamp design uses fibre-optics so that only one high intensity discharge bulb is required for both main- and dipped-beam operation. Light conducted from the upper (main beam) source, by the optical fibre at the rear of the lamp, serves as a source for the lower, dipped-beam unit, a shutter mechanism controlling the amount of light passed. The arrangement is also said to improve colour distribution between each source, give better balance of luminous intensity and thus better forward vision.

Body system innovation in the fuel tank region

Fig 53: Pedestrian protection device patented by Concept Mouldings Ltd (GB2316371A)

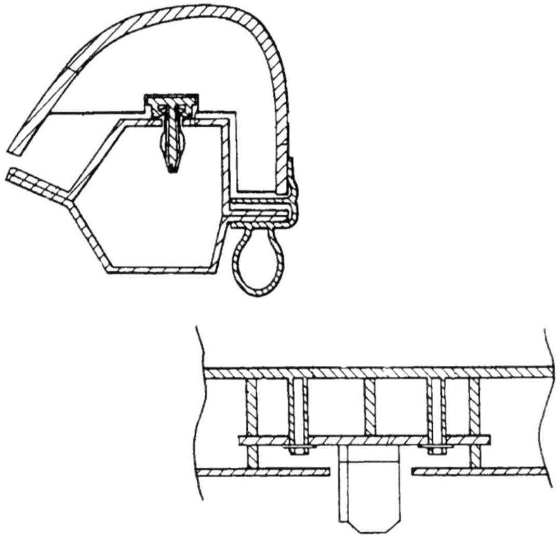

Fig 54: Controlled-collapse trim by Nissan (GB2318551A)

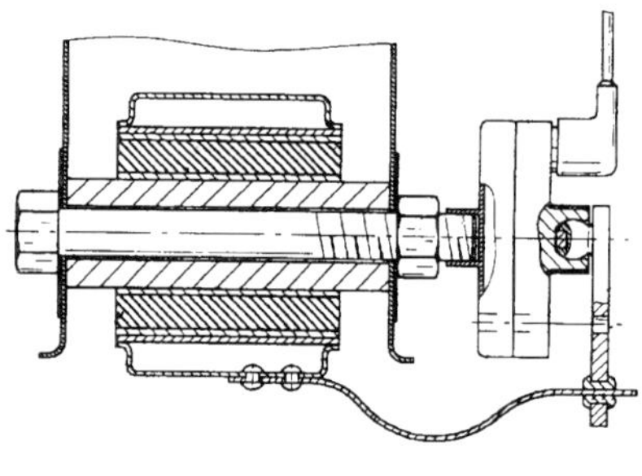

Fig 55: Simplified level-sensing by Mercedes-Benz (GB2292614A)

includes the fuel delivery system of Fig 57. Simplified assembly and test can take place of this fuel distributor system, which is an add-on unit to a fuel tank, rather than fabricated within it. The carrier casing, which is mounted to the fuel tank, has a non-return valve, bottom right, and an ingenious jet-pump arrangement, for filling. The jet-pump, left side of carrier, involves excess fuel flowing back through a duct from the engine.

An improved fuel tank, overall, is covered in the patented design of Fig 58. Prevention of overfilling on refuelling, and avoidance of any bad effects of vapour formation on vehicle acceleration, are the objectives of this fuel tank design. Two separate ullage spaces are formed by means of an inverted canister suspended from the roof of the tank, which forms a vapour reservoir. An acceleration-sensitive valve controls the opening between the two ullage spaces which are also interconnected by a small bleed valve.

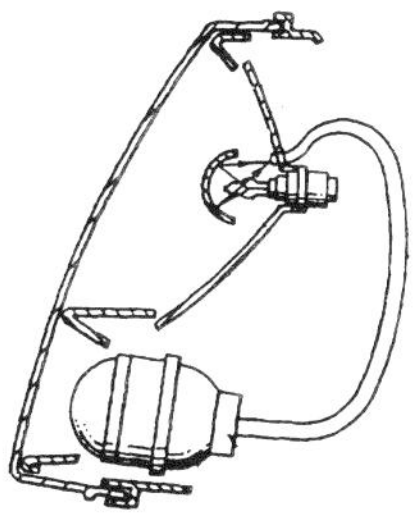

Fig 56: Fibre-optic light division system by Fuji Jukogyo (GB2315538A)

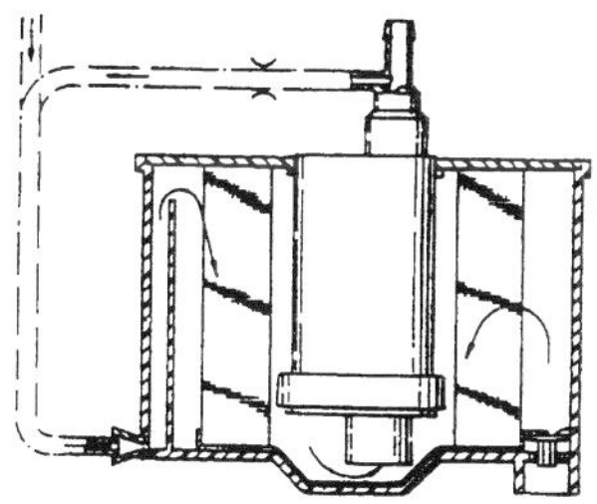

Fig 57: Fuel delivery system by Mercedes-Benz (GB2304821A)

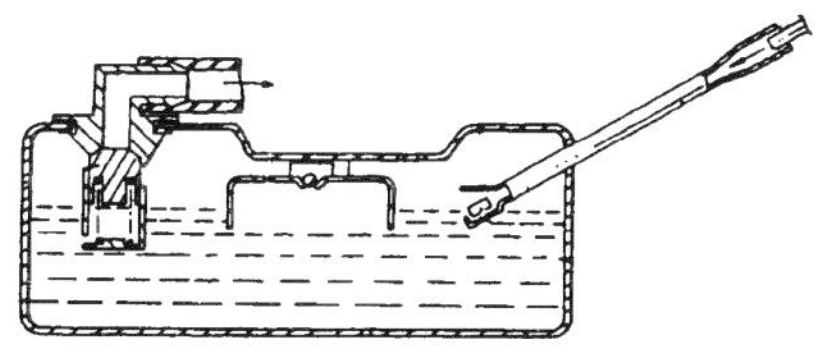

Fig 58: Improved fuel tank layout by Ford GB2306952A

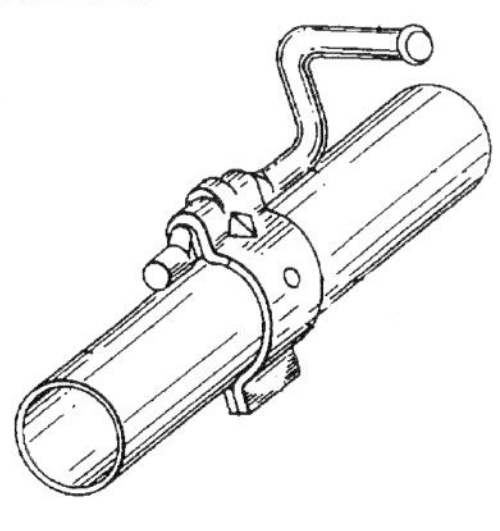

Fig 59: Exhaust fixing by Calsonic Corp (GB2313581A)

An exhaust clamp without bolts is an interesting piece of underbody innovation, Fig 59. Problems of looseness in an exhaust pipe assembly, following the fixing clamp nuts becoming untightened in service, are overcome in this relatively easy-to-assemble design. The ubiquitous bolted exhaust pipe clamp is replaced by a spring-clip-shaped member which attaches to a support arm by notched engagement with its flattened end. The rod is hung from a flexible mount in the usual way.

Aerodynamics and weight saving

Attention to underbody panel shape is also important in extracting the last mph of speed. Dr Dominey of Durham University[3] points out that the stability of regulations has led to a convergence of design which means that competitive circuit performance now depends on fine tuning of aerodynamic design. As ground clearance beneath the front wing is reduced flow can no longer tolerate the pressure gradient and the aerofoil stalls in the limiting condition. The limited permitted span of the aerofoils makes it necessary to control the associated tip effects. The end-plate solution aimed at maintaining approximate two-dimensional flow increases loading at the wing tips but does not eliminate strong tip vortices — for which semi-tubular guides along the lower edges of the end-plates has been one of the ways of containing them. To overcome the compromising effect of the front wing on engine cooling performance, the use of flapped aerofoils has been introduced — with chord and camber at the outer end of the flap greater than at the chassis end. Combining this with taller and narrower engine air-intakes ensures that only the wake from the inner section of the wing impinges on the intake. With rear-wing aerofoils the requirement for high downforce to be generated from already disturbed airflow has necessitated the use of highly cambered multi-element configurations which are inefficient in terms of drag. The trailing edge of the wing is governed by regulations while leading edge position is chosen as a compromise between a high one better exploiting airflow and a low one allowing a larger aerofoil to be used. The undertray of the car is the other important factor, Fig 60, air flowing beneath the car leading to a high-pressure stagnation point on the upper surface and a separation bubble below the entry lip. Addition of a diffuser creates a further zone of low pressure and its design is crucial to obtaining marginal advantage.

How the influence of a vehicle underbody affects aerodynamic drag was also discussed by Rover engineer J.P. Howell[4]. He showed, Fig 61a, at (a) the contributions made by different elements of the body on the overall aerodynamic drag of a saloon car. By fairing the complete underbody of an 800 saloon it was found possible to improve drag by factors of 0.39 and 0.46 for a notchback and hatchback configuration respectively. So as to gauge the effect of progressively adding fairing panels (b), wind-tunnel tests were made with successive increase in the numbered panels to obtain the results in (c). The extremities of the floor gave the largest gains but there is evidence of interactions between different sections of the floor. Also obtained were interesting results due to the progressive alteration of rear-end shape between the limits of notch- and hatch- (3) and of adding on and progressively increasing the depth of a front spoiler (6).

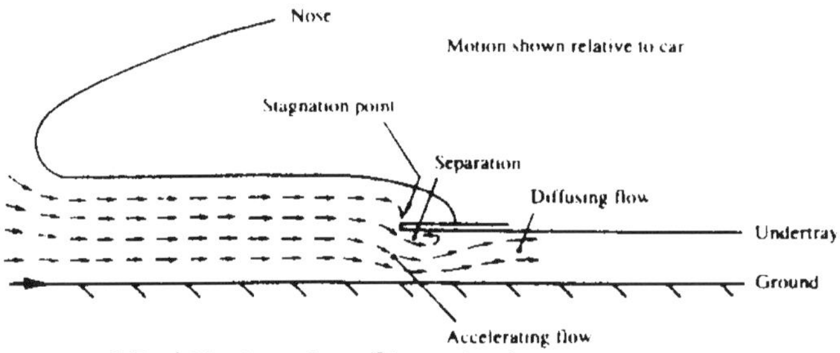

Fig 60: Aerodynamic under-layer

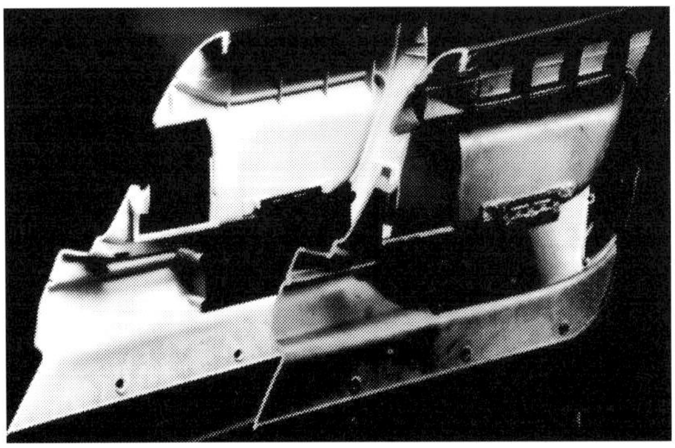

Fig 62: GM Corvette doors: front in RRIM rear view in LD-SRIM

An innovative body construction material is low density structural RIM. An LD-SRIM system has been trailed as a RRIM replacement for production of automotive interior trim panels via the Strappazini process. Trials at United Technologies Automotive confirm that ICI's MDI-based polyurethane LD-SRIM system has exhibited advantages over the commercial RRIM system. The Strappazini process is a patented method of moulding in place multiple covers with well finished joints and seams. The process is useful in moulding complex automotive interior trim components, including interior door panels. 'RIMline' SL-87339, a system containing an internal mould release (IMR), was tested in a Corvette RRIM door panel tool at UTA. In a closed mould, with only slight variations in mould temperatures and throughput of the current production equipment, the 'RIMline' system was poured directly onto a glass mat using the Strappazini process. Multiple releases were demonstrated with ICI's new IMR technology. During initial evaluation, 15 releases were achieved. Based on previous production experience, it is expected that in a plant environment at least 30 to 40 additional releases between external mould release applications can be achieved. In comparison, the current RRIM system is not available with IMR technology.

The LD-SRIM part, Fig 61b, weighed 20 percent less than the same door part moulded from the RRIM material. This was accomplished without optimization of part thickness or density. An additional 20 to 25 percent weight reduction can be achieved by reducing part thickness due to the superior strength properties of low density SRIM composite. RRIM requires substantially thicker sections to meet stiffness and thermal stability requirements as well as adequate impact resistance.

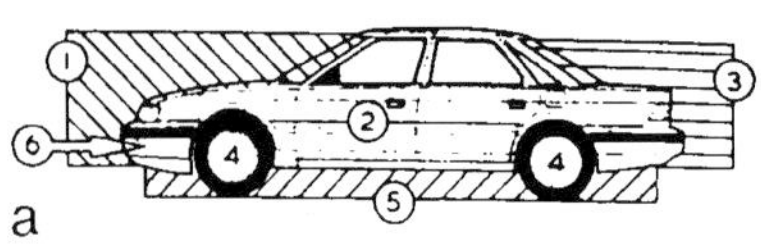

c

	ΔC_D
1. Front	0.00
2. Sides/Roof	0.05
3. Rear	0.14
4. Wheels/Wheelarches	0.06
5. Underfloor	0.05
6. Cooling	0.02
	0.32

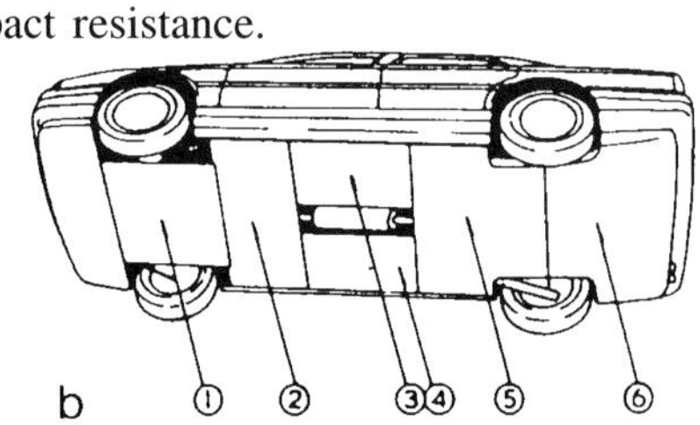

Fig 61: Drag factors involved

CV systems

CV chassis-cab configuration

Advanced-concept trucks are the trendsetters in truck design; DAF's future concept vehicle (FCV), Fig 62, is a fully driveable working unit whose features all have prospects, individually, for eventual production fitment. The chassis-frame is a fabricated rectangular section aluminium alloy box beam; the box girder design gives high torsional rigidity necessary to mount the independent suspension assemblies. Outboard of it are also hung fuel tanks and air reservoirs, while inside houses the major elements of the drivetrain. The apertures in the box tube are for access to various parts of the drive train, the front rectangular one giving engine access. A flat-bottom sump is fitted to the engine which is installed as low as possible within the box-tube, forward of the front wheels centre-line.

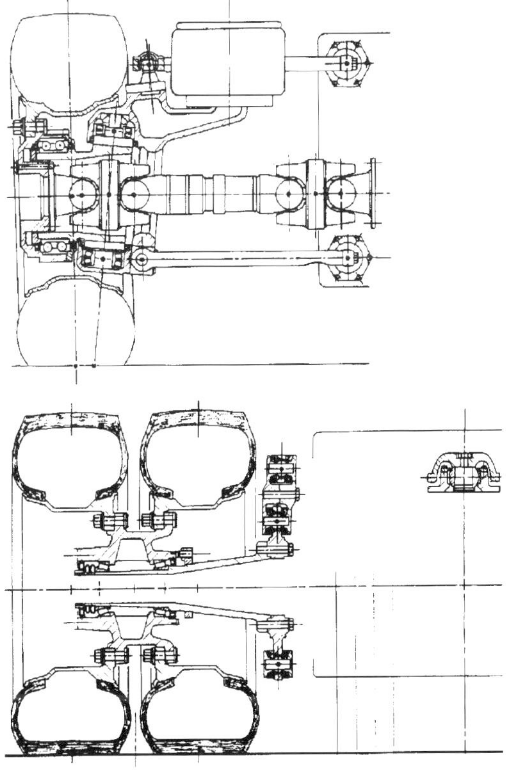

While engine, semiautomatic transmission and prop-shaft are otherwise conventional, the rear wheel drive is novel, in the form of a 'double drive differential system,. A torque apportioning device transmits two-thirds of it through a straight-through shaft to the second axle, the other third passing, via an annular gear drive, to the input pinion of the lead axle. Being a hypoid gear, its pinion offset allows the through-shaft to pass beneath the cross-shaft of the front differential. A short propshaft then connects the drive to the second axle. Independent suspension is fitted to both front and second 'axles', the latter both driving and steering. Special low-profile air-springs are employed which bear on saddles on the stub-axles. Lower links of the wishbone suspension, whose roll-centre is at ground level, are cranked downwards to clear the inboard disc brakes. In the second axle, drive extends outwards to the wheel hubs via a ring-type kingpin housing. Suspension travel is 80 mm to bump and 160 mm to rebound stops. For the third axle, twinned wheels on special two-flange hubs are supported by stub axles on a rectangular frame. Its

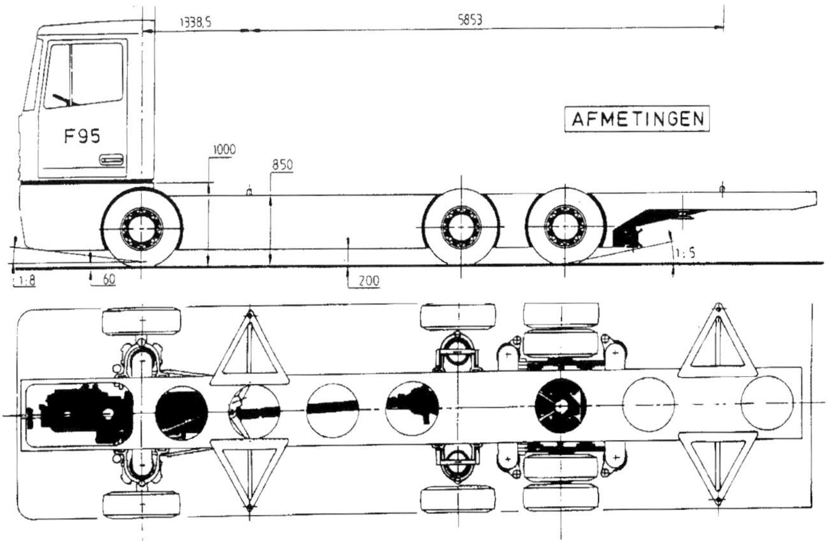

Fig 62a: DAF concept truck, with suspension and drive detail above

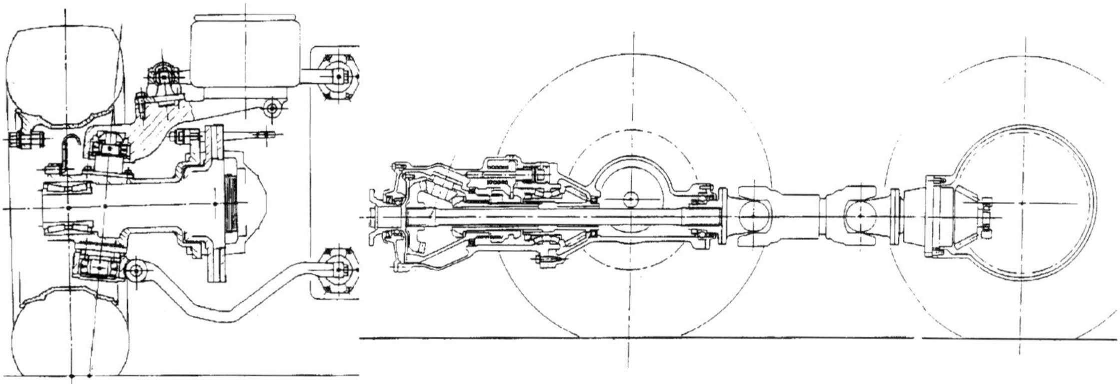

Fig 62b: Concept truck suspension and drive detail

slender, deep sidemembers have brackets extending outwards, beyond the inner road-wheels, to give as wide a spring-base as possible.

Concept has become production in the case of the important breakthrough which has been made by Dennis Specialist Vehicles. It involves a dedicated space-frame chassis for the new Rapier fire-appliance vehicle, Fig 63. Dramatic improvement in ride and handling is claimed over the conventional ladder-frame and leaf-spring layout — with the availability of a more rigid frame to mount a suspension of greater sophistication. Over a tonne of weight has been pared from the conventional structure and a low deck height has also been achieved — with a particularly low-slung engine. The latter has permitted the cab to be mounted 250 mm lower than usual, removing the need for entry steps. The entire crew cab tilts in one piece and the lightweight body is made in corrugated aluminium alloy sheet over an extruded section frame — incorporating water tanks made from plastic. Substantial coil springs suspend both front and rear wheels and, at the rear, wheels are independently suspended by a double wishbone system. Body roll, under 0.6 g lateral acceleration, has been reduced to 3.1 degrees compared with 8.5 degrees for a comparable conventional vehicle.

Fig 63: Dennis Rapier space-frame chassis

Cab/body fittings

Recent patent specifications reveal a number of the detailed improvements which have been made to truck systems. An example is the suspended cab-tilt pivot design in Fig 64. Here a truck cab suspension which is sympathetic to the cab-tilt system, without causing discomfort to the occupants, is proposed. The cab itself hinges on the uppermost pivot of the

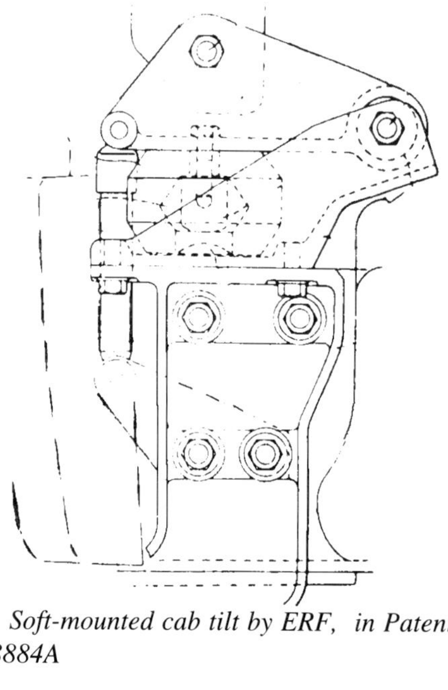

Fig 64: Soft-mounted cab tilt by ERF, in Patent GB2188884A

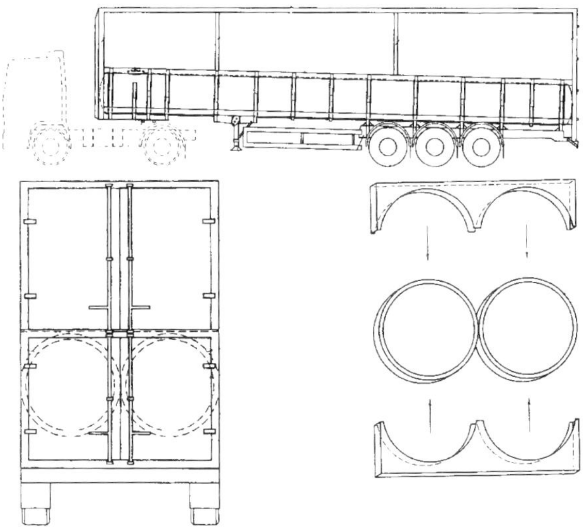

Fig 65: Return-load tanker by Clayton Commercials (GB2298830 A)

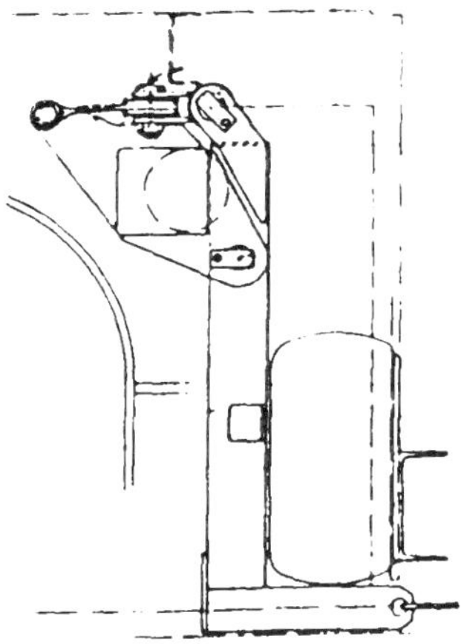

Fig 66: Variable cube body by York Trailers in Patent GB2198091A

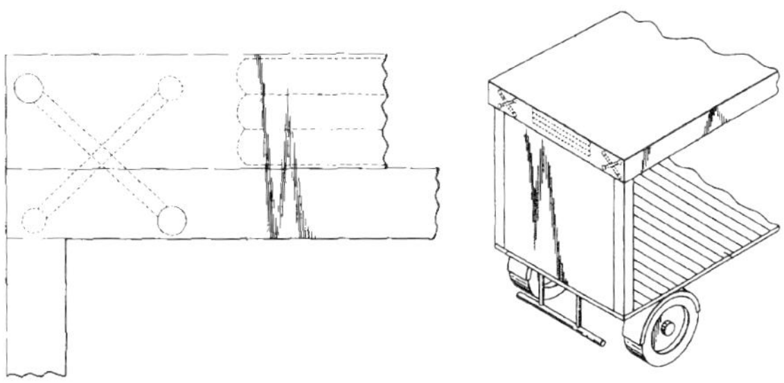

Fig 67: Air-bag system vari-cube system by JR Cramp, covered in Parent GB2299311A

mechanism shown while beneath it a second pivot forms the scissor linkage of the suspension mechanism — based on a rubber 'doughnut' compression spring and hydraulic damper.

A return-load tanker is the subject of the interesting patented design in Fig 65. The classic difficulty is tackled here in the operation of tankers and bulkers, not being able to fill the tanks with a return load, which might contaminate the principal product carried. According to the claimants, existing efforts to build dual purpose vehicles, carrying both dry goods and flow products on flat-topped tankers, have involved overweight structures. Here the lower portion of the vehicle, for flow products, comprises typically two large cylindrical vessels which form substantial structural members of the vehicle. Cradles around the vessels support the upper floor for carrying dry goods as well as proving a mounting for the running gear of the vehicle.

A number of attempts have been made to produce 'variable cube' bodies. A means of, literally, raising the roof of this van body is proposed by the design in Fig 66. An air-inflatable bag is positioned between the body wall and a lever which compresses the bag. Inflation of the bag moves the lever and pulls cables to operate the roof-raising device. Shown in this example mounted to the step of a step-frame trailer, cables from the pulleys shown pass beneath the chassis to operate jacks in the body corner posts. Another proposal also concerns raising the roof with airbags. Previous efforts have made been made to lift the roof of curtain-sided vehicle by an amount corresponding to the depth of the projecting border which covers the curtain tracks, to allow easier loading of tall cargo items. Because such efforts have thus far involved cumbersome mechanisms, the design of Fig 67 is aimed to simplify the approach by the use of airbags. A scissor linkage is associated with the air-bag, positioned at the top front/rear corners of the vehicle body. The compressed air supply would normally be reservoir tanks fed by the vehicle's air-compressor and air-lines would run through the roof support uprights.

Improved cargo retention is the object of the design in Fig 68. Problems of complexity, and associated proneness to wear, of existing cargo retention devices for goods vehicles can be reduced by the adoption of this mechanism in the design shown. Channels on each side of the deck anchor bracing beams, by means of a socket beneath each channel. The beams are either stowed within the channels or tilted upwards to brace the load. The beam is released from the load-retention position by raising it towards the vertical and then displacing it axially, from where it is pivoted to the stowed position. The interior of the beam encloses a restraining strap which is withdrawn to secure the cargo, the strap retracting into the beam automatically when not in use.

The object of the Fig 69 design is to provide an easy-access curtainsider. Means to avoid complex slider mechanisms for the body-side support pillars of a curtainsider are provided in the arrangement shown. In order to move the pillars to ease side-access of bulky loads, the pillars are mounted on a hinged linkage.

Hinged corner-posts for a curtainsider are covered by the design of Fig 70. The need to provide front-corner posts, of curtain sided CV bodies, which are faired in plan view to achieve aerodynamic efficiency, results in a dead space behind the section profile when loading the container body by fork truck from the side. In order to avoid a secondary loading operation, to enable pallets to fill the dead space, hinged post sections are suggested. When the flaps are in their closed position the side-curtain is wrapped around them and secured in socket catches prior to tensioning.

Considerably reduced effort to open and close the side-curtains of a curtainsider truck-body is claimed in the interesting roller arrangement of Fig 71. The roller track section is profiled to allow tilting of the rollers as they roll along the track. The rollers comprise ball bearings with convex sectioned tyres of hard plastic over their outer races. This set-up overcomes the problem with conventional systems of

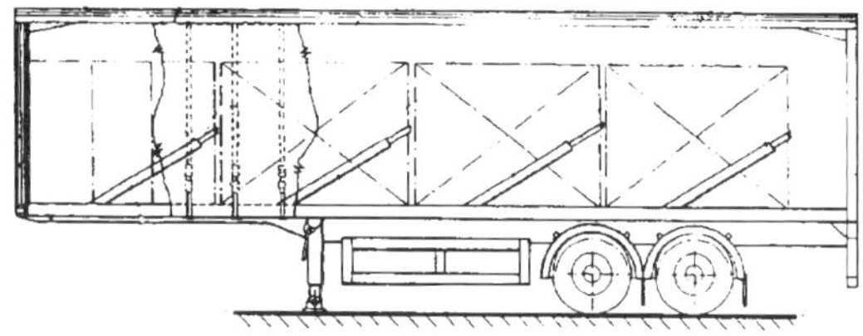

Fig 68: Improved cargo retention system suggested by Boalloy Industries, covered in Patent GB2301067A

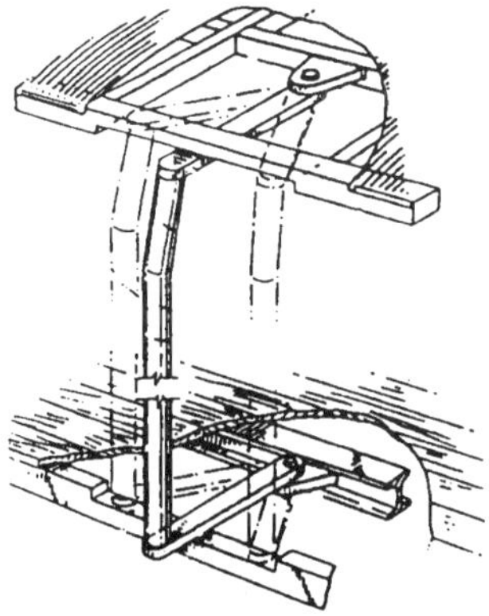

Fig 69: Easy-access curtainsider by Cartwright Freight Systems (GB 2185715A)

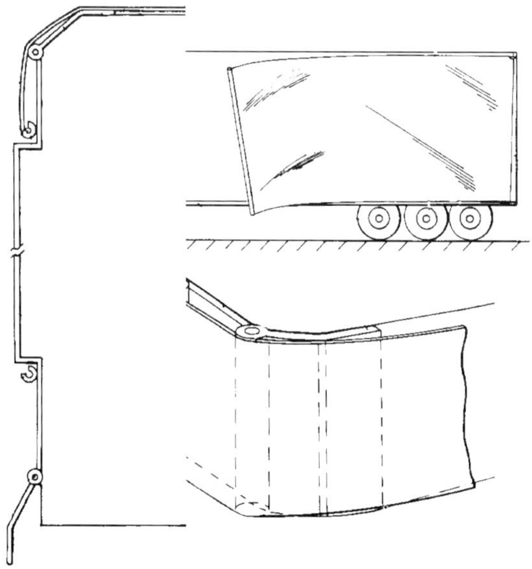

Fig 70: M & G Tankers & Trailers have suggested hinged post sections, in Patent proposal, GB2293144A

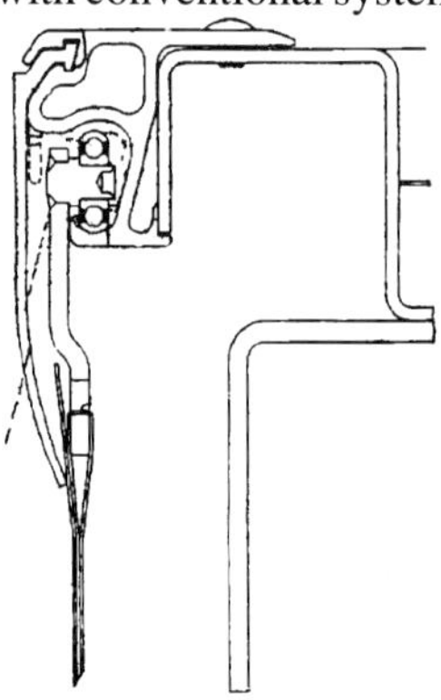

Fig 71: Improved curtain-sider by Montracon covered by Patent GB2291095A

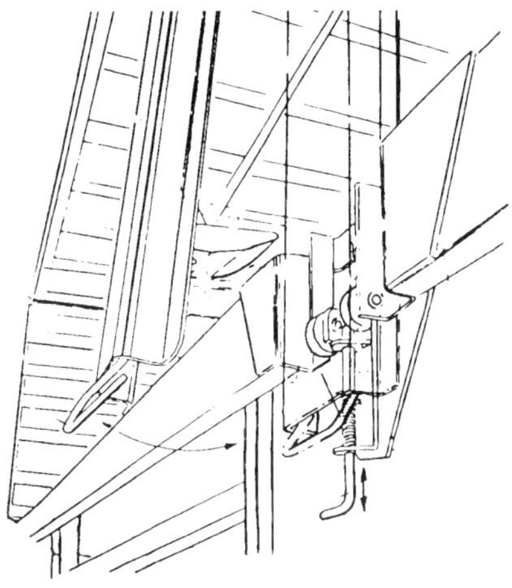

Fig 72: Simplified curtainsider support post in Patent GB2209712A by John J Cameron

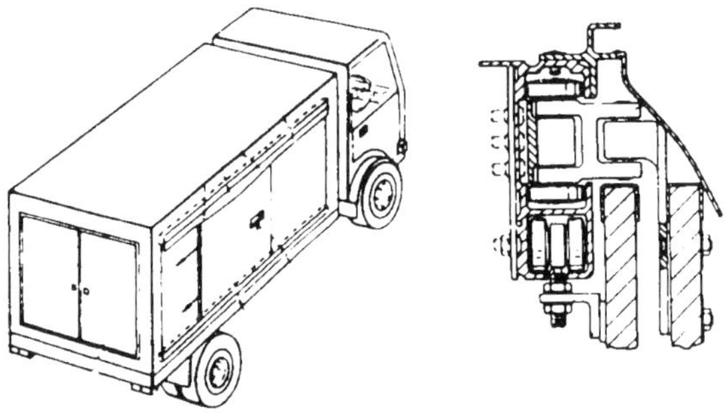

Fig 73: Compact sliding door gear by Bedwas Bodyworks, in Patent GB2203184A

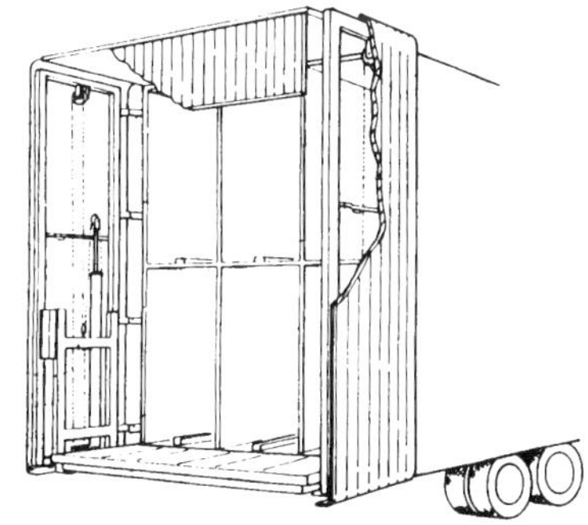

Fig 74: Tail-lift in door by Tidd Strongbox, covered in Patent GB2207112A

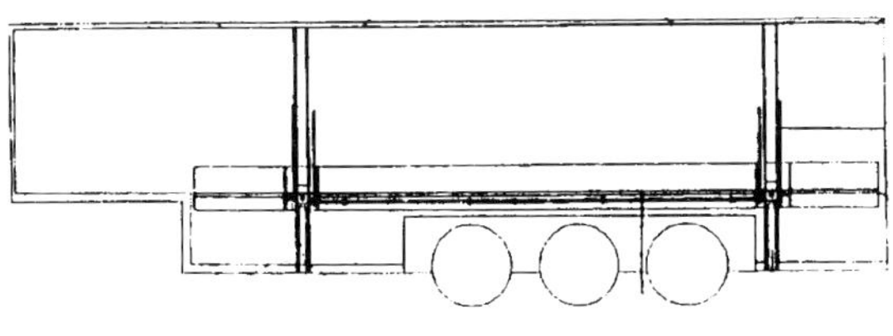

Fig 75: Upper-deck levelling achieved by the UK Lift Company, covered in Patent GB2299791A

binding of the rollers when the curtain is displaced sideways by the operator during pulling. Each hanger bracket is supported by two or more rollers.

A simplified support post for curtainsider CV bodies is proposed in the design of Fig 71. The cumbersome toggle linkage provided at the lower end of the column is dispensed with. The upper end slides on the cant rail, as normal practice, but the lower end locks to the side-rave, by means of a simple spigot linkage shown here.

A compact sliding door gear system is offered in the design of Fig 73. The difficulty is overcome here of accommodating two ISO pallets lengthwise across the platform of a goods vehicle — within the maximum legal vehicle width and the clearance envelope of the door gear. This design uses a 'master' plug-and-slide door suspended from a carriage and swing-arm assembly.

Tail-lift in a door is the result of the ingenious design of Fig 74. A tail-lift platform which doubles as a part-closure to prevent cold air fallout from refrigerated vehicles is raised and lowered by carriages in the rear doors of the body in this design. The body comprises a number of refrigerated compartments and a separate folding canopy can temporarily close the upper ones.

Levelling a lifting deck is the object of the Fig 75 design. Existing mechanisms used to raise the decks of goods vehicle bodies, to achieve two-tier stowage of the payload, suffer from the moving deck becoming unlevel as it is raised on jacks. In this design movement of the deck is synchronized by the interaction of two toothed racks mounted on the body and toothed wheels mounted on the moving deck. Connection of the wheels by shaft or endless chain ensures the level movement of the deck regardless of ram motions.

Folding catwalk rail is covered by the design of Fig 76. Added safety for road tanker operators is provided by this pivoting catwalk rail. When access to the top of the tanker is not required the rail lays flat across the catwalk to prevent ingress by unwanted intruders. When legitimate access is required hydraulic rams raise the rail to an upright position while interlocks ensure the vehicle's brakes stay on, in this position.

Perhaps the ultimate in cab access is offered by the Fig 77 design. Overcoming the ingress and exit restrictions for driver and operators of crew-cabbed commercial vehicles is the objective, particularly suited to vehicles with front overhead loaders. The driver's seat and controls are centrally mounted while those for the crew are mounted above and behind, on either side of the cab and over the wheel housings. Doors of 1.9 metres in height open inwardly to prevent pavement obstruction. The floor is inclined upwards at the front and level at the back and provides cross cab access ahead of the engine compartment.

Advanced bus/ambulance design

Rethinking the ambulance configuration is behind the design of Fig 78. Emergency transport of accident victims in heavy traffic was the impetus behind an interesting vehicle to handle urban emergencies discussed at a recent Autotech congress. Authors Da Silva and Miranda of Lisbon Technical University explained that the key design factor was provision of an efficient life condition support for the patient in a vehicle with exceptional traffic mobility. That involved a highly compact layout yet a comfortable posture and space for medical assistance to the patient in transit. The short journeys inherent in the duties of the vehicle make it an attractive proposition for electric traction and in the proposed design the patient would be positioned above the battery container. Twin rear wheels would be mounted on an oscillating axle, the 'banking' of which would be controlled by the driver, or 'rider' in a semi motorcycle situation, sitting behind the single steering front wheel of the vehicle. Retractable outrigger wheels would also be used to provide cornering stability in less dense traffic conditions. Translucent panelling around the patient compartment would reduce the claustrophobic effect as would careful colour selection for the interior.

A joint venture between MAN and Voith has resulted in the NL 202 DE low floor concept city bus, Fig 79, designed to carry 98 passengers at a maximum speed of 70 km/h. No steps are involved at any of the entrances, which lead directly to a completely level deck height of between 317 to 340 mm. The rear-mounted horizontally positioned diesel engine allows fitment of a bench seat at the rear of the bus; it drives a generator with only electrical connection

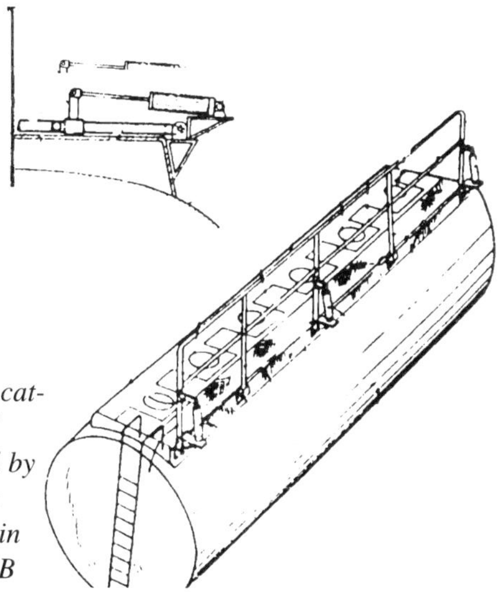

Fig 76: Pivoting catwalk rail proposed by Safewalk Railings in Patent GB 2203390A

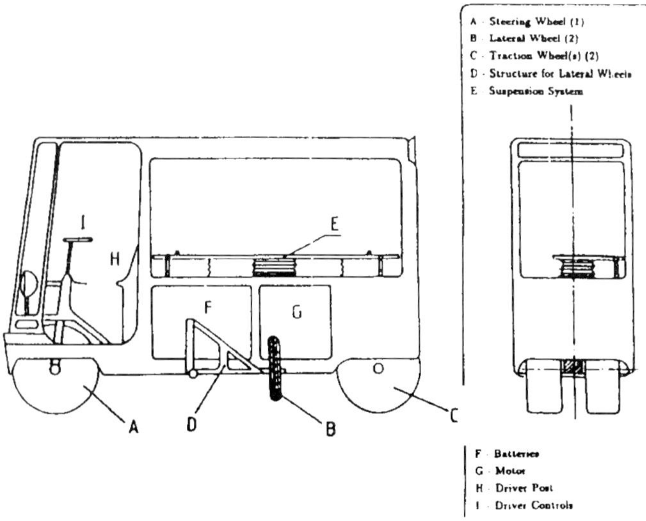

Fig 77: Short-haul ambulance

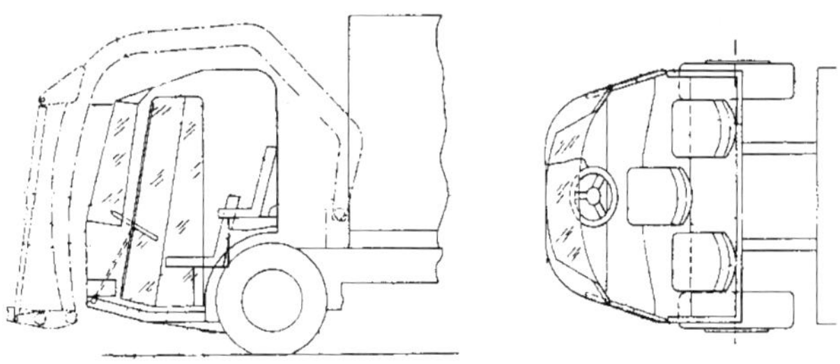

Fig 77: Easy-access cab by Britannia Trucks (GB2297951A)

Fig 79a: NL 202 DE concept MAN bus with propulsion system detail below

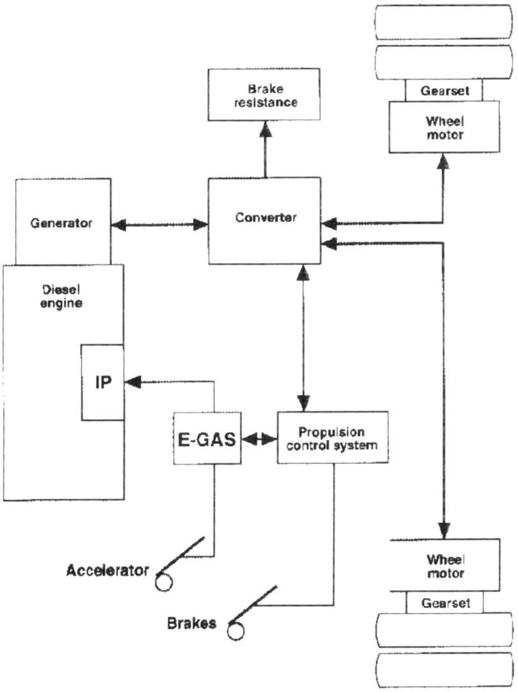

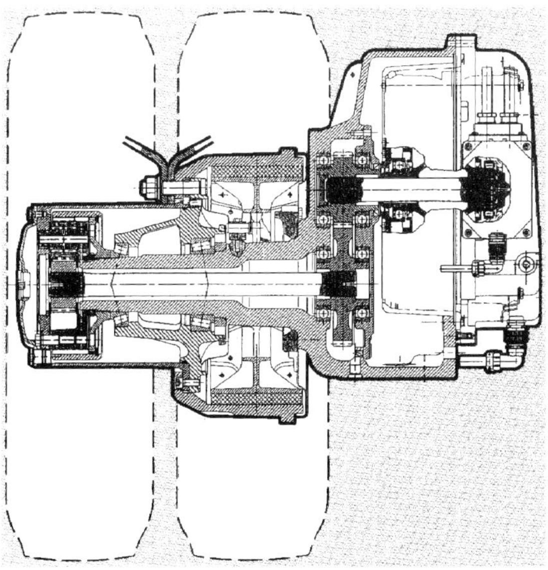

to the Voith transverse-flux wheel motors, which drive the wheels through two-stage hub-reduction gear sets. The diesel is rated at 127 kW and the generator at 135 kW; the controller is of the IGBT converter type and also developed by Voith. It provides a differential action to the wheel motors on cornering. Permanent magnet synchronous wheel-motors are rated at 57 kW and have a maximum speed of 2500 rev/min; see Table 1 in Fig 79a. The bus is 12 metres long and has water-cooling for its generator, converters and wheel motors. As well as providing virtually jerk-free acceleration, the drive system is seen by MAN as providing the possibility of four-wheel drive on future articulated buses to improve traction and stability in slippery road conditions. The term 'transverse flux motor' refers to the means used to guide the magnetic flux in the stator; this is new to inverter-supplied PM types and involves a novel collector configuration. Double-sided magnetic force generation is also new and involves a patented double air gap construction having high idling inductances and force densities up to 120 kNm/m^2, with relatively low losses.

A new control process permits operation of the motor in a field-weakening type mode, in spite of PM excitation. The generator is almost identical in concept but involves no field weakening. Each has concentric construction of permanent magnets, rotor/stator soft-iron elements and stator winding; see Fig 79a. Armature elements are U-shaped cut strip-wound core sections, embedded in the ring-shaped supporting structures of inner and outer stators. Each core surrounds the windings and forms a stator pole with its cut surfaces facing the rotor. The latter is pot-shaped and positioned between poles of the outer and inner stators. In the stator pole region it comprises magnet and soft iron element while in the winding region a ring of GRP serves as the connecting

	TFM wheel motor	TFM generator
Power	57 kW	135 kW
Rated speed	735 rev./min.	1,750 rev./min.
Approx. max. speed	2,500 rev./min.	2,400 rev./min.
Max. fundamental frequency of stator	1,350 Hz	–
Rated torque	740 Nm	740 Nm
Approx. max. torque	1,050 Nm	740 Nm
Approx. torque conversion	1:3	
Power/weight ratio	1.8 kg/kW	0.9 kg/kW

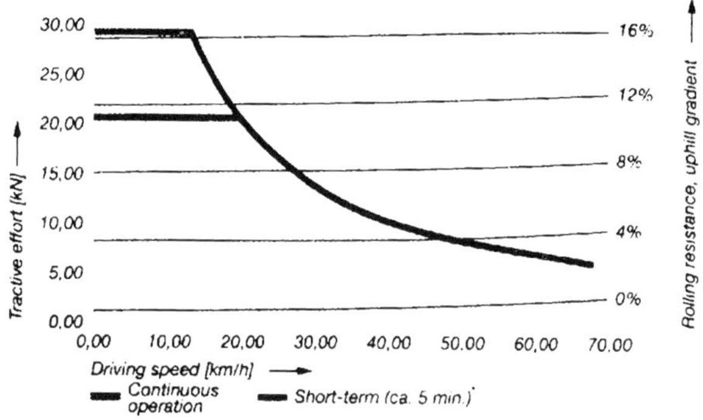

element. The inverters supply the motors with sinusoidal currents and voltages until the nominal operating point is reached; operating frequency is 10 kHz. In field-weakening mode the induced voltage exceeds intermediate circuit voltage and only square wave voltages are supplied to the motor. Power output then remains constant and the operating frequency equals the fundamental motor frequency. A large speed ratio, 1.5:1, is thus possible.

An electric city-bus designed for low drag is seen in the Centro concept midibus from Capoco Design, Fig 80. It is intended to show the practicality of 'available now' electric drive technology for city centre operation. The 7.3 metre long vehicle would operate at up to 40 kph and have a range of 40 miles. Energy consumption is reduced by the vehicle's drag coefficient of just 0.31 and low rolling-resistance tyres of modest section profile. A lightweight structure comprises mechanically fastened aluminium

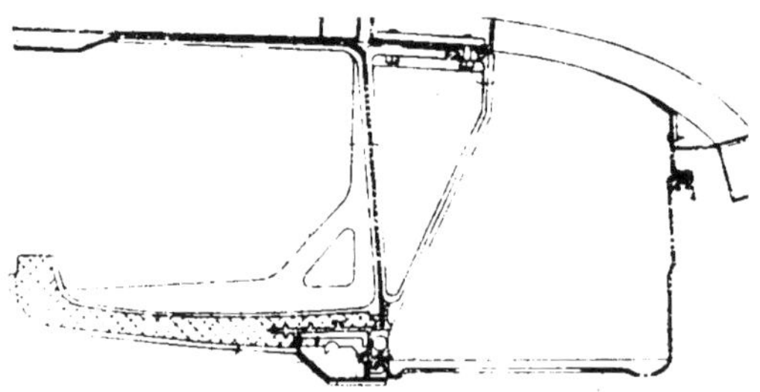

Fig 81: Aero-style luggage rack designed by Ikarus covered by Patent GB2209140A

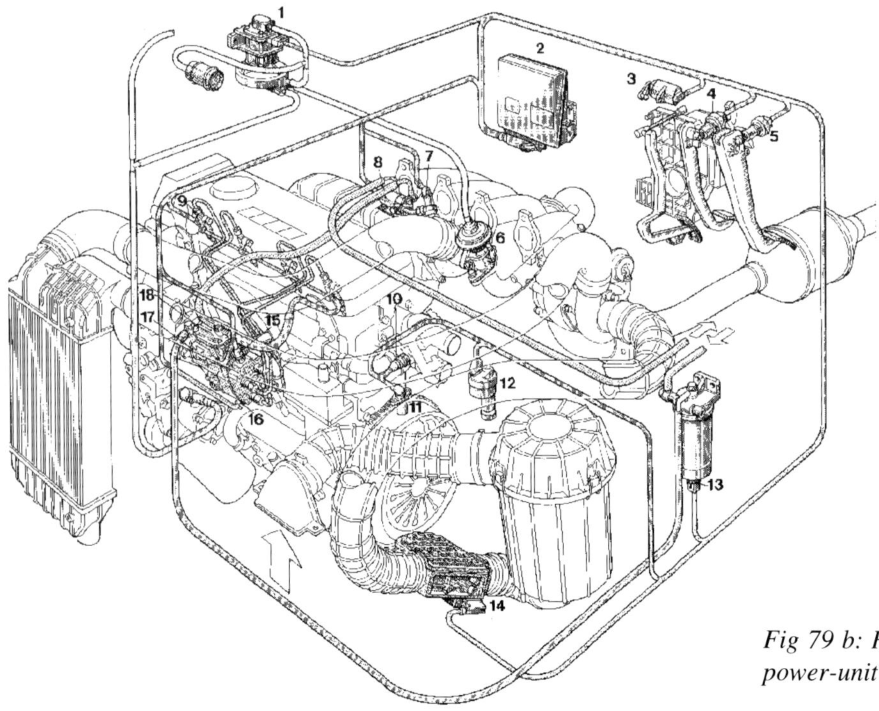

1) EGR electromagnetic control valve
2) Electronic control unit
3) Accelerator position reinforcement
4) Brake pedal switch
5) Clutch pedal switch
6) EGR valve
7) Thermostarter
8) Thermostarter feed valve
9) Instrumented injector
10) Water temperature sensor
11) Engine speed sensor
12) Vehicle speed sensor
13) Water in fuel sensor
14) Water flow and temperature sensor
15) Engine stalling solenoid
16) Engine advance adjustment
17) Fuel flow and temperature adjustment
18) Injection pump.

Fig 79 b: Performance characteristic and power-unit environmental control system

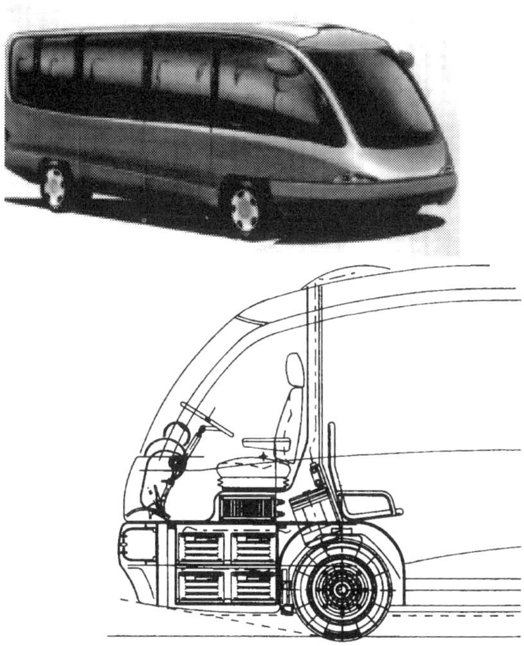

Fig 80: Capoco Centro electric bus

alloy extrusions covered by a single-piece roof moulding. In full standee form up to 56 passengers can be carried and deck height is just 290 mm. Shown also in Fig 80 is the economical packaging of batteries and drive motor, with near vertical drive axis. Width of the vehicle is 2380 mm and height 2550 m, on a 4050 mm wheelbase.

An aero-style bus saloon assembly follows from modular elements used in this patented roof liner and luggage rack assembly, Fig 81, for touring coaches. The arrangement involves zed-shaped brackets screwed to the roof frame which secure headlining, ventilating ducts and luggage shelves. The shelf elements incorporate a profiled channel, for housing light fittings, which is closed by a snap-fit translucent cover.

References

Chapter 1

1 Sears, K., Automotive engineering: strategic overview, *Steel Times*, Vol. 2. No.1, 1997

2 Matthews and Davies, Precoated steels development for the automotive industry, Proc. I.Mech.E., Vol. 211., Part D, 1997

3 Rink and Pugh, The perfect couple — metal/plastic hybrids making effective use of composites, IBCAM Conference, 1997

4 Merrifield, R., Instrument panel structural concepts which integrate functions utilizing injection moulded plastic components, Paper C524/120/97, Autotech Congress, 1997

5 Ashby, M., Material Property Charts, Performances Indices, Section 4, Material Selection and Design,20th edition , ASM International, 1997

Chapter 2

1 Bass *et al*, A system for simulating structural intrusion, Proc. I.Mech.E., Vol. 211., Part D, 1997

2 Yim and Lee, Design optimization of the pillar joint structures using equivalent beam modelling technique, SAE Paper 971544, part of P-308

3 Hardy, M., Automotive modelling and NVH seminar, Proc.I.Mech.E., 1997

4 Hargreaves, J., I .Mech.E .Vehicle noise and vibration conference, 1998

5 Lawrence and Hardy, Development and use of pedestrian impactors to reduce the injury potential of cars, paper C524/049/97, Autotech 1997

6 Coleman and Harrow, Car design for all, IMechE seminar , 1997

7 Delphi Automotive, Adaptive restraint technologies, *Vehicle Engineer*, December 1998

Chapter 3

1 Evans and Blaszczyk, A comparative study of the performance and exhaust emissions of a spark ignition engine fuelled by natural gas and gasoline, Proc.I.Mech.E., Vol. 211, Part D, 1997

2 Perkins and Penny, Design options and performance characteristics for 0.25–0.3 litre/cylinder HSDI diesel engines, Euro 4challenge: future technology and systems, IMechE Seminar, 1997

3 Sadler *et al.*, Optimization of the combustion system for a direct-injection gasoline engine using a high speed in-cylinder sampling valve, Euro 4challenge: future technology and systems, I.Mech.E. Seminar, 1997

4 Zhou and Qian, Development of a modified diesel engine cycle, Proc. I.Mech.E., Vol 212, Part D, 1998

5 Cains *et al.*, High dilution combustion through axial and barrel swirl, I.Mech.E. Seminar Publication: Automotive Engines and Powertrains, 1997

6 Bassett *et al.*, A simple two-state late intake valve closing mechanism, Proc IMechE, Vol 211, Part D, 1997

7 Woods and Brown, Flow area of multiple poppet valves, Proc IMechE, Vol 210, Part D, 1996

8 Ohashi *et al.*, Honda's 4 speed all clutch to clutch automatic transmission, SAE Paper 980819

9 Abo *et al.*, Development of a metal belt-drive CVT incorporating a torque converter for use with 2-litre class engines, SAE Paper 980823

10 Ahluwalia *et al*, The new high torque NVT-750 manual transaxle, SAE Paper 980828

11 Turner and Kelly, A transmission for all seasons, Advanced vehicle transmission and powertrain conference, I.Mech.E., 1997

12 1998 SAE software based on book: Gillespie, G, Fundamentals of vehicle design, SAE, 1992

13 Gadola and Cambiaghi, MMGB: a computer-based approach to racing car suspension design,*ATA*, Vol. 47., no. 6/7., 1994

14 Ellis, J, Vehicle handling dynamics, Mechanical Engineering Publications, 1994

15 Potter *et al.*, Assessing 'road-friendliness': a review, Proc. I.Mech.E., Vol. 211., Part D, 1997

Chapter 4

1a. Bodoni-Bird, C, What can seamless electro-mechanical vehicles learn from Nature?, 96C001,

1b. Miyata et al, Engine control by ion density analysis, 96C003

1c. Pinkos and Shtarkman, Electronically controlled smart materials in active suspension systems, 96C004

1d. Hatanaka and Noro, New approach for intelligent steering system development, 96C007

1e. Milburn, S, Integration of advanced functions into electric drivetrains, 96C050

...from the bound volume of proceedings of the 1996 Convergence Transportation Electronics Association conference

2a Kuragaki *et al.*, An adaptive cruise control using wheel torque management technique, 980606

2b Olbrich *et al.*, Light radar sensor and control unit for adaptive cruise control, 980607
...from the bound volume of proceedings of the ITS Advanced controls and vehicle navigation systems seminar at the 1998 SAE Congress
3 Bolton, W., Essential mathematics for engineering, Heinemann
4 Jones *et al.*, HYZEM — a joint approach towards understanding hybrid vehicle introduction into Europe, Proceedings of the IMechE Combustion engines and hybrid vehicles conference, 1998
5 Friedmann *et al.*, Development and application of map-controlled drive management for a BMW parallel hybrid vehicle, SAE Special Publication SP1331, 1998
6 Nagasaka *et al.*, Development of the hybrid/battery ECU for the Toyota Hybrid System, SAE Special Publication SP1331, 1998
7 Saito *et al.*, Super capacitor for energy recycling hybrid vehicle, Convergence 96 proceedings
8 Origuchi *et al.*, Development of a lithium-ion battery system for EVs, SAE paper 970238

Chapter 5
1 Bowsher , G, Braking and traction at supersonic speeds, *Special Vehicle Engineer*, January 1998

Chapter 7
1 SAE papers 870147/8
2 I.Mech.E .Autotech paper C427/40/108
3 Proc. I.MechE., Vol 206, Aerodynamics of Grand Prix cars
4 I.Mech.E .Autotech Paper C427/6/032

Index

Index

A

B

C

D

E

T

U

V

W

Z